Civil Engineering Construction Materials

For Students of B.Tech, B.E., AMIE, Engineers, Architects and Builders

DR. S. K. SHARMA
B. Tech (Hons) I.I.T. Kharagpur,
M.Sc. Engg. (Distinction) Ph.D., FIE.
Former Professor and Head of Civil Engineering
PEC University of Technology, Chandigarh

KHANNA BOOK PUBLISHING CO. (P) LTD.
Publisher of Engineering and Computer Books
4C/4344, Ansari Road, Darya Ganj, New Delhi-110002
Phone : 011-23244447-48 **Mobile:** +91-9910909320
E-mail : contact@khannabooks.com
Website : www.khannabooks.com

Dear Readers,

To prevent the piracy, this book is secured with HIGH SECURITY HOLOGRAM on the front title cover. In case you don't find the hologram on the front cover title, please write us to at contact@khannabooks.com or whatsapp us at +91-99109 09320 and avail special gift voucher for yourself.

Specimen of Hologram on front Cover title:

Moreover, there is a SPECIAL DISCOUNT COUPON for you with EVERY HOLOGRAM.

How to avail this SPECIAL DISCOUNT:

Step 1: Scratch the hologram
Step 2: Under the scratch area, your "coupon code" is available
Step 3: Logon to www.khannabooks.com
Step 4: Use your "coupon code" in the shopping cart and get your copy at a special discount
Step 5: Enjoy your reading!

Price: ₹ 425.00

ISBN: 978-93-82609-84-1

Copyright_Civil Engineering Construction Materials
by S.K. Sharma

First Edition: 2016
Reprint: 2021, **2024**

To view complete list of books, Please scan the QR Code:

Published by:
Khanna Book Publishing Co. (P) Ltd.
CIN: U22110DL1998PTC095547

Visit us at: **www.khannabooks.com**
Write us at: **contact@khannabooks.com**

Printed in Bharat.

Preface

The subject of Engineering Materials is of basic importance to Civil Engineering students, engineers, architects and builders. Judicious selection of materials can be done only if the user has proper knowledge and understanding of the sources, production techniques, properties, performance and relative economy of various materials.

In all technical courses the study of engineering materials commences from the initial stage when the student learns about the various materials. The emphasis is mostly on traditional and conventional materials in common use. However, the study should be continued to know more about the new and improved materials developed as a result of research and technological advancements. This is because materials play a key role and modern advancements can be attributed to great strides in the development of new materials. Moreover, on account of more exacting demands as regards quality and performance of the materials, right selection and use of materials have become matters of vital importance. On this is dependent, to a great extent, the cost of engineering works.

The intention of bringing out the present volume is to provide an introductory treatment of the subject in the light of above observations. An attempt has been made to describe in greater details some aspects of availability, production and use of conventional materials such as stones, bricks, lime, cement, concrete, steel, timber etc. Basic information on new materials like structural clay products, metals and non-metals, varieties of cements and concrete, wood products, plastics, has been presented.

In bringing out this volume reference has been made to a number of publications, including books, articles in journals, research reports, standards, trade literature etc. from which useful information has been sifted and suitably presented. Special attention is made of the research publications of the Central Building Research Institute,Roorkee, various technical publications of the National Buildings Organisation and the standard specifications and codes of practices brought out by the Bureau of Indian Standards. The usefulness of these publications is gratefully acknowledged. It may be mentioned that one of the objectives sought to be achieved by this volume is to bring to the readers, specially the students, the knowledge and information made available by the above national research and development organisations, through the medium of text book. Another feature of this book is that it is written entirely in Metric System.

The author feels, obliged to Shri Mukul Seth, Director, Khanna Book Publishing Co. (P) Ltd. for taking special interest in bringing out the book in a nice form.

Though every care has been taken to present up-to-date information on the subject, suggestions for improvement are always welcome.

January 2016

Dr. S. K. Sharma

Brief Table of Contents

Detailed Table of Contents

STONES

1.1 STONE AS A BUILDING MATERIAL

Stone has been used throughout the world as a building material from very early times. Important examples of its use in India are the Hindu and Budhist Temples, the Taj Mahal at Agra, the Red Fort at Delhi, Victoria Memorial in Kolkata and many other massive buildings throughout the length and breadth of the country. The pyramids of Egypt, the excavated ruins of Troy, the remains of Greecian and Roman structures, etc. illustrate the popularity of stone as a building material practically in all parts of the world.

Where permanancy of structure is required, building stone continues to enjoy superiority over bricks and other rival materials. When properly selected, it lasts long and does not involve huge maintenance cost. These qualities favour its use in permanent engineering works, a few examples of which are given below:

(*i*) Different parts of buildings such as foundations, walls, columns, roofs, etc.

(*ii*) Structural members in buildings such as slabs, etc.

(*iii*) Face-work to brick walls for reduction in cost of maintenance.

CLASSIFICATION OF STONES

Stones are derived from rocks forming the crust of earth's surface. Rocks yielding stones may be classified as follows :

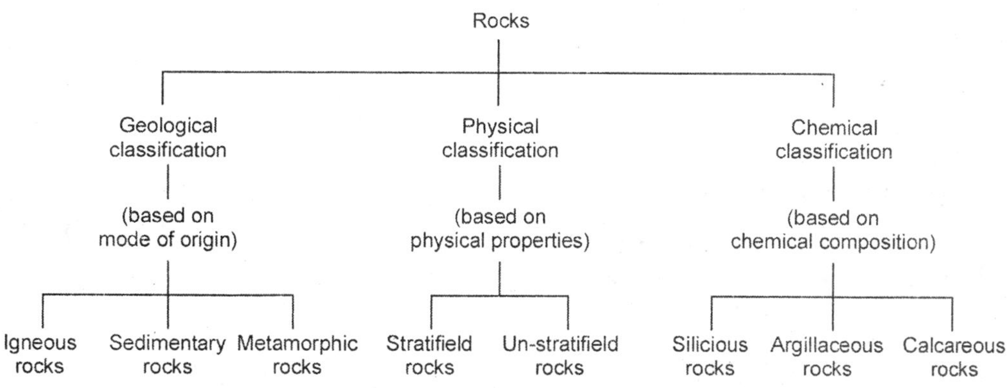

These are described in the following paragraphs:

1.2 IGNEOUS ROCKS

Igneous rocks are formed by the cooling and solidifying of the molten material called magma erupted from inside the earth. They are volcanic in nature and are generally strong and durable. Granites, traps and basalts belong to this group. Granite indicates the formation of large crystals due to extremely slow cooling of the molten material, which is due to a thick covering on the top. Trap and Basalt are formed with a non-crystalline or amorphous and glassy texture due to the sudden cooling of the molten matter as it is forced to the surface.

1.3 SEDIMENTARY ROCKS

Sedimentary or aqueous rocks are formed by gradual deposition or sedimentation under water. Most of the sedimentary rocks are formed by water depositing the materials which it once conveyed in suspension and in solution. These materials are subsequently cemented together by lime, silica, potash, etc. and result in limestone and sandstone, which are typical examples of sedimentary rocks.

1.4 METAMORPHIC ROCKS

Metamorphic rocks are igneous or sedimentary in their origin, which when subjected to intense pressure or heat or both, undergo structural changes in their structure, texture and colour. They are generally hard and durable and difficult to work with. As a result of metamorphic action, granite, trap, basalt and limestone change into gneisses, chists, laterite and marble respectively. Similarly sandstones turn into quartzite and mud-stone and shale change to slate.

1.5 STRATIFIED ROCKS

Rocks which show distinct signs of layers *e.g.*, limestone, sandstone, slate, etc.

1.6 UNSTRATIFIED ROCKS

Rocks which do not show any sign of strata and cannot be easily split into thin slabs, *eg*. granite, trap, marble, etc.

1.7 SILICIOUS ROCKS

These rocks have silica (sand, granite, flint) as its main constituent. Silica is very hard and durable and is unaffected by weathering agencies. Chief types of silicious rocks are sandstone, trap, granite, etc.

1.8 ARGILLACEOUS ROCKS

These rocks have clay or alumina as its base. They are hard and durable but not elastic as they are brittle and do not absorb shock. Slates, laterites form an important group of argillaceous rocks.

1.9 CALCAREOUS ROCKS

These rocks have calcium carbonate as the main ingredient. They are acted upon by dilute hydrochloric acid. The principal varieties of these rocks are marble, limestone, etc.

COMPOSITION OF ROCKS AND THEIR CONSTITUENT MINERALS

Chemically, rocks are composed of the following main constituents.

(*i*) Mineral Earths	:	Silica (SiO_2)	Alumina (Al_2O_3)
		Lime (CaO)	Magnesia (MgO)
(*ii*) Alkalies	:	Soda (Na_2O)	Potash (K_2O)
(*iii*) Acids	:	Carbonic acid	

In addition to the above mentioned ingredients, there are sometimes very small percentages of iron oxide and manganese oxide present.

Physically, rocks are composed of minerals formed by one or more of the above mineral earths, held mechanically together by some cementing material. The most predominant minerals are described below :

1.10 QUARTZ (SiO_2)

Silica is very abundant in sandstone, granite and certain varieties of gneiss. It is white but is often coloured in various shades due to the presence of a small amount of metallic oxide. Common sand is a form of quartz. Flint, chalcedony, agate and jasper are different varieties of coloured quartz. In purer crystalline conditions quartz is usually transluscent. Its crystals are in the form of regular six-sided prisms terminated by a pyramid summit. Hardness and specific gravity are 7 and 2. 6 respectively.

1.11 FELSPAR (K_2O, Al_2O_3, 6 SiO_2)

This is silicate of aluminium with silicate of sodium or potassium or a mixture of the two. Sometimes K_2O is replaced by Na_2O or CaO. Felspar is an important mineral present almost in all igneous rocks. It is often coloured due to the presence of a small quantity of iron oxide and this factor is responsible for the colour of red granite. When felspar -contains large proportions of calcium, potassium or sodium, it is liable to cause decay quickly . Felspars containing potassium are known as Orthoclase and are flesh coloured. Those containing sodium as their base are grey such as Oligoclase and Albite felspars. With calcium as a base, the colour of felspar is whitish grey, as in the Labradorite varieties. Hardness and specific gravity are 6 and 2. 6 to 2- 75 respectively.

1.12 MICA

This is silicate of aluminium with potassium (sometimes with magnesium in addition) and occurs in thin transparent plates full of very thin laminae, in which it readily splits. Mica flakes shine with metallic lustre. There are two varieties. Muscovite which is white and Biotite, which is dark brown or black. Mica is a source of weakness as it disintegrates quickly and slotes containing over 2% mica are considered unsuitable for structural purposes. Hardness and specific gravity are 2 to 3 and 2'7 to 3-2 respectively.

1.13 HORNBLENDE [5 (Ca, Mg). 6 SiO_2]

This is silicate of calcium and magnesium and occurs in crystals, fibre and grains and has a glossy lustre. This is dark green, brown or black in colour. It is very heavy, strong and durable mineral. It is present in a number of igneous rocks in the form of fibres or four sided prisms. It weathers fairly well. It is brittle but strong and durable. Hardness and specific gravity are 5 to 6 and 2.9 to 3.0 respectively.

1.14 AUGITE (Ca, Mg) SiO_2.

This is very much similar to but heavier than hornblende. Only the angles of crystals are different. It is found abundantly as a constituent of Deccan trap. It is also called Pyroxene. Hardness and specific gravity are 5 to 6 and 2.3 to 3.6 respectively.

1.15 CALCIUM CARBONATE($CaCo_3$)

This is the basis of limestone, chalk and calcareous stones. It may occur as a cementing material of sandstones and shales. Hardness and specific gravity are 3 and 2.72 respectively.

CHARACTERISTICS OF STONE

The following characteristics should be looked into while selecting stone for use in a building and other structural purposes:

(*i*) General structure
(*ii*) Durability
(*iii*) Strength
(*iv*) Appearance and colour
(*v*) Facility for working
(*vi*) Fire resistance
(*vii*) Natural bed
(*viii*) Cost

1.16 GENERAL STRUCTURE

This relates to the manner or mode in which the particles and the mass, which go to form a stone are arranged. Principally there are three types of structures namely unstratified (granular), stratified and foliated.

(*i*) *An unstratified structure.* It relates to the stone obtained from the rocks of the igneous origin. They are formed by the process of cooling of the molten rock called magma. Depending upon the time taken to cool down, they are either of crystalline or of amorphous variety. They yield strong building stones in general.

(*ii*) *A stratified structure.* It is a characteristic feature of sedimentary rocks indicating the manner in which they are formed, *viz.* by the deposition of the materials produced by weathering of originally formed rocks. Layer after layer of these materials are deposited, which give rise to stratified structure. These rocks also have thin laminae interposed between their layers of stratification and this give rise to laminated structure. Such rocks can be easily split along these planes of stratification which are called planes of cleavage.

(*iii*) *A foliated structure.* Metamorphic rocks have foliated structure *i.e.,* parallel arrangement of minerals having a tendency to split in definite directions. They are not formed by the process of deposition but are developed in the existing rocks as a result of metamorphic action of heat and pressure.

1.17 DURABILITY

Durability of a stone depends upon :

(*i*) Chemical composition of stone

(*ii*) Texture of stone

(*iii*) Resistance to atmospheric influences.

1.17.1 Chemical Composition of Stone

Ordinary buildings of limestone and weak sandstone can be considered to have a life of about 50 years, while the estimated life of granite, gneiss and good sandstone buildings is over 200 years. Air and water containing carbon dioxide seriously affect limestone, while the disintegration of alkaline silicate of alumina in stone is mainly due to the solvent action of water which becomes chemically active and acidulated when it passes through decaying vegetation and thus gets charged with different types of organic acids. The presence of mica, even in proportions less than 2 to 3 percent, makes the stone unsuitable for building purposes. All the varieties of mica are soft in themselves and are readily decomposed by exposure to atmosphere.

Stones with silicates as the cementing material will weather better than those with calcareous or ferrugenous binding material. If the grains and the cementing .material are both lasting, the stone will be very durable. If one of these is weak, it will soon disintegrate.

1.17.2 Texture of Stones

The texture of a stone indicates the arrangement of constituent minerals in it. Stones with crystalline or homogeneous texture are close grained and compact. They have better weather resisting qualities than the open textured variety. Generally igneous rocks and metamorphic rocks such as granites, traps, gneisses and schist are practically impervious.

Sandstones and limestones belonging to sedimentary rocks are porous. Their texture could be described as granullar crystalline. Some of the metamorphic rocks also have the same texture.

Amorphous rocks have a glassy or fused texture. They have no grains or crystals and look like glass, flint etc. They are hard and compact.

Fracture denotes the appearance of the surface obtained by breaking the rock in a direction other than that of cleavage if it exists and indicates its texture. The fracture may be of the following types :

(*i*) *Even Fracture.* It shows the surfaces of both the sides of a fracture as more or less flat and shows crystalline texture.

(*ii*) *An uneven Fracture.* It shows sharp projections with rough or irregullar texture.

(*iii*) *Concoidal Fracture,* It represents a curved, shell like appearance and corresponds to glassy texture.

Crystalline and compact varieties always indicate an even, broken surface. Good building stones should always represent an even fracture. But for the construction of road surfaces and for concrete work, an uneven or concoidal fracture indicated by sharp and angular bulging or hollow faces is desirable to secure proper interlocking and solidity of work. The fracture of a stone should be bright, clean and sharp, with grains well cemented together. A dull, chalky and earthly appearance clearly indicates signs of early decay.

1.17.3 Resistance To Atmospheric Influence

Stones used on exposed faces of structure are continuously acted upon by atmospheric agencies such as wind, rain, frost, sun and also by vegetable life and organisms. These agencies set up disintegration and decay and subsequent separation of the particles of the stone. This action is known as *weathering*. The process of weathering, is associated with the transportation of the separated particles, usually from a higher region to the lower ones. So powerful are these agents that even the hardest varieties of stones are worn out and crumble to pieces and also to powder form under unfavourable conditions by the continuous process of weathering. A stone which has an ability to resist this destructive action is said to weather well.

Different atmospheric influences to which structural stone is subjected are :

(*i*) *Alternate heat and cold.* Heat induces tension stresses and causes the stone to contract The greater the difference in the temperatures, the earlier the stones wear out.

(*ii*) *Alternate wet and dry.* Wetness is caused by rain, dew, snow etc. and the sun dries it. Under this action, stones wear out on the face which, is most exposed to the direction of the prevailing wet wind.

(*iii*) *Wind.* A piercing wind forces the moisture or rain into its pores. It causes the moisture on the surface to evaporate and while doing so, induces cold. Wind carries grit, dust and decomposing organic matter in its grain and each of these have a distinct adverse effect on the surface of the stone.

(*iv*) *Chemical agencies in the atmosphere.* These are :

(a) *Rain.* Rain water acts both, (*i*) physically, and (*ii*) chemically, as an agent in the weathering and decay of stones. The former action is due to its erosive and transportation powers and the later mainly due to its solvent capacity, decomposition, oxidation and hydration. The strongest of the rocks crumble down to powder under their joint action.

Physical action of rain-water. The alternate soaking of stones and subsequent drying causes disintegration and loosening of their particles. The showers of rain strike on disintegrated exposed faces of stones, thereby causing the loosened particles to be washed away. This action of weathering is purely a mechanical one and if there are structural defects in a stone such as open joints, cracks etc, it reaches their interior section also. If a stone is porous, moisture will travel to its interior and the decaying process is rapid.

Chemical action of rain water. Rain, in its passage through the atmosphere to the surface of the earth, get contaminated with gaseous impurities such as oxygen, nitrogen and carbon dioxide. Alkaline materials like lime, magnesia, soda, potash and iron are readily dissolved and carried away in solution by water containing carbon dioxide.

When rain passes through the acid laden atmosphere of industrial localities, it gets charged with H_2S, HCl and SO_2 present in the atmosphere; SO_2 which is particularly present in industrial towns, where coal is burnt, form a weak acid, *i.e.*, sulphurous acid which reacts on calcium carbonate in stone and forms a thick black deposit, commonly called "soot", which is $CaSO_4$. Even calcareous sand stones, immediately behind this black skin or blisters of soot, crumble on account of the crystallisation of $CaSO_4$ in the pores and the crust of the soot flakes off, exposing a fresh surface of the stone to the action of the acid. The carbonate of lime, which is dissolved in rainwater, also crystallises out from the solution.

In addition to the decomposition effects and solvent action of aciduated rain water, the chemical reactions which disintegrate stones are mainly oxidation and hydration. This is due to the oxygen and water entering into chemical combination with the rock minerals. Iron is present in almost all rocks in one form or the other. Rocks containing iron compounds in the pentoxide, sulphide or carbonate form are readily oxidised and hydrated. Thus petroxide *i.e.*, FeO is converted into ferric oxide, *i.e.*, Fe_2O_3 which combines with water to form $Fe_2O_3.nH_2O$. This chemical combination is accompanied by an increase in volume and results in a physical change manifested by the liberation of the neighbouring minerals composing the rock. Similarly iron sulphide readily oxidises to lemonite and liberates sulphur, which in its turn unites with oxygen and water to form sulphuric acid and finally to sulphates. Similarly iron from iron carbonates or sinderite is also oxidised to lemonite.

(*b*) *Frost.* During frosty weather, the water absorbed by the stones freezes in the pores and at freezing point it increases in volume by about 10 percent and exerts considerable bursting pressure. Porous stones break up and disintegrate very easily by this action. The action of frost is naturally limited to very cold countries.

(*c*) *Movement of chemicals.* If limestone and sandstone are laid close to each other in a structure, the chemicals formed in the lime by the reaction of the atmospheric gases enter the sandstone and causes disintegration in it.

(*d*) *Vegetable life and organisms.* The decaying of stones due to these agencies occur to a much less extent in buildings when the lower forms of vegetable life such as mosses grow on the surfaces of stones and in the open joints or cracks in buildings. The roots in the joints of stones cause mechanical and chemical adverse effects. They secrete organic acids, which bring about chemical decomposition of rock minerals. The vegetable matter also attracts moisture and keeps the stones under damp condition, which is favourable to augment decay.

Living organisms play their own part in promoting disintegration. A certain category of worms bore holes in stones, thereby weakening them. These organisms also secrete organic acids which have a corrosive action on stone minerals.

Retardation of decay : In order to retard the decaying action of the weathering agencies, the following points should be kept in mind :

(*i*) For important buildings, only compact silicious stones should be used for external surfaces. These stones must have a dense crystalline texture. The use of sandstones cemented with silicious binding material should be made. Limestones or calcareous sandstones with open texture, should not be used for external surfaces in industrial towns.

(*ii*) The freshly quarried stones should be seasoned for a sufficient period by exposing them before use. This will reduce the quantity of quarry sap which accelerates decaying action.

(*iii*) Large size stones should be used as far as possible to minimise the number of joints which are signs of weakness and through which ingress of water takes place.

(*iv*) Use of well dressed, finished and polished stones should be preferred to rough stones.

(*v*) For ordinary buildings, external renderings such as pointing or plastering should be given at the time of construction. Cement plaster should be applied, using pure silica sand.

(*vi*) Stones should be placed on their natural beds.

(*vii*) All the joints in the stone masonry should be completely filled in. No hollow or cavity should be left. The aim should be to have a sound and solid structure.

(*viii*) The exposed stone surface should be kept as clean as possible. This can be achieved by washing it with water at regular intervals.

1.18 STRENGTH AGAINST CRUSHING

Building stones are usually subjected to direct compressive load when used in ordinary buildings and as such their strength against crushing is to be determined. However when subjected to transverse loads as in the case of beams, lintels, corbels etc, the strength of stone is also considered from the point of view of bending stresses.

The compressive stresses on stone are never sufficient to cause .any danger of crushing in the case of ordinary buildings. In a three-storeyed residential building having 400 mm thick walls, the maximum compressive stress works out to be about 0.435 N/mm^2 (4.35 kg/cm^2). per square centimeter which is a very safe figure even for stones of the weakest varieties. Stones are however, subjected to high compressive stresses in case of dams, high retaining walls and arches of major bridges. While designing structures in stone, the compressive stresses should not exceed one tenth of the crushing strength.

The texture and specific gravity of a stone greatly affect its crushing strength; the more compact grained and heavier a stone, the stronger it is. A dense and compact stone has also very few or no pores and thus does not absorb water. Stone of igneous origin is stronger than that of sedimentary formation. A stone having a crystalline texture is superior to a non-crystalline one and the finer the crystalline texture, the stronger it is. If the texture of stone is even and its specific gravity is not less than 2.7, the stone can be considered to be quite safe against crushing, when used even for fairly heavy residential buildings.

1.19 APPEARANCE AND COLOUR

These properties mainly indicate the fitness of a stone to produce the desired architectural effect on the face work of a structure. Better architectural effects can be produced by using stones of suitable colours to match with the surroundings for aesthetical reasons. Appearance depends on the ease with which a stone can be dressed, rubbed or polished, qualities for which even a slightly softer stone is selected for bringing out an architectural effect. Deep colours in sedimentary rocks are due to the presence of oxides of iron in the cementing material. Such stones should not be used because they are liable to be disfigured from unsighty rusts and stains caused by oxidation of iron upon exposure to the atmospheric influences.

For high class buildings great importance is attached to the colour of the stone for the face work. Good weathering stone should also be free from flaws, holes and mottled colour. For face work of important buildings, harder varieties of crystalline stones having a dense texture are preferred because they take a finer polish. Granites, marbles, slates and compact varieties of limestones can be polished well. Polished stones are especially used for pillars, balusters, cornice, pedestals, paving stones etc.

1.20 FACILITY FOR WORKING

Stones have to be shaped by dressing before they are set in a wall. For facility of dressing, a stone must be comparatively soft, yet durable, compact grained and homogeneous in texture, rather than crystalline and must be free from veins and planes of cleavage. The soft stones can be dressed more economically, while hard stones are costly and more difficult to dress. For purposes of carving, close-grained stones with distinct beds such as a large variety of beautiful sandstones are more suitable, while hard stones can easily be dressed in the shape of handy blocks for rubble masonry. It is easier if the stones are worked upon quickly, at the site of quarry after they have been quarried.

1.21 FIRE RESISTANCE

Fire brings about rapid destruction of stone structures by disintegration. The disintegration of stone can be caused due to unequal coefficients of expansion of different rock forming minerals. Traps and basalts resist fire action very well but crystalline varieties of igneous and metamorphic rocks *e.g.* granites disintegrate quickly when subjected to high temperature under fire. Free quartz which is present in the harder varieties of igneous rocks like granites etc. is a dangerous substance, when under fire, because even at a temperature of 600°C quartz expands suddenly and splinters freely causing destruction of the stone. Even timber is able to withstand a much higher temperature. Limestones remain unaffected upto about 800°C, but thereafter they quickly decompose into CaO and CO_2. When extinguishing fire, water applied to the burning stone further causes slaking of CaO. The latter then crumbles to powder and also increases in bulk. Sandstones are very good in resisting fire because they have silicates as binding material which are not easily affected by fire. Claystones are also good fire-resistant but they are not used as building stones because of their poor strength and durability.

1.22 NATURAL BED

The original bed or plane on which a sedimentary stone is formed is known as the natural bed of stone. In sedimentary rocks, the natural bed can be easily observed as it lies along the planes of stratification. For igneous rocks, natural bed is of little importance and is also difficult to determine.

Stone should be placed in a building component with its natural bed at right angles to the direction of load or pressure, which the member is called upon to bear. This is technically known as placing the stone on its natural bed. Placed in this position, the stone will not easily split under load nor get affected by moisture and frost between the layers. In arches, the laminations should be parallel to the central line of the voussoirs and at right angles to the face of the arch. In the case of string courses, cornices, etc. the laminations should be placed vertically.

1.23 COST

Stone is very heavy and its transport to far off places is costly. Moreover, there is also the cost of quarrying and dressing it to suitable sizes and shapes. In plains, use of stones is uneconomical due to its non-availablity and exhorbitant transport cost. Hence it finds favour in hilly tracts.

QUARRYING OF STONES

Quarrying is the process by which stones are taken out from their natural rock beds for use in general building works. The exposed part of such a natural rock bed is termed as a quarry. The operations involved in quarrying are open to the sky. The difference between quarrying and mining is that the latter is essentially a deep underground operation and the work is not carried out in an exposed condition. Quarrying in preference to mining is often recommended, since at a comparatively low cost much of the materials could be obtained easily and the stone always admits of inspection in-situ. The factor of cost in cutting and transporting stones from the quarry is the chief consideration in the selection of a quarry site.

1.24 QUARRYING SITES

Quarrying is usually started from a side and goes deep, it is because of the difficulty experienced in the lifting and hauling of the stones loosened and broken up from the bottom of a quarry. An inclined track is made for the haulage of material by trucks running on a light railway track. For a large scale operation, it is important to arrange a proper layout to ensure economic and safe working of the quarry. Such a layout should show the sites for quarrying, dressing and stacking of stones. The disposal of spoil heaps becomes an important problem for consideration in major quarries.

Stones to be quarried are usually found covered with a thick covering of soil or small loose stones, which have to be removed before the stones can be extracted. In order to remove the loose earth, sometimes small charges of explosives have to be resorted to, to break up the covering layer and enable the spoil to be easily removed. Often the work on a quarry commences on an open rock face and the stone is removed by working on the natural joints, fissures or lines of cleavage that can be easily located.

The location of a quarry depends on the following factors :

(*i*) Availability of the desired building stone at or near the surface of ground.

(*ii*) Nearness of the quarry to the existing lines of communication such as roads, railways and waterways.

(*iii*) The possibility of arranging an adequate drainage for the quarry pit at a very low cost.

(*iv*) The haulage facilities of the quarried materials to outside of the quarry for further conveyance.

(*v*) Availability of suitable sites for the location of power station, crushers, stone dressing and stone cutting machinery etc.

(*vi*) In the case of quarry where blasting has to be resorted to, the quarry should be sufficiently distant from the site of building works and other permanent structures.

1.25 METHODS OF QUARRYING

Depending on the structure and nature of the rock and the purpose for which the stone is intended to be used, there are principally two methods of quarrying in use:

(*i*) Quarrying by hand tools such as crowbars, jumpers, pick-axes, feathers and wedges, channelers etc.

(*ii*) Quarrying by blasting with the aid of explosives.

Some tools used for quarrying and blasting are shown in Fig 1.1.

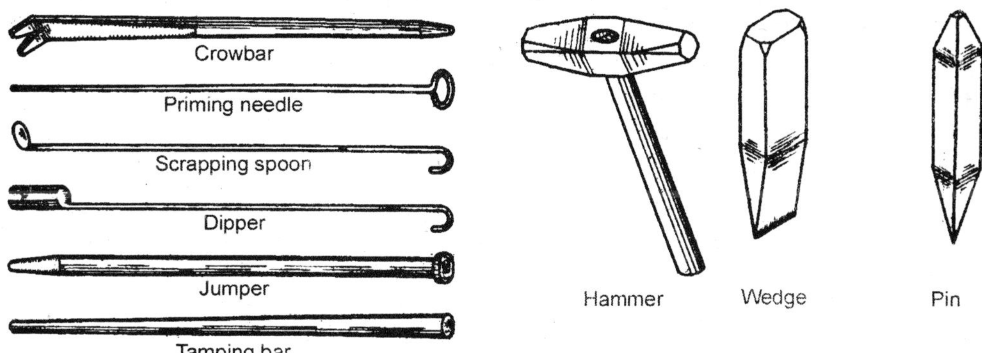

Fig. 1.1. Quarrying tools

In either case, quarrying may be carried out by manual labour or with the aid of machinery or by both. Sometimes heating is resorted to for splitting a rock, particularly if large blocks are not required. Fuel is collected on the exposed surface of rock and a fire is burnt steadily for several hours. The rock gets heated, expands and separates out from the cool mass with a dull noise. The portion, so detached, is removed with steel wedges and hammers and crowbars. Only stone which is to be crushed into small size aggregate, road metal or railway ballast may be quarried by heating.

In stratified rocks, there are joints or fissures at which the stone has little or no natural adhesion and therefore, can be easily split up into blocks. If the rocks is not very hard and is distinctly stratified and fissured, the stone can be easily removed with pick-axes, crowbars, quarrying hammers and chisels. If the blocks between the natural joints and fissures are not very big, it is possible to loosen and separate them by driving steel wedges into the fissures under the strokes of a sledge hammer or mask hammer and applying leverage by long steel crow- bars. When the blocks are very large or when natural fissures do not exist, holes are first bored or grooves cut in the upper surface of a bed at regular short intervals. The holes may be about 12 mm in diameter, 100 to 150 mm apart and 150 to 200 mm deep. Blunt wedges of steel are then inserted into the holes or conical pins of steel are driven with a hammer, until a block is removed from the compact mass of rock. The splitting of rock is sometimes accomplished by wooden plugs driven in the holes and soaked with water, as the plugs swell, the rock gets split. Instead of conical pins, plugs and feathers are sometimes used, as shown in Fig. 1.2. The plug is a conical wedge and the feathers are flat wedges with upper ends slightly bent. The plug, is placed between two feathers and the three are slipped into a cylinderical hole. If the plug is now driven in with hammer blows, it exerts a great force. If these plugs and feathers are placed a few mm apart in a row of holes and all driven at the same time, the stone will be cracked along the line of the holes even if it is comparatively thick. When the stone is hard, a pneumatic drill is used to make holes for the plug and feathers.

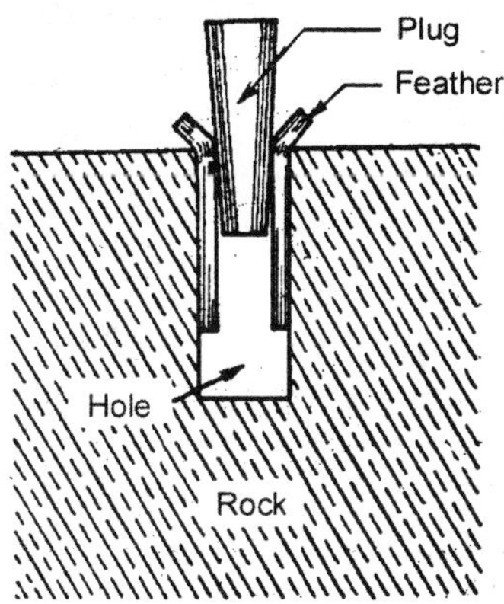

Fig. 1.2. Plug wedging

Igneous rocks are very hard and compact and are quarried in blocks by either cutting grooves or boring holes at regular intervals on the surface of the bed. Blunt wedges in grooves or conical pins in holes are inserted and hammered hard to cause the block to split off from the layers. Sometimes splitting is caused by heating, as described above.

1.26 BLASTING

Metamorphic rocks are highly compact and solid and are difficult to be quarried with hand tools. These are, therefore, usually quarried by blasting.

Blasting is the operation of disintegrating or quarrying rock with the help of an explosive or with an electric current. It must be remembered that the object of blasting is to shake up and break up the mass of the rock, so as to render it easy to work and handle by tools; and not to blow the material violently out and split all the good stone into useless small pieces. Blasting is also, sometimes, resorted to convert the larger blocks into smaller, convenient and handy sizes. This process is known as secondary blasting.

1.26.1 Different Operations in Blasting

The operations which constitute the entire process of blasting are :

(*i*) Boring the holes

(*ii*) Charging with explosives

(*iii*) Tamping and

(*iv*) Firing.

1.26.1.1 Boring the holes

The holes are made with the help of a jumper which is an iron bar of variable length according to the depth required, for the blast. Its two ends are tipped with steel, formed like knife edges, their length across depending upon the diameter of the hole for which they are to be used.

In using the jumper, one man sits down on the rock in a convenient posture, and holding the lower part loosely, he guides it that it may strike fair in the hole. Another man stands upright, and raising the jumper, perhaps 300 mm. above the surface of the rock, he drives it forcibly home. At intervals, little water is poured in, by which the rock is to a certain extent softened, and the dust formed by the jumper is made of a pasty consistency, and thus more easily removed. For the purpose of doing this, a metal spoon or scraper is used consisting of an iron rod having one of its ends beaten out and curved so as to form a groove, like a hollow cylinder. The under part of this is closed by a semicircular disc on which the mud, filling the groove, rests. The hole is then cleaned and dried by inserting and rotating a rag tied to the end of a small iron rod. The size of the hole usually varies from 20 to 40 mm diameter and 1.25 tc 2.50 metres deep.

When large quantities of stones are to be blasted, drilling of the holes is done with machines and the cleaning operation carried out by a blast of compressed air.

Great care should be exercised in locating the exact position of holes as the efficiency of blasting will depend upon the judicious selection of the holes. Position of holes for different conditions of rock blasting is shown in Fig. 1.3.

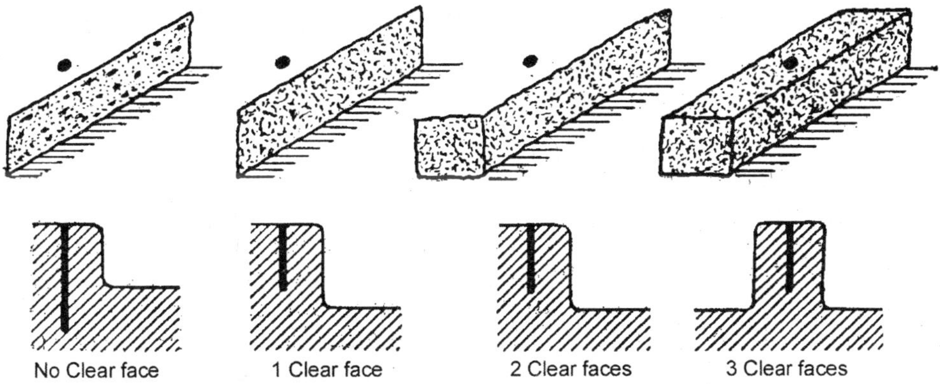

Fig. 1.3. Rock faces with location of bore holes

1.26.1.2 Charging with explosives

After the hole has been sunk to the proper depth, the charge is deposited in requisite amounts. The various types of explosives used are :

Blasting powder or gun powder. It is a chemical compound of 60-70 percent by weight of salt petre (KNO_3); 15 to 25 percent of sulphur and 10 to 20 percent of charcoal. This is a weak explosive and is useful for quarrying large blocks in soft rocks.

Dynamite contains about 75 percent of nitroglycerine mixed in about 25 percent of fine sand to form a thick paste. Due to its semiliquid form, it can be easily tamped into the hole. This is six times stronger than gun powder and can be used in wet condition and under water. This also does not require careful tamping and is useful for heavy work in tunneling and mining. It is available in catridge form also, which is more suited under wet conditions.

Gun-cotton or Nitrocellulose consists of nitric acid, HNO_3, and saturated cotton. This is more powerful than even dynamite and great care has to be exercised in its preservation and use, especially in moist conditions.

Blasting gelatine contains 95 percent of nitroglycerine, $CH_2OHCHOHCH_2ONO_2$, and 5 percent of guncotton.

Cordite is made of nitroglycerine and is a very powerful smokeless explosive.

Gelignite. Consists of 63 percent of blasting gelatine and 37 percent of absorbing powder and is a very strong explosive. It is used in water.

Liquid oxygen is used these days to a great extent as an explosive.

Blasting fuse is a device consisting of a length of slow burning composition, which delays the firing of an explosive charge sufficiently to allow the person firing it to get out of danger.

Detonator is a substance which initiates explosion, generally used to set off high explosives such as gun cotton, gelignite and dynamite. The ignition of the charge in the detonator is often carried out electrically. These are extremely sensitive to explosion and decompose from very slight excitation. They usually comprise of fulminates of silver and mercury, chlorides and iodides of nitrogen etc.

Precautions to be observed while handling explosives

These are covered under the Explosives Rules, published by the Government of India. The general precautions to be observed are :

(1) The floor of any place or any carriage or vessel on which any explosive is to be laid should be carefully examined to check that there is nothing likely to endanger the safety of the consignment. Further it should be thoroughly cleaned of all grit and foreign matter and swept before and after use.

(2) The packages or casks containing the explosives should not be thrown or dropped down or rolled or pulled along the ground or floor but should be passed from hand to hand and carefully deposited and stored.

(3) Detonators, fuses, and percussion caps should be kept in separate packages and stored in a separate building, away from other explosives.

(4) Where a package is to be slung, due precautions should be taken to sling it in such a manner so as to prevent the possibility of a fall.

(5) No person should smoke and no fires, lights or articles of an inflammable nature or liable to spontaneous ignition or to cause or communicate fire or explosion, such as sulphuric acid, petroleum, carbide of calcium or compressed gases should be allowed at any time in proximity to a place where the explosive is handled or stored.

(6) No person on, in or near any place, where explosives are handled should : *(a)* wear boots or shoes with iron nails or shod or strengthened with iron unless such boots or shoes are covered with leather, India rubber, felt etc.

(7) No iron or steel tools should be used in handling explosives. Only wooden, copper or brass levers, wedges and mallets should be made use of.

(8) Explosives should be stored in a well designed building, far away from roads, dwelling houses, factories, public places, railway lines etc. The minimum clear distance specified is about 50 metres.

(9) The room or magazine, where the explosives are stored, should have an efficient lightning conductor, which should be tested at least once in every year.

(10) The electrical wires should, as far as possible, be concealed in the walls. In case there is no electricity, use of torch should be made.

(11) Before repairs are done to any room or magazine or part there of, the same should, as far as practicable be cleaned by the removal of all explosives.

(12) No unauthorised person should be allowed to come near the premises where explosives are stored.

(13) No child under 16 years of age and no person who is in a state of intoxication should be employed on the loading, unloading, transport and handling of explosives.

(14) A broom or brush should be kept for cleaning the magazine on each occasion of opening, receipt, delivery and inspection.

(15) In the case of any explosive which is liable to be dangerously affected by water, due precautions should at all times be taken to prevent water from coming into contact with such explosives.

LINES OF LEAST RESISTANCE

The effect produced by blasting of an explosive depends upon the resistance that the rock offers to the escape of gases produced during the explosion. Generally these gases escape into the atmosphere through the shortest route and across the portion of the rock which offers them the least resistance.

The line of least resistance is defined as that line by which the explosion of the powder will find the least opposition to its vent in the air. This may be the shortest distance from the centre of explosion to the nearest face of the rock, as shown in Fig.1. 4. In case the rock contains fissures or other weak spots, then the line of least resistance is the least distance from the centre of disturbance to the nearest fissure, provided it is less than the distance to the face of the rock. In case tamping is not done effectively, then the line of least resistance is simply the depth of the hole. To avoid this, the depth of hole is normally kept as 1-5 times the line of least resistance.

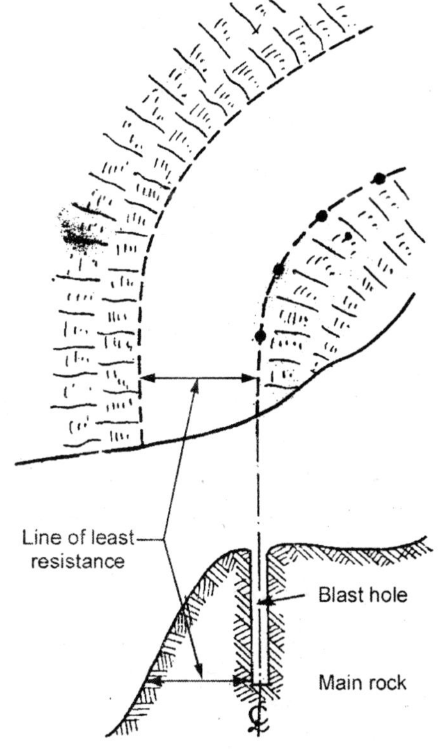

Fig. 1.4. Line of least resistance

QUANTITY OF BLASTING CHARGE

The blasting charge or the quantity of explosive material depends upon the following factors:

(*i*) power of the explosive
(*ii*) nature or toughness of the rock
(*iii*) mass of the rock to be removed with each blast
(*iv*) depth of hole
(*v*) the number of exposed faces of the rock and
(*vi*) the line of least resistance.

As a rough guide for estimating the quantity of blasting charge, the following relation may be adopted.

Blasting powder = 0.15 (Line of least resistance in metres)^2 in kg. For example, if the line of least resistance is one metre, the quantity of blasting charge required is 0-15 kg or 150 g. This applies to blasting powder only. Precise quantity should be determined by a few actual experiments on the particular rock to be worked and the explosive to be used.

1.26.1.3 Tamping

Tamping is the process of filling of a hole containing explosive charge with clay or other stemming material to confine the force of explosion and prevent explosion along the hole itself. The tamping or filling in again, is done so that a due resistance may be offered to the explosive force of the powder and that this force may be directed through the line of least resistance.

The best material for tamping is burnt clay. When small fragments of the rock in which the blasts are being prepared, are used the process is never free from the danger of striking fire. An hill earth forms an excellent tamping material; failing this clay or wet sand can be used. Dry sand is the worst, as it offers least resistance to the force of the powder.

Tamping is done by means of a tamping bar, which is a heavy brass rod of a diameter a little less than that of the hole in which it is to be used, and slightly tapering at the extremities. At each end, an open groove is formed along the side, to admit the bar being used with facility, while the priming needle is placed in the hole. Brass is selected for tamping bars, to avoid the risk of accidents from the charge being ignited by sparks which would be struck if iron or steel were to be employed. Thus iron or steel should never be used for making tamping bars. In using the bar, the tamping material is put into the hole, in small quantities, sufficient to fill from 20 to 40 mm at a time, and each is well and firmly rammed home by successive blows. The time required varies with the material used.

Through the tamping, it is necessary that a communication with the charge, for the purpose of priming should be made. This is done by means of the priming needle, which is a thin metal rod having a loop handle at one extremity, and pointed at the other. Copper is a bad material for priming on account of its softness, those of iron about 1-5 mm in diameter answer better, and to guard against accident by sparks being struck by them, they may be tipped with brass. The needle is placed in the hole and the tamping rammed round it. It is necessary to grease it well before the tamping is commenced, and to turn it frequently during the process, since, if not done, the friction ultimately becomes so great, as to make it difficult to withdraw.

The priming needle is removed after the tamping is over. The space occupied by this needle is filled with fine gunpowder and on the top 'Brick fords' fuse is introduced above the top of the hole.

1.26.1.4 Firing

The fuse, as mentioned above, is lighted by a match. Its length is such that the person igniting the fuse can retreat to a safe place before the blasting charge explodes. Brick fords fuse usually burns at the rate of about one metre per minute. Two or more holes may be coupled to a common fuse for firing simultaneously. In the case of multiple blasting one or more groups of holes could be fired each time. A tremendous explosion with a sound of a muffled character occurs when the powder gets ignited and the gases formed due to the explosion try to escape along the line of least resistance. In case proper tamping is not done, explosion takes place along the line of the hole without blowing off the rock. It must, therefore, be ensured that the tamping is very hard so that the gases find it easier to blow out the rock and escape through the openings.

Electrically operated igniting devices, using detonators are employed these days to create the spark needed for explosion. These are very safe, convenient and number of explosions either simultaneously or with a time lag can be controlled from one place.

COMMON BUILDING STONES

1.27 GRANITE

Formation and constituents. This is a crystalline igneous rock consisting essentially of quartz and felspars (orthoclase, microcline and sodic plagioclase), mica and some accessory minerals such as hornblende, pyroxene, amphiboles etc. It is an acid type rock containing 60 to 80 percent of silica, which acts as a cementing material. Silica is more in the form of felspar than quartz.

Granites are plutonic or deep seated rocks with large over burdens. They develop crystalline structure because of the slow cooling of the molten magma at great depths. Granites get exposed to the earths' crust due to the weathering away of the overlying strata.

Properties. Granties have a variety of shades of colour, varying from while to light grey, red, pink, dark blue, green and black depending upon component minerals. Generally they present a some what mottled appearance and the colour imparted to them depends upon the colour of their felspar. The texture of granites varies from coarse grained to fine grained but the individual crystals of the different minerals are large enough to be recognised by a naked eye. Their specific gravity ranges from 2.6 to 2.7. Granites have high compressive strength varying from 77 to 130 N/mm^2 (770 to 1300 kg/sq. cm) IS code 3316 : 1974 (R 2003) also specified a minimum compressive strength of 100 N/mm^2 (1000 Kg/cm^2) specific gravity 2.6 and maximum water absorption of 0.5 per cent. The durability of granites depends upon the relative proportion of quartz, felspar and mica present in them. Quartz is unaffected by the weathering agencies but felspar decomposes and falls to powder rapidly, especially the potash or the orthoclase variety. Mica is also subject to rapid disintegration and is a source of weakness. Therefore, the most durable form of granite is one which contains a large proportion of quartz, oligoclase felspar and very little of mica.

Uses. On account of the difficulty in working it to shapes required, and high cost of dressing, use of granites is restricted in building construction except at places where no other building material is available. It is suitable for works of importance such as bridge abutments, piers, harbour and dock works, light houses etc. where weight and durability are essential. It is not suitable for carving.

Fine grained granties takes and preserves high polish and is suitable for commercial and monumental works. Granites having uniformity of colour and texture, works to a fine smooth polished surface and looks very beautiful in massive architecture. Such stones are used for ornamental works, polished columns, pedestals, statues, slabs for lintels, etc. Physical properties of polished granite, as per I.S. 14223 (Part I): 1995 (Reaffirmed 2001) are given in Table 1.1.

Table 1.1: Physical Properties of Granite

Characteristic	*Requirements*	
	Pink granite	*Multi-coloured and grey granites*
Moisture content (percent)	0.15, Max	0.15, Max
Dry Density (Min)	2.58 to 2.63	2.60 to 2.68
Specific gravity (Min)	2.75	2.75
Water absorption	0.50, Max	0.50, Max
Porosity (percent)	1.02 to 2.50	1 to 2
Compressive strength (kg/cm^2)	1.000-1.500	1.300-2.200
Teasile strength (Min)	90 kg/cm^2	90 kg/cm^2
Shear strength (kg/cm^2)	280-425	300-540
Hardness (mohs)	6 to 7	6 to 7
Hardness (Schmidt) No.	80 to 100	85 to 110
Hardness (shore) No.	50 to 60	46 to 61
Ultrasonic pulse velocity	5.000 Min	5000 Min
Resistance to wear	Not greater than 2mm(avg)	Not greater than 2mm (avg)

The quarry refuse can be used for road paving and the crushed stone as road or railway ballast.

Availability. These are available in Assam, Maharashtra, West Bengal, Bihar, Andhra Pradesh, Himachal Pradesh, Kashmir, Tamil Nadu, Odisha, Mysore, Gujarat, Punjab, Kerala and Rajasthan.

1.28 TRAP AND BASALT

Both these are amorphous varieties of igneous rocks, being basic in character. They are formed on the surface of the earth's crust by the sudden cooling of the basic lava issuing out from the interior through the fissures. They contain 40 to 60 percent of silica. Trap contains felspar and hornblends while basalt contains felspar, pyroxene with or without intersitial glass and having individual crystals too fine to be easily distinguished. When olivine is present, the rock is termed as olivine basalt.

Properties. These are usually of dark colour varying from bluish or greenish grey to even light red and yellow. These stones are very hard, tough and compact. Their specific gravity is 3.0. Their compressive strength varies from 1530 to 1890 kg/sq cm. Due to their amorphous texture and hardness, they are seldom quarried in blocks.

Uses. These rocks are expensive to work to regular sizes. They are used in foundation of structures usually, but may be used for superstructure also. Being heavier, stronger and durable, they are most suitable as road metal, paving stones and flag stones. Where they are abundantly available, they have been used in rubble masonry work also. These stones are not suitable for carved work and mouldings.

Availability. These stones are found in Maharashtra,West Bengal and Madhya Pradesh.

1.29 SANDSTONES

Sandstones are stratified (often false bedded) sedimentary rocks. They consist essentially of rounded grains of quartz compacted and cemented together by siliceous, calcareous, argillaceous and ferruginous cementing materials. The cementing materials are carried by the percolating waters charged with these materials and deposited in the voids of the sand-grains. After evaporation, the mass becomes compact. According to the nature of the cementing material, the varieties of sandstone may be distinguished as siliceous sandstones, calcareous sandstones, argillaceous sandstones, ferruginous sandstones. Sandstones sometimes contain mica also.

Properties. The colour of sandstones depends upon the cementing material. It varies from red to brown, earthy to buff, white and sometimes yellow or greenish. The presence of iron, even in small traces is responsible in imparting colour. The texture of sandstones is also fine to coarse grained depending upon the size of the grains of which they are made. The grains may be angular, sharp and of varying sizes and if the interspaces are completely filled by silica, a very solid, compact, and hard rock is formed, which is termed quartzite. If the interspaces are not completely filled, open textured and porous sandstones are formed.

Silicious sandstones are very durable. If the cementing material is lime, clay or iron, the decay of the sandstones depends upon the quality of the cementing material. Sandstone containing carbonate of lime as a cementing material does not weather well. A porous sandstone, sandstone containing clay and sandstone deposited near the sea-coast and impregnated with iron decays more quickly and are of poorer quality. The specific gravity of sandstones varies from 2.65 to 2.95 and their compressive strength is 65 N/mm^2 (650kg/cm^2)

The water absorption, as per I.S. 3622 : 1977 should not be more than 2.5 per cent by mass, transverse strength not less than 7 N/mm^2 and resistance to wear not greater than 2 mm on the average.

Sandstones can resist heat well, except for those containing mica, clay and pyrites. They are easier to work and dress in general, unless they are of the silicious and more harder varieties.

Sandstones can be easily quarried and are available in large blocks. Freshly quarried stones contain a certain amount of quarry sap and have to be seasoned completely by exposure, before they are put to use. The drying of quarry sap should be affected from the interior of the stone.

Uses. The better, harder sorts are most excellent building stones, sufficiently strong to last, and yet soft enough to work. Such rocks are more suitable for ashlar masonry, other jobs of architectural importance. Care must be taken to lay these stones along their natural bedding planes i.e. with their stratification normal to the direction of the load ; else they will flake of in thin layers and weather very rapidly.

For massive works such as monumental buildings, dams, bridges, docks etc., the stones should be used in thick blocks showing faint traces of stratification. Use of sandstone should be made with caution for such works if the cementing material is weak or contains lime. For ornamental carvings, fine grained sandstones capable of being worked with a chisel are tc be preferred. Sandstones are not suitable as a road metal or railway ballast as their resistance to abrasion is not well.

Availability. Sandstones are available in Andaman Islands, West-Bengal, Bihar, Andhra Pradesh, Himachal Pradesh, Kashmir, Tamil Nadu, Madhya Pradesh, U.P., Rajasthan, Punjab, and Gujarat.

1.30 LIMESTONES

Limestones are also sedimentary rocks and consist chiefly of grains of calcium carbonate (calcite) cemented together by a matrix of the same material with small proportions of magnesium carbonate, and of siliceous, argillaceous, ferruginous, phosphatic and bituminous varieties. Limestones are classified as follows, based on their texture :

(*i*) *Kankar* : These stones are found either in a layer, from a few mm to 450 mm thick which are quarried in blocks or in separate nodules disposed in a horizontal layer at various depths underground in the alluvial deposits of older formations. The nodules vary in sizes from 10 to 100 mm. These are composed of nearly pure compact carbonate of lime, but their upper surface consists of mixture of calcium carbonate, sand and clay. Kankar is also formed in large quantities in the stream beds, where it is deposited by water holding it in solution. Such stones contain about 70% calcium carbonate and 30% clay and sand.

Kankar limestones are grey coloured, open textured, and like a sponge in appearance. They do not weather well on exposure to atmosphere. Kankar containing less clayey material are hard and tough.

These are chiefly used as a road metal or railway ballast and for burning of lime and manufacture of cement. Argillaceous stones, where found in a solid layer, make a good durable stone for rough work, as in foundations.

(*ii*) *Granular limestones*: These in appearance are like sandstones, but composed of grains of limestones cemented together by some natural cement. They have got open and porous texture. These are available in variety of shades of colours ranging from light white, pink, blue, light brown to even black. These are often quite soft when first taken out of rock, but hardens when exposed to air. Granular limestones are fairly strong and durable.

These, on account of the ease with which they can be worked to any shape, and also that the good sorts are very durable though not strong to withstand great pressure or any wear by friction, are most valuable stones for ordinary architectural purposes.

According to their composition, limestones are classified as :

(i) *Argillaceous limestones* or clayey limestones. These are usually found as Kankar.

(ii) *Silicious limestones.* These are similar to argillaceous lime-stones except that they contain silica instead of clay and as such they are better suited for building purposes. These stones are more compact, weather well and have granular texture.

(iii) *Magnesium limestones or dolomites.* Carbonate of calcium and magnesium are present almost in equal proportions in this variety of limestone. These stones are more compact and heavy. The presence of magnesium make such stones more resistant to fire and these stones are superior to ordinary limestone.

Limestones have a specific gravity of 2.0 to 2.75. They have a compressive strength of 55 N/mm^2 (550 kg/sq. cm). The transverse strength is specified as 7 N/mm^2(70 kg/cm^2). The water absorption should not exceed 0.15 percent by weight. On the whole the weathering quality of limestones is good, although this depends upon the condition in which they are used. These should not be used in exposed work in areas laiden with smoke and acid, such as industrial towns. Dolomites and silicious limestones are better resistant to weathering action.

The standard sizes of limestone slabs and tiles as prescribed by I.S. 1128: 1974, is given in Table 1.2

Table 1.2: Standard Sizes of Limestone Slabs and Tiles

Length (1)	*Breadth* (2)	*Thickness* (3)
150 to 600 mm in stages of 50 mm	150 to 600 mm in stages of 50 mm	15 to 95 mm in stages of 10 mm
600 to 1000 mm in stages of 100 mm	300 to 1000 mm in stages of 100 mm	-do-
1000 to 1500 mm in stages of 100 mm	300 to 1000 mm in stages of 100 mm	25 to 95 mm in stages of 10 mm

Availability. These are found in Andaman Islands, West Bengal, Bihar, Andhra Pradesh, Himachal Pradesh, Madhya Pradesh, U.P., Rajasthan, Punjab and Gujarat.

1.31 LATERITES

Laterite is a sedimentary rock and consists of a mixture of hydrated ferric oxide, with hydroxide of aluminium in various proportions, frequently it also contains maganese dioxide, titanium dioxide and free silca. This is formed on the surface of rocks due to their alteration in tropical countries. The original minerals of the rock, namely alkalies, lime, magnesia and silica are removed in solution. Silica is removed very slowly and is, therefore, present even at the last stage of alteration. The formation of laterite takes place in a free movement of water at the top and a good drainage from below. Bacteria causes decomposition and disintegration of the original rock surface. The sub-soil water acts upon subsequently, thereby allowing the movement of iron oxide to the top and silica and other minerals to the bottom in the solution.

Properties. Laterites have a brownish red colour due to the presence of oxide of iron. Its texture is cellular and the structure is bedded. It is a soft rock when quarried and is cut into rectangular blocks. It hardens on exposure to some extent. It has a specific gravity of 2 and a very low compressive strength. IS 3620 : 1979 has specified a minimum compressive strength of 3.5 N/mm^2. (35 kg/cm^2)

Uses. Laterite is used for rough stone masonry work.

Laterite is also found in another variety, known as nodular laterite which on crushing yields very good road metal.

Availability. These stones are found in Maharashtra, Bihar, Andhra Pradesh, Tamil Nadu, Mysore, West-Bengal and Kerala.

1.32 SHALES

Shales are also stratified (sedimentary) rocks consisting of a mixture of compacted clays with calcareous, ferruginous or corbonaceous matter. They possess a finely laminated structure by virtue of which they are fissile and break into parallel sided fragments easily. Their lamination is parallel to the bedding planes. The texture of shales is plastic with grain size less than 0.01 mm. These are available in a variety of shades of colours *viz.* grey, greyish, white, reddish, pink, blue, brown and black.

1.33 MARBLES

Marble is a metamorphic rock. The term strictly applies to a granular crystalline limestone, but in a loose sense it includes any calcareous or other rock of similar hardness that can be polished for decorative purposes. These are formed by deposits originally but afterwards crystallised by intense heat and pressure. Limestones get converted into marbles when heated to redness without the escape of carbon dioxide and on subsequent cooling. Only the texture gels altered. No change in the chemical composition takes place. All the original minerals such as silicious, argillaceous, hornblende, mica, pyroxene etc. are present in the metamorphosed limestones.

Properties. Marble is available in variety of shades of colour namely, white, grey, black, pink, red, brown, yellow, blue, green and combination of these. This has a jointed structure and a crystalline texture. Marble is compact, close-grained and shining. It can be easily sawn and carved. Marble takes a fine polish, but it cannot be retained for long if exposed to weathering action. It has a specific gravity of 2.65. Minimum specified is 2.5 and compressive strength of 72N/mm^2 (720 kg./sq. cm) Although very hard yet it can be worked to the most delicate shapes and surfaces.

Uses. On account of its cost it is used only for high class ornamental and superior type of building work. It is used as blocks, slabs and tiles. It is suitable for floor finishes, dados, veneer work, steps, balusters, mosaic floor tiles etc. The quarry waste and impure variety are used for making lime.

Availability. U.P., Rajasthan, Madhya Pradesh, Andhra Pradesh Mysore and Gujarat.

1.34 SLATES

Slate is a metamorphic rock formed from fine grained deposits such as shale, mudstones and volcanic ashes. The metamorphic action is carried out to such an extent by long continued pressure that the original planes of stratification are completely obliterated and new planes of splitting called cleavage planes are developed in the rock. These are well defined and enable splitting of stone into very thin sheets, 5mm or more thick. The original mineral grains are flattened out. Sometimes new micaceous minerals are also formed. The cleavage planes develop at right angles to the force of meta-morphism and are at various angles to the planes of original bedding.

Properties. The colour of slates is highly variable from purple to greenish black, dark grey, greenish grey and purplish grey. It has a slaty structure *i.e.* the property of easy splitting along planes independent of the original bedding. It is compact and closely grained such that the individual minerals cannot be recognised by the naked eye.

Slate containing mica, lime and carbonaceous matter, does not weather well. It has a specific gravity of 2.89 and compressive strength varying from77 to 210 N/mm^2 (770 to 2100 kg/sq. cm.)

The physical properties of slate tiles, as per IS. 6250: 1981 is given in Table 1.3

Table 1.3: Physical Properties of Slate Tiles

Sl.No.	*Characteristic*	*Requirement*
(*i*)	Water absorption	(a) Maximum average – 2 percent by mass. (*b*) Variation should not exceed 20 percent between individual sample.
(*ii*)	Modulus of rupture	60 N/mm^2 (dry), Min 40 N/mm^2 (wet), Min
(*iii*)	Depth of softening	0.05 mm, Max
(*iv*)	Permeability	No water shall ooze from the bottom
(*v*)	Sulphuric acid immersion (see Note)	Shall show no sign of delamination along the edge or swelling. Softening, flaking of the surface and shall not exhibit gaseous evolution during immersion.
(*vi*)	Wetting and drying	Shall show no sign of delimination or splitting along the edge nor flaking of the surface.

Note: This requirement is related to the conditions of atmospheric pollution and the slate tiles be subjected to this requirement only if required by the purchaser.

Uses. Slate is not much used as a building stone, but is good enough for ordinary work. It is mostly used as a roofing and flooring material. When the cleavage is into fine thin layers, it makes the best roofing material, being fairly light and practically everlasting. Harder varieties of slate are used for dado work and steps of stairs.

Availability. U.P., Bihar, Rajasthan, Tamil Nadu, Mysore and Madhya Pradesh.

1.35 QUARTZITE

Silicious sandstones under the effect of metamorphic action yield quartzite. Quartz is the important and most predominant mineral. Felspar may or may not be present. Quartzite also contains various accessory minerals like zircon, garnet, apatite, iron ore etc.

Quartzite is available in white, milky or grey colours. It has stratified structure, compact beds and breaks up into irregular shapes. It is very dense and strong stone. Quartzite is difficult to dress and work.

Uses. Quartzite is used for rubble masonry, paving, and curb blocks and as an aggregate for concrete. Judged from the attrition tests, quartzite stands high in the list of building stones.

Availability. These are available in West Bengal, Andhra Pradesh, Himachal Pradesh, Madras, U.P., Mysore, Gujarat, Punjab and Rajasthan.

1.36 GNEISSES

These are metamorphic rocks which generally (but not necessarily) have the composition of granites. They are formed by the metamorphism of granites. The rocks are of coarse grains, characterised by a mineral banding, in which the light minerals (quartz and felspar are separated from dark ones (mica and hornblende). The layers of dark minerals are foliated, while the light bands are granulitic.

They are available in white to light grey, pink and greenish black colours and have a specific gravity of 2·69. Their compressive strength is 210 N/mm^2(2100 kg/sq. cm) ·. They are very durable stones, which contain lesser percentage of mica and are better building stones than granites.

Uses. These are used for paving slabs and rough stone masonry work. Since they have a banded structure which does not represent a clean appearance, they are not used in important structures where stone facing is desired. When crushed, gneiss stones provide a good aggregate for concrete and artificial stone. They are also used for stone pitching, revetments and aprons.

Availability. Gneisses occur in Maharashtra, West Bengal, Bihar, Andhra Pradesh, Tamil Nadu, Gujarat and Mysore.

1.37 SCHISTS

It is a name given to a group of metamorphic rocks which have a tendency to split on account of the presence of folia of flaky and elongated minerals, such as mica, talc and chlorite. These rocks are formed from original igneous or sedimentary rocks by the action of combined heat and pressure. They have a schistose (foliated) structure and can be easily split in layers. This is due to the large quantity of mica present.

Schists have variable shades of colours ranging from purple to greenish black, dark grey, light-grey and greenish pink. They have properties similar to gneisses.

Availability. These stones are available in Madhya Pradesh, U.P., Rajasthan and Gujarat.

1.38 GRAVEL, SHINGLE AND CONGLOMERATE

Gravel is a rounded water-worn pebble of any size. This is found in river beds and in alluvial tracts. It is commonly used as an aggregate for concrete and for road surfacing.

Shingle denotes larger water-worn pebbles available near the sea shore.

Conglomerate is formed when gravel and shingle pebbles are bound together by some kind of cementing material such as sand, clay, lime etc. These stones have a porous granular texture and are irregular in shape. These are not suitable for building purposes and are used for revetments and aprons.

DIMENSIONS OF NATURAL BUILDING STONES

IS 1127: 1970 (Reaffirmed 1993) has recommended the dimensions of natural building stones used for various types of stone masonry as given in Table 1.4.

Table 1.4 : Dimensions of Natural Building Stones

S.No.	Type of Masonry	Length mm	Breadth mm	Height mm
(1)	(2)	(3)	(4)	(5)
(*i*)	Stones for ashlar	597	297	297
		697	347	347
		797	397	397
(*ii*)	Stones for block in course	394	194	194
		494	244	244
(*iii*)	Stones for square rubble	90	90	90
		140	90, 140	90, 140
		190	90, 140, 190	90, 140, 190
		240	90, 140, 190,	90, 140, 190
		290	90, 140, 190, 240	90, 140, 190, 240
		390	90, 140, 190, 240, 290	90, 140, 190, 240, 290
		440	90, 140, 190, 240, 290	90, 140, 190, 240, 290
		490	90, 140, 190, 240, 290	90, 140, 190, 240, 290
		590	90, 140, 190, 240, 290	90, 140, 190, 240, 290
(*iv*)	Stones for random rubble	May be of any size and shape but not less than 150 mm in any direction		
(*v*)	Stones for Sills and lintels.	(*a*) 890, 990 1090, 1190 1290	90, 190, 290, 390, 490	90, 140, 190
		(*b*) 1390, 1490, 1590, 1690, 1790	190, 290, 390, 490, 590	140, 190, 240, 290
(*vi*)	Stones for arches, domes and circular moulded work	The dimesions depend on the particulars of the curve		
(*vii*)	Coping stones	190, 290, 390 490, 590, 690 790	200, 300, 400, 500, 600	100, 150, 200
(*viii*)	Kerb stones	390, 490, 590 690, 790	100, 200, 30	300, 400, 500

ARTIFICIAL OR CAST STONES

Where durable natural stone blocks are not available at reasonable cost, processes have been invented for manufacturing cast stones. One advantage with cast stones is that they can be moulded to most intricate forms and, in this respect, they are more economical than carved natural stones. Cast stones are made with good setting mortar or concrete, covered or impregnated with a preservative. These can be cast into one monolithic piece of any size and are suitable for face work since all grooves, rebates etc. can be easily cast. The strength of such stones can be increased by providing steel reinforcement.

The varieties of these stones are :

(*i*) *Simple Cement Concrete Blocks.* They are made at site for the construction of piers or cast in moulds for steps, window sills etc. Artificial paving slabs and stones composed of cement concrete and sometimes treated with sodium silicate solution also come under this class.

(*ii*) *Terrazzo.* It is an artificial stone made by mixing marble chips with cement and some pigment. It is either precast or laid in-situ and used for face of concrete or brick walls or for floors. Ornamental patterns are invariably incorporated when laid in-situ and after initial set, the surface is ground and polished well.

(*iii*) *Cement Concrete Subjected to some Hardening Process eg.* Victoria stone, Imperial stone, Indulated stone, etc.

For Victoria Stone, a concrete of four parts of crushed granite and one part of Portland Cement is allowed to set for 3 or more days into a hard block of the required shape. It is then immersed in silicates of soda for 7 to 8 weeks. This is chiefly used for carvings, window sills, copings, stairs etc.

(*iv*) *Chemical Stone or Ransonie's Patent Artificial Stone.* This is made by mixing dry sand with silicate of soda. The mixture, while in a plastic state, is poured into moulds and the castings on withdrawl are immersed in a solution of $CaCl_2$, $CaSiO_3$, which cements the particles of sand. This stone weighs 1920 kg/m^3 and has a compressive strength of 32N/mm^2 (320kg/cm^2). This can be dressed and carved like natural stone.

(*v*) *Reconstructed Stone.* This is made by taking the debris of limestone quarries, crushing them into grit, mixing with lime made from dolomite, heating in a closed retort upto 980°C to drive off CO_2, slaking the powdery residue of CaO and MgO, mixing with water and consolidating under great pressure into blocks. They are then dried and CO_2 is admitted until the carbonization of the hydrate of lime blocks is complete.

(*vi*) *Bituminous Stone.* Diorite and other granite stones are often impregnated with prepared or refined tar to form bituminous stones. They are used for a durable, noise, wear and dust resisting stone surfaces.

(*vii*) *Artificial Marble.* This can be either pre-cast or made in situ from sand and either Portland or gypsum cement. The pre-cast marble is made on glass-face plates with a thin and uniform coating of water-clear oil. The slab is removed after 3 days, treated with a liquid fluorate of magnesia solution on the fifth day, then washed and wrapped in paper for another day and again treated with the liquid. After 3 to 5 weeks, polishing is done by rubbing emery over the surface with a linen rag ball dipped in a mixture of lime water and silicate of potash and repeating without emery. This stone is used for external work.

For in-situ work the mix laid on canvas is applied to the surface and a brass trowel is worked over the canvas till the facing is 1-5 mm greater than the required thickness. The canvas is then peeled off from bottom upwards, the surface rubbed over and air holes filled with mix. Grinding is done by hand or machine. In the former, grinding is done with fine-grained pumice till the surface is smooth, then with fine-grained carborundum under running water, then rubbing with a polishing stone using a little water and finally rubbing with a ball of wool moistened with alum water.

DRESSING OF STONES

Dressing means preparation of the surface of stone to obtain plain edges or to obtain stones of required shape and size. Well dressed stones are required to have thin and fine mortar joints in masonry work. Their use provides good appearance with neat vertical and horizontal joints and proper bedding. Dressing is of special importance for preparing stones for ornamental constructions such as for voussoirs and skew backs for arches, for columns and copings to masonry pillars or walls etc.

Tools Required. The use of different types of tools is required depending upon the type of rock and dressing desired. The common tools used by masons are : Hammers which may be club, mash, sledge etc. types, jumpers, chisels, which may be pitching, point broad, claw etc. ; drawing equipment for marking such as trammel and scriber, set squares, sinking square, double sinking square, bevel, campasses etc, saws of frame, hand, fillet, cross, whip etc. types.

1.39 TYPES OF MASONRY

The following types of masonry construction are adopted. Some types are shown in Fig. 1.5

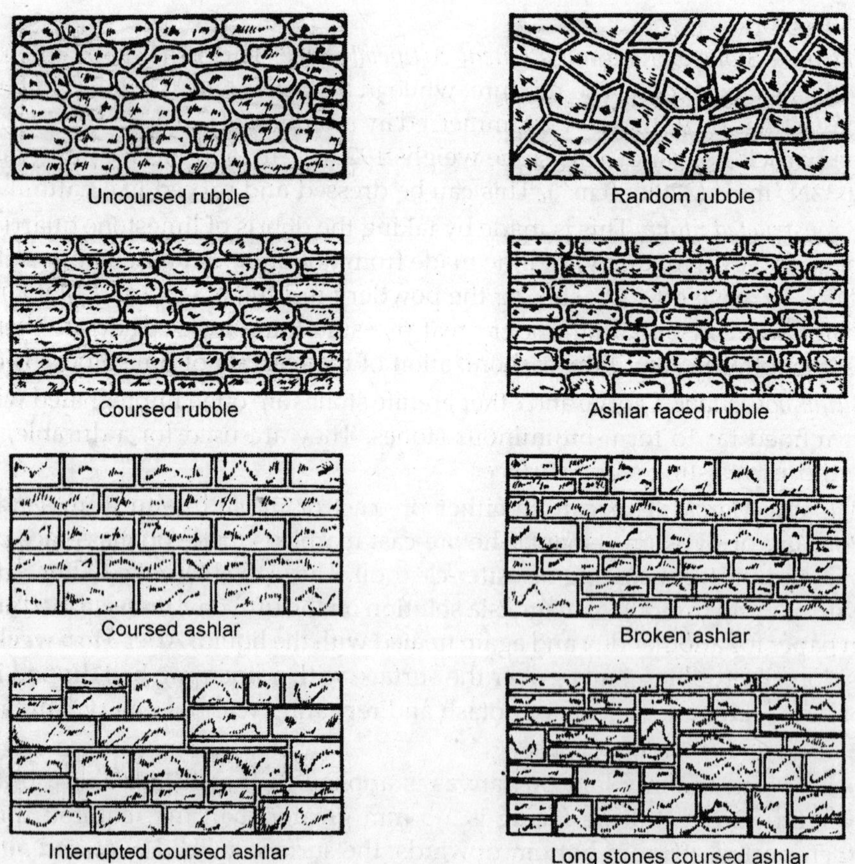

Fig. 1.5. Different types of stone masonry

(i) *Riprap Masonry.* This consists of masonry which is done by piling together uncut stones without using any mortar between them.

(ii) *Uncoursed Ordinary Field Stone Wall.* This is built from rock-faced stones of various sizes and shapes.

(iii) *Random Rubble Wall.* In this the stones are built in the wall to any shape or pitched to fit adjacent stone.

(iv) *Coursed Bubble Wall.* The stones are roughly squared and laid in courses for building course rubble wall. Another variation of this type of masonry is obtained when stratified squared stones with quarry face are used.

(v) *Ashlar.* For this type of work the stone is finally dressed and worked to fit in the general face of the wall. The type of ashlar depends upon the nature of dressing given to face of the stone.

Ashlar could be of courses, broken, interrupted coursed, and coursed (long stones).

(vi) *Cut-stone.* This is also called finished stone work. In this the stone used is generally costly. Each stone is dressed accurately to shape on all faces in accordance with the drawings. This type of masonry is employed for high class work.

1.40 VARIETIES OF DRESSING

Stone dressing can be done in several ways depending upon the type of masonry face desired and the quality of the work. Some dressed stone faces are shown in Fig. 1.6

Rock face

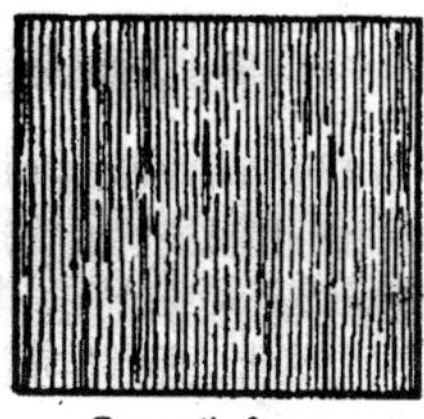
Smooth face

Pointed finish

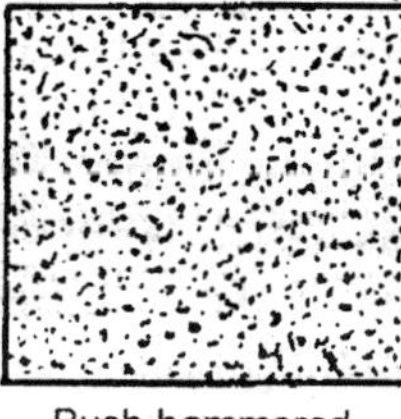
Bush hammered

Fig. 1.6. Dressed Stone Faces.

(i) *Hammer faced-quarry* or quarry-pitched rustic rock, or pitch-face, means a freshly fractured rock surface. This type of dressing is done by just knocking off the sharp and irregular corners of stones obtained by blasting. This helps in proper bonding of individual stones. This method of dressing is generally done for random rubble masonry. To obtain uniform joints, sometimes the joint faces and bed faces of stones are fully dressed or dressed only for a width of 70 to 100 mm from the edges of the exposed face or faces so as to obtain reasonably uniform joints. Such joints are obtained in coursed rubble masonry.

(ii) *Chisel-drafted.* This type of dressing is done in the same manner as above; except that in addition margins 30 to 50 mm wide depending on the size of the block are worked around the face of the stone. The central portion of the stone face is either kept rough, or it is 'punched' or 'picked'.

(iii) *Bush hammered.* The rough finish of the stone face is obtained when the outer skin of the stone is removed.

(iv) *Boasted.* The finish of the stone face done with a mason's 50 mm boaster in fairly even tool marks gives this type of dressing.

(v) *Rusticated.* The stone work projects from the wall and is finished in several ways *e g.*, chamfer, chamfer-fillet, rebated, etc.

(vi) *Broached or Punched.* Lines of punched marks are made vertically, horizontally or diagonally across the face, usually between chisel drafted margins.

(vii) *Rubbed.* Rubbed fiinish is obtained by means of a rotating steel disc using water and sand or carborandum.

(viii) *Pitched face.* The beds and joints are worked true and the lines for the face of the wall marked on the beds and joints. The lines are then pitched with a hammer and pitching tool. The rough surface of the stone should extend over the surface, no punch or chisel marks being shown.

(ix) *Reticulated.* The surface is worked true with series of sinkings cut into the stone about 10 mm., the sinking being separated by bands of regular width.

(x) *Vermiculated.* The bands separating the sinkings of the reticulated are irregular in width and form giving a worm eaten appearance.

(xi) *Sand blasting.* It is used to get quickly the effect of carving upon the stones. The polished surface is covered with a molten glue like compound which hardens to a tough elastic consistency. The designs are imprinted on this covering and the coating is removed from all parts that are to be cut below the surface. A stream of a powdered carborandum is blown against the design by compressed air when the exposed hard stone is cut, leaving the surface coating unaffected. The coating is then removed and the polished stone with sunk carving is obtained.

Polishing of stones. For obtaining very smooth and plane surface the stones like granite, marble etc. are polished and used for ornamental work, and flagstones are polished and used for paving in floors. The stones are polished by holding them firmly on the top of a revolving table to which some abrasive material like chilled shots or sand or crushed carborandum is led. The final polishing is done by using putty and finishing with rubbers of felt. Hard polishing is done by rubbing.

Rough tooled dressing. This method of dressing is used when the masonry has to be built with fairly regular shaped stones with more or less plane faces. In one line dressing the projections are knocked off with chisels from the surfaces, testing now and then with the mason's square. There may however, remain shallow depressions on the plane surfaces. The corners and edges are rendered square and true.

To do away with the depressions in one line dressed stone and to obtain even surfaces the surfaces are dressed with chisels at closer intervals and to a depth over the depressions. In this the chisel marks are left over the whole surface. This method is called two line dressing.

Three line dressing of stone is done for obtaining finer surfaces the ridges or projections of the chisel marks being quite tiny.

Fine tooled dressing. For further finish, the three lined dressed stone is fine tooled with many pointed, serrated or saw like chisel and hammer. Like this all the projections caused by dressing are removed. A fine tooled surface is then polished. It is only the exposed faces of the stone blocks that are two line or three line dressed or fine tooled. The beds and side joints are generally one line dressed.

Where large quantities of dressed stones are required expeditiously, dressing of stones is done with machinery run by electricity or other power.

PRESERVATION OF STONE

In order to prolong the life of a structure against the decaying action of the weathering agencies and to prevent the face work from becoming unsighty, stone work should be treated suitably to retard its deterioration. The preservation of stone, for ordinary building may be done by either pointing or plastering at the time of construction. For important buildings, use of preservatives is made which act in the following ways :

(*i*) fill up the voids in the stone

(*ii*) provide a coating to the stone surface and prevent the ingress of moisture ; and

(*iii*) react with the constituents of stones and provide a hard and durable surface.

Various proprietary paints, oils and chemical solutions are available in the market which can be used for a preservative treatment to stones. These treatments are not ever lasting and have to be renewed regularly say after 5 to 10 years. The choice of a particular type of preservative will depend upon the following factors :

(*i*) Physical and chemical properties of the stone

(*ii*) chemical composition of the preservative ; and

(*iii*) nature of the surroundings.

However, before giving any preservative treatment, it must be ensured that :

(*i*) the surface to be preserved has been scruplously cleaned with a wire brush to remove dust etc. and for proper bond of the stone and preservative

(*ii*) the surface has been washed with a jet of water under pressure; and,

(*iii*) the surface has been dried sufficiently.

1.41 TYPES OF PRESERVATIVES

The various types of preservatives used are :

(*i*) *Paints.* Paint is applied to the exposed stone surface with a brush. To give the desired colour, a pigment may be added. For deep penetration, paint may be applied under pressure. Use of paints, as a preservative on stone masonry, is seldom recommended as it never gives the desired shade of the stone. Moreover it has to be renewed frequently ; which increases the cost.

(*ii*) *Boiled Linseed Oils* are brushed on the surface in one or two coats. This provides a greasy surface on the stone, thereby allowing the rain water to drain of rapidly. They do not affect the original colour of the stone but evaporate readily in hot climates. A coat of dilute ammonia in hot water over linseed oils helps to make uniform the discoloration produced by the oil.

(*iii*) *Coal Tar or Bitumen.* An application of one or two coats of bitumen or coal tar provides a very effective preservation treatment over stones. However, they are rarely used especially on face work,, because of their black colour.

(*iv*) *Paraffin.* Melted paraffin or paraffin dissolved in naphtha is applied with a brush and forced into the pores of the stone by heating the surface to a temperature of about 60°C. Paraffin discolours the stones. Addition of some creosote to paraffin prevents organic growth and fungus on the stone.

(*v*) *Ransome's Process.* First sodium or potassium chloride solution is filled into the pores of the stone by repeated applications with a brush. As the surface dries up, a coat of calcium chloride is applied. This results in the formation of insoluble calcium silicate.

(*vi*) *Alum Soap Solution.* Sometimes solution of alum and soap in water is applied to the stone surface. However, they have to be frequently renewed.

(*viii*) *Baryta Solution.* In industrial towns, the atmosphere is usually charged with sulphurous or carbonic acids. These acids react with the calcium present in the stones to form calcium sulphate. Soot and dust adhere firmly to $CaSO_4$. Also it flakes of in due course of time and exposes fresh surfaces for further action.

To safeguard against this, use of solution of barium hydrate [$Ba(OH)_2$], knowr as 'Baryta' is made. This is applied in three coats. Baryta reacts with $CaSO_4$ to form barium sulphate, which is insoluble in water and is not affected by the acid charged atmosphere. The calcium hydroxide, formed during the above reaction absorbs carbon dioxide from the atmosphere to form calcium carbonate.

(*viii*) *Szerelmey's Stone Liquid.* Comprises of sodium or potassium silicates in combination with calcium and barium chlorides. They are applied in either two or three coats. Generally a coat of some bituminous liquid is given before the application of szerelmey's stone liquid.

TESTING OF STONES

A number of tests are to be carried out to check the suitability of stones for structural use. The various tests that are usually made are:

1.42 CRUSHING STRENGTH TEST

The crushing strength should be tested if the stone is to withstand very high pressure, as for instance if the stone is to be used in the bottom layers of a very high tower. Testing blocks 100 mm cube, truly cut and finely dressed are used. It is essential to make their bottom and top surfaces absolutely flat and horizontal. Plaster of Paris may be used to achieve flawless top and bottom surfaces. The load must be applied axially. The test blocks are tested in a compression testing machine to destruction and the changes in the blocks recorded for the corresponding loads.

1.43 CRYSTALLISATION TEST

This test is meant to ascertain whether the specimen possesses a cohesive property and the particles of stones are well cemented together, thereby assessing the weathering quality of stones. The decay of stone by weathering is mainly due to crystallisation of salts, particularly $CaSO_4$, in the pores of stones. The test samples, 10mm cubes are first weighed, then immersed in a 14% solution of sulphate of Soda at 20°C for 2 hours and dried at 100°C. They are again weighed. The process of weighing, immersing in salt solution, drying and reweighing is repeated 10 to 15 times. Each time the weight is recorded and the resulting difference in weight noted. A stone which is not much affected has very good weathering properties.

1.44 ABSORPTION TEST

To ascertain the relative qualities of different stones they are immersed for 24 hours in water and the amount absorbed by each specimen is noted. Greater absorption of water by stone means that it is porous and cannot be relied upon to weather well.

1.45 ABRASION TEST

The test is useful for determining the hardness of a stone as a material of construction. Resistance to abrasion is tested by putting a cylinder of stones 25 mm diameter and 25 mm long in the abrasion testing machine shown in Fig. 1.7 and revolving it at about 2000 times per hour. Silica sand is fed continually upon the revolving test specimen for half an hour. The weight lost by the stone measures the amount of abrasion loss. The more resistant stone will show comparatively small loss.

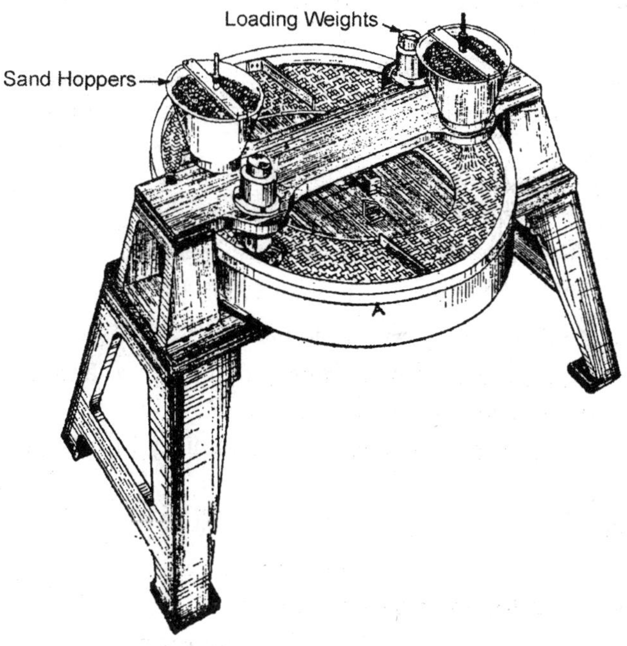

Fig. 1.7. Abrasion testing machine

1.46 ACID TEST

The test is carried out to ascertain the resistance of stone to the acid conditions in the atmosphere in commerical and industrial towns. The stone is soaked for some days in 1 percent solution of sulphuric and hydrochloric acids. The solution is frequently agitated. If the sample has still maintained its edges and corners as sharp as before, the stone will weather well. If, however, they are rounded and there are loose particles of deposit on the surface, it indicates presence of $CaCO_3$ in it, which detracts considerably from its usefulness.

1.47 FIRE TEST

When a stone building is under fire and the face of the stone is subjected to high temperature, a number of stresses are set up due to different coefficients of expansion of its constituent minerals and also due to low heat conductivity of stone. The stress conditions get accentuated when water is used for fire extinguishing. This results in **the** cracking of stone and sometimes parts of the stone falls off.

The fire resisting quality of stone can be tested satisfactorily by building a small hut or a wall panel of it and subjecting one side of the structure to a high temperature of about 950° C and studying the behaviour of stone under fire. If the cracks developed are deep, is should not be used for important buildings.

QUESTIONS

1. List five structures in India, where stone has been used as a building material. Give at least eight different parts of a building where stone can be used.
2. What are the different classifications of rocks yielding stones? Briefly describe them.
3. Discuss the geological classification of rocks. *(JNTU. A.P. 2014)*
4. What are the main constituent minerals which are present in rocks; and what properties are imparted by them. Give examples of rocks containing these minerals.
5. Name the characteristics which should be considered while selecting stone for a building/structure. Briefly discuss them.
6. On what factors depend the location of a quarry for stones? Briefly state the methods of quarrying stones and draw sketches of the tools used.
7. Discuss, in detail, the different operations involved in quarrying by blasting.
8. What precautions will you observe while handling explosives for blasting of stones?
9. Define line of least resistance with the help of a diagram. What factors govern the quantity of explosive material used in blasting of stones.
10. Give the constituents, properties, uses and availability in India of the following stones:
    (*i*) Granite
    (*ii*) Sandstone
    (*iii*) Marble.
11. Briefly describe the following types of stones:
    (*i*) Trap and basalt (*ii*) Limestones
    (*iii*) Laterites (*iv*) Slates
12. Write short notes on:
    (*i*) Shales (*ii*) Quartzite
    (*iii*) Gneisses (*iv*) Schists
    (*v*) Artificial or cast stone.
    List any four commonly used building stones and state their suitability in constructions.
13. What do you understand by the term 'dressing of stones? Illustrate the different varieties of dressing, with the help of sketches; wherever possible. *(VTU Belgaum Jan 2011)*
14. Why is it necessary to preserve a stone? What steps will you ensure before applying a preservative? List the various types of preservatives used.
15. Name the tests which are carried out to check the suitability of stones for structural use. Explain the method of testing for determining the compressive strength of stones.
16. Briefly describe the following types of tests for stones:
    (*i*) Abrasion test (*ii*) Crystallisation test
    (*iii*) Absorption test (*iv*) Acid test; and
    (*v*) Fire test.

17. Briefly discuss about quarrying and dressing of stones. *(JNTU Dec. 13)*
18. Give the structural requirements of various building stones based on their applications and utility services. *(JNTU/A.P Dec. 2013)*
19. What are the qualities of a good building stone? What are the various uses of stones? What is meant by dressing of stone? Describes its various varieties. *(RGTU Bhopal M.P. June 2014)*

OBJECTIVE TYPE QUESTIONS

1. Which of the following rocks does not fall under geological classification

(a) Sedimentary rock (b) Argillaceous rock
(c) Igneous rock (d) Metamorphic rock

2. Which of the following is not a startified rock

(a) Sandstone (b) limestone
(c) marble (d) slate

3. Select calcareous type of rock from the following:

(a) marble (b) granite
(c) Basalt (d) gneiss

4. Which of the following stones is less durable (about 50 years)

(a) Granite (b) Limestone
(c) gneiss (d) Good sandstone

5. Which of the following minerals is alkaline :

(a) Silica (b) Alumina
(c) Lime (d) Potash

6. Calcium carbonate has a hardness of :

(a) 6 (b) 7
(c) 2 (d) 3

7. The specific gravity of quartz is :

(a) 2.6 (b) 3.2
(c) 2.9 (d) 2.72

8. Stratified structures is a characteristic of :

(a) Igneous rock (b) Metamorphic rock
(c) Sedimentary rock (d) None-of these

9. The compressive strength of granite varies from

(a) 40 to 70 N/mm^2 (b) 70 to 130 N/mm^2
(c) 130 to 175 N/mm^2 (d) 175 to 225 N/mm^2

10. The specific gravity of stone should normally be not less than

(a) 2 (b) 2.25
(c) 2.5 (d) 2.75

11. The water absorption of sandstone should not be more than (percent by mass) :
    (a) 2.5 (b) 5
    (c) 10 (d) 20
12. The transverse strength of limestone is specified as
    (a) 3 N/mm^2 (b) 5 N/mm^2
    (c) 7 N/mm^2 (d) 9 N/mm^2
13. Laterites should have a minimum compressive strength of:
    (a) 0.5 N/mm^2 (b) 1.5 N/mm^2
    (c) 2.5 N/mm^2 (d) 3.5 N/mm^2
14. Which of the following stones has a highest compressive strength
    (a) Marbles (b) Basalt
    (c) Gneiss (d) Shales
15. Slate tiles are used in a building :
    (a) Foundation (b) Walls
    (c) lintels (d) roof covering
16. Sandstone comprise :
    (a) Rounded grains of quartz (b) Silica
    (c) quartz and felspar (d) Mica
17. Which one of the following is an example of silicious rock :
    (a) Gneiss (b) Quartzite
    (c) Granite (d) all of them
18. The most powerful explosive used in blasting of rocks is :
    (a) Dynamite (b) gun cotton
    (c) cordite (d) blasting powder
19. The weathering property of stone is determined by :
    (a) Abrasion test (b) Acid test
    (c) Crystallisation Test (d) Crushing strength test
20. Abrasion test on stone is carried out to determine its:
    (a) transverse strength (b) hardness
    (c) cohesive property (d) compressive strength

Answers

1. (*b*) 2. (*c*) 3. (a) 4. (b) 5. (d) 6. (d) 7. (a) 8. (c)
9. (b) 10. (c) 11. (a) 12. (c) 13. (d) 14. (c) 15. (d) 16. (a)
17. (d) 18. (b) 19. (c) 20. (b)

FURTHER REFERENCES

Bureau of Indian Standards, New Delhi

IS Codes 1127; 1128; 3316; 3620; 3622; 6250; 14223 (Part I)

BRICKS, TILES AND OTHER CLAY PRODUCTS

2.1 USES

Bricks, tiles and other clay products are important building materials which have been used in our country since earliest times.

Building bricks are the oldest structural unit and are used because of their strength,, durability and pleasing appearance.

Clay tiles are commonly used for roofing, specially in the country-side. Clay-tiles and paving bricks provide cheap and durable flooring.

A variety of structural clay products like hollow bricks, perforated bricks and hollow clay tiles are coming in vogue, particularly vitrified high strength bricks called engineering bricks are used for road and air-field pavements and for the construction of bridge piers etc. Vitrified clay tiles called quarry tiles provide hard and wear resistance flooring. Terracotta and wall tiles of clay are popular for decorative use.

Refractory bricks are used for lining of furnaces and at places where high temperatures are to be resisted.

Earthenware and stone-ware pipes are used as sewer pipes and for draining of rain water.

2.2 BUILDING BRICKS

The use of building bricks is most common for the construction of walls-both of load bearing and non-load bearing types, columns, pilasters etc. The method of manufacturing bricks is simple and the production can be done locally at cheap cost. However, due to traditional methods of brick making it has not been possible for the brick industry to cater to the huge demand which has come about. The present annual production of bricks has been roughly estimated to be around 300,000 millions ; whereas the demand is of the order of 500,000 millions. It has, therefore, become necessary to modernize the brick industry to improve the quality of bricks and to step up the production considerably. To achieve these scientific evaluation of clays, partial mechanisation of manufacturing process and improved and economical method of burning bricks have to be adopted.

Bricks of various sizes are being produced and the quality of bricks also varies considerably. The need has been greatly felt to standardise the size of bricks and to produce bricks of uniform quality. The Bereau Standards has specified 190 × 90 × 90 mm as a standard brick size, which with mortar joints would mean a brick of 200 × 100 × 100 mm size. This size conforms to the modular dimension based on 100 mm module and the manufacture of standard sized bricks is said to be feasible and its use has many advantages.

BRICK MAKING SOIL

2.3 CONSTITUENTS OF BRICK MAKING SOIL

Bricks are made from clays or shales, extensive deposits of which are available in almost all parts of our country, particularly in the plains. Residual, glacial and sedimentary clays are the most satisfactory raw materials. The chief constituents of the brick making soil are :

(*i*) *Kaolin.* A principal constituent of igneous rocks is felspar which may constitute upto 60 percent of the rock. Of this family of minerals, orthoclase felspar, $K_2O\ Al_2O_3\ 6SiO_2$ is mainly responsible for yielding kaolin. When igneous rocks are subjected to chemical weathering under the action of atmosphere and water, orthoclase felspar decomposes to give kaolinite. This, with the addition of more water yields a clay deposit called kaolin.

(*ii*) *Lime and Magnesia.* Igneous rocks on decomposition also yield silicates of calcium and magnesium, such as hornblende, pyroxene and lime felspar which contribute to the lime-magnesia content of the clay. These are usually in the carbonate form ; but silicates of lime and magnesia are also commonly found.

(*iii*) *Iron.* Iron oxide is also present in clays as a result of the disintegration of rock minerals contained in it. These minerals vary in size from colloidal to nodular dimension.

(*iv*) *Uncombined Silica (SiO_2) or free sand.* Sand exists in the form of silicates and in addition also in a free state when it is called flint or sand. Free sand is produced from quartz, SiO_2 contained in the igneous rocks. This is a result of the disintegration of the rocks. Sand varies in size from 0.05 mm to 1.00 mm and depending upon it the sand is termed as very fine, medium and coarse. Silt is an intimate mixture of clay and very fine sand.

(*v*) *Other minerals.* Minerals of sodium, manganese, chromium, etc. are also present in clay in a very small percentage.

2.4 FUNCTIONS OF VARIOUS CONSTITUENTS

The functions performed by the various constituents of a brick making soil are described below :

Kaolin. This is the principal constituent of all kinds of clays. Kaolin is a plastic material when wet and it sets when burnt. Thus it imparts plasticity to the clay which is essential for the purpose of moulding. Excess of kaolin causes shrinkage and warping.

Silica or free sand. Free silica, if not present in excess quantities, produces resistance to heat, preserves the shape of brick at higher temperature, prevents shrinkage, warping and undue hardness. Further, though sand alone is infusible, it slightly fuses at kiln temperature in the presence of iron and lime and serves as a hard cementing material in the burnt brick. Thus it also imparts strength and durability.

Lime. It serves two purposes. Firstly, it reduces shrinkage in drying and secondly it acts as a flux during the burning process enabling the silica particles to melt and thus binding the particles of bricks together. Lime should be present in a very finely divided state so that it is completely hydrated or fluxed in the process of the manufacture of bricks. The presence of lumps of unhydrated lime is the cause of unsighty defects on brick walls called 'lime pops'.

Oxide of Iron and Magnesia. Oxide of iron acts as a flux and adds greatly to the hardness and strength of brick. The colour of brick depends very much upon the content of iron and ranges from light yellow to red ; the colour increases in intensity with the increase in the iron content in the clay. If iron oxide is present in large quantities i.e. 8 to 10 percent, on burning of bricks in kiln, the red iron oxide gets converted into black oxide, which on combining with silica produces dark brown colour.

The presence of manganese in the clay changes this colour into bluish black or dark black. Iron, in the presence of small quantities of manganesia, makes the brick yellow. If manganesia is present in excess it causes the brick to decay.

Lime mixed with iron gives cream or brown coloured bricks.

2.5 HARMFUL CONSTITUENTS OF BRICK MAKING SOIL

(*i*) *Pebble of Stone and Gravel :* These do not allow the clay to be mixed uniformly and thoroughly and result in weak and porous bricks. While preparing the soil for moulding bricks these should be removed.

(*ii*) *Alkalies.* Alkalies have very injurious effect on bricks. They lower the fusion point of clay. If present in excess, they cause the bricks to fuse, twist and warp during burning and thus tend to disfigure them. Alkaline salts, if present in masonry structure, act like hygroscopic substances. They absorb moisture from the atmosphere in due course of time and create damp conditions. The moisture on drying leaves behind a greyish white deposit. This it termed as *efflorescence* on account of which the appearance of the building is spoiled.

The common salts met with are the sulphates of calcium, sodium, potassium and magnesium. These are either present in the original clays or are formed during the burning process as a result of reaction between the sulphur compounds of coal and bases in the clay. The presence of '*Reh*' or '*Kallar*', consisting of sodium sulphate with more or less of sodium carbonate and sodium chloride render the clay utterly unsuitable for brick making. It does not allow bricks to be burnt properly. Even after burning, the salts crystallise and appear in the shape of irregular and unsighty white patches. This causes the surface of the bricks to peel off layer by layer and gradually crumble away. Presence of '*reh*' or *kallar*' can be easily detected by the efflorescence on the sides of a fresh excavation, if the soil is moist, but it would be appropriate in all cases to moisten the clay with water and subject it to evaporation and check for efflorescence.

(*iii*) *Limestone and Kankar.* This should be present in a very finely divided state because even small particles of the size of a pin-head cause flaking of the bricks. Limestone should never be present in lumps and in excess quantity. $CaCO_3$ on burning gets converted into an anhydrous substance (CaO). On coming in contact with water it gets hydrated, swells and increases in volume two to three times. This causes the flaking of the bricks *i.e.* to split and crumble to pieces.

(iv) Iron pyrite. If present in the clays, it decomposes, oxidises-and finally crystallises in the brick. This causes the brick to split.

(v) Vegetation and Organic Matter. Clays containing vegetation and organic matter, if not thoroughly burnt, will produce porous bricks. This is due to the evolution of gases during the burning of the carbonaceous matter resulting in the formation of small pores. These gases should be thoroughly driven out during burning, if good bricks are to be prepared.

2.6 CHARACTERISTICS OF SUITABLE BRICK SOIL

The mechanical composition of the soil, suitable for brick making, should preferably conform to the following requirements :

(i) Clay 20 to 35 percent
(ii) Silt 20 to 35 percent
(iii) Sand 35 to 45 percent

If a single soil is not suitable, two or more soils may be mixed so as to conform to the above requirements. The soil should be free from gravel, coarse sand (particle size greater than 2 mm.), lime and kankar particles, vegetable matters and roots etc.

The chemical composition of soil may roughly conform to the following requirements :

(i) Silica (SiO_2)	Not less than 60 percent
(ii) Alumina (Al_2O_3)	Not less than 15 percent
(iii) Ferric Oxide (Fe_2O_3)	Not less than 3 percent
(iv) Lime (CaO or $CaCO_3$)	Not less than 3 percent
(v) Magnesia (MgO or $MgCO_3$)	Not less than 3 percent
(vi) Alkalies (K_2O+Na_2O)	Not more than 4 percent
(vii) Total water soluble	Not more than 1 percent
(viii) Loss on ignition	Not more than 7 percent

The *physical* properties of a good soil should vary as follows :

Liquid limit	25 to 38%
Plasticity index	7 to 13%
Volumetric shrinkage	15 to 25%

2.7 FIELD TESTS FOR DETERMINING THE SUITABILITY OF BRICK EARTH

Generally the soil which contains four parts of clay and one part of sand is considered to be suitable for brick making. Experienced people are able to determine the suitability of the soil by looking at the soil and by rubbing in the hand.

The scientific method for testing the suitability of the soil in the field is as follows :

Sampling. Collect samples of soil from different places from the site selected for brick making. The samples are either taken from the surface or by removing some layers of top soils. If 15 kg of soil is taken for each sample then the work of testing is done very conveniently. For identification, the bricks made from each sample and the respective place from which the sample has been collected should be similarly marked.

For determining the suitability of the soil, the following four field tests should be made :

2.7.1 First Test—The Test for the Consistency of the Soil

The above test can be divided into the following steps :

(*i*) First of all remove stones, kankar etc., which cannot be ground.

(*ii*) After this grind every soil sample to a fine powder.

(*iii*) Then mix sufficient quantity of water gradually with the soil so that it becomes plastic and can be moulded with hands. So long as the soil does not become plastic, water should be added in small quantities and the soil mixed with hands. It must be remembered that no more than the required quantity of water should be added, otherwise it would not be possible to mould bricks.

(*iv*) From this plastic soil take a handful of soil sample and add little water to it and form a ball by pressing it. Keep this ball on the ground in the sun for drying.

(*v*) When the ball has dried, find out by inspection if the ball has lost its shape and examine the cracks which have occurred on the surface of the ball.

(*a*) The ball which does not get deformed on drying and which shows very little cracks on its surface, shows that the soil used for it is suitable for brick making.

(*b*) The ball which gets deformed on drying and crumbles easily shows that the soil from which it is made contains more of sand than required. From such soil, bricks cannot be moulded.

If some soil which contains less sand is available, then by mixing this soil with sand, the above soil can be made suitable for brick moulding. It is far easier to add sand to the soil which contains less of sand in order to make it suitable than to mix two types of soils lor the purpose.

(*c*) The soil of the ball which after drying has become hard and shown wide cracks on the surface contains less of sand. Such a soil can be improved by adding the required quantity of sand. The easiest way for this is to add a small quantity of sand gradually and form balls which should be examined after drying to know the suitable sand proportion.

(*vi*) By mixing different proportions of two types of soils, form small balls and keep them for drying. After drying examine the shape of the balls and the cracks formed on the surface. The ball which shows least deformity and cracks has the correct soil proportion for brick moulding.

2.7.2 Second Test—Testing The Soil For Moulding

It is necessary that the soil which has been found to be successful in the first test should be used for brick moulding.

Generally, for moulding of the bricks properly, some additional quantity of water is required to be added.

(*i*) The soil, used for the ball which after drying does not loose shape and shows very little cracks is considered to be suitable for the second test. In the mixture, the proportion of water and sand should be the same as the one used for making the ball.

(*ii*) The soil mixture should be kneeded very well. If the soil mass has mixed properly then it would be possible to form thick threads from the soil.

(*iii*) This mass of soil should be put into the mould and brick moulded. The brick should be examined to find out whether it is properly moulded.

(*iv*) If the edges and corners of the moulded brick are not well formed, then the mixture contains less quantity of water. In this situation by adding more quantity of water, the soil mix can be made suitable for brick moulding. By adding water gradually to the mix, brick should be moulded and this process should be repeated until well formed bricks could be moulded.

(*v*) The same process should be repeated with the sample of soils which have been found to be successful in the first test.

(*vi*) The bricks which have been made from the same soil, sand and water mix should be given similar mark.

(*vii*) Leave all moulded bricks for drying on the ground for four days.

2.7.3 Third Test—Determining the Shrinkage and Deformation if any in the Bricks after Burning

From this test, the extent of deformation and shrinkage that occurs to the sample brick is determined. The test is done as follows :

(*i*) The sample brick should be burnt in the ordinary potter clamp for three days and four nights, until the bricks become red hot. As a matter of fact, the time for burning of bricks depends upon the size and shape of the clamp but normally it is two times, the time taken for baking the earthen pots.

(*ii*) While setting the bricks in the clamp, the bricks should be so arranged that some space is left in between them so that they may burn properly.

(*iii*) The sample bricks should be put here and then in the clamp in such a manner that all the bricks from the same sample may not get over burnt.

(*iv*) Great care should be taken while arranging the bricks in the clamp.

(*v*) When the bricks have been burnt, allow all the bricks to remain in a lot until all of them get cooled.

(*vi*) When the bricks have cooled, remove them from the clamp and these should now be examined for shrinkage and deformation.

(*vii*) The bricks which have lost their shape and have completely been deformed should be sorted out. Those bricks which have shrunk evenly and do not show any bad effect due to burning should be selected. These should be tested for strength and quality.

2.7.4 Fourth Test—Test for Determining the Strength and Quality the Bricks

The strength of the bricks should be determined in the following manner :

(*i*) The soundness of the brick is found out by striking the flat surface of a brick with that of another. If this sound is dull, then the bricks are not of good quality and the burning has not been proper.

(*ii*) The strength of bricks is also found by soaking them in water. For this the bricks are completely immersed in water for 24 hours. After this, examine them. The Bricks which become soft by immersing in water should be broken and again immersed in water after which the extent of their becoming soft should be found out. Good bricks do not absorb more than $\frac{1}{2}$ to $\frac{1}{6}$th of their own weight of water when immersed in water for 24 hours.

(*iii*) By dropping on the sample bricks a wooden log from a height of 100 mm., the strength of the brick is determined. By dropping the log of wood from different heights, the extra strength of the brick samples can also be found out.

The compressive strength of very good bricks is not less than 7 N/mm^2 (70 kg./sq. cm)

2.8 FOR BETTER RESULTS—WHILE PREPARING THE SOIL FOR TESTS THE FOLLOWING POINTS SHOULD BE KEPT IN MIND

(*i*) It is better if the sample bricks could be burnt in bigger kilns. By doing so the burning is proper and results are more satisfactory.

(*ii*) It is better to have a heavy and dense soil sample.

(*iii*) If the burnt bricks break easily then they are deficient in two respects. They are either not properly burnt or the soil from which they have been made contains too much of sand.

(*iv*) The method of making the bricks which have been found suitable after successfully carrying out the four tests should be noted so that the same method could be adopted for making bricks on a large scale.

Wherever possible particularly for large-scale production of bricks the physical and chemical properties of the soil used for brick making should be determined by carrying out scientific tests in the laboratories. From the tests, the suitability of the soil for brick making is determined and it is also known if for improvement of the soil composition, certain materials are required and if so the proportions in which they are to be added. However, following general suggestions are made for the improvement of certain types of soils :

(*a*) The black cotton soil which shrinks very much on drying and expands considerably when watered can be made suitable for brick making by the addition of coal ash in certain proportions and by other methods.

(*b*) If the soil for brick making contains lime nodules then the bricks made of such soils get cracked which is called lime bursting. For checking lime bursting in bricks, a solution of 0.5 to 0.75 percent sodium or potassium chloride (table salt) in water and 5 to 10 percent of coal ash should be added to the soil. The brick formed form this soil should be burnt under light fire. The burnt bricks on cooling should be dipped in water. This further checks lime bursting.

MANUFACTURE OF BRICKS

The following steps are required in manufacturing bricks :

(*i*) Selection of site for brick making ;

(*ii*) Preparation of the soil by tempering of earth ;

(*iii*) Moulding of bricks ;

(*iv*) Drying of bricks ; and

(*v*) Firing or burning of bricks.

2.9 SELECTION OF SITE FOR BRICK MAKING

Such a site should be selected for the manufacture of bricks where suitable quality of soil is available in sufficient quantity. Mostly such sites are found along the river banks, in the valley or in the plain where soil brought by the river beds gets deposited.

First of all from experience and investigation of the site, the suitability of the available soil is estimated. When manufacturing of bricks is to be done on a large scale, it is necessary to carry out scientific tests to determine the suitability of the soil, as described above.

At the same time it should be kept in mind that the soil for brick making should be easily available ; otherwise unnecessary labour and cost for digging and transportation of the soil would be involved.

Generally, for making the soil suitable for brick making, certain materials are required to be mixed with the soil and therefore the availability of such materials near the site for brick making is of great importance. It is necessary that the water and the fuel like coal or wood should be easily available in sufficient quantities.

2.10 PREPARATION OF SITE FOR BRICK MAKING

Site for brick making should be levelled. When the ground is uneven, there is a danger of the shape of bricks being spoiled. At the site, there should not be any vegetation, plants or trees. It would be better if the pits which are made due to the digging of the soil could be utilised for some useful purpose. Such pits when dug in the bed of a dam could increase its storage capacity and if these pits are near the field, these could be used as manure pits. By digging the soil from the higher portion of the ground, the ground could also be levelled.

2.10.1 Preparation of the Soil for Brick Making

First of all the soil for brick making should be dug and brought to the site where bricks are to be moulded and burnt.

Generally after sprinkling some water on the heap of the soil, it is left over for some days. After this necessary quantity of sand, if required, and water should be mixed with the heap of the soil very well. The quantity of sand and water required for mixing is found from the first test which has been described earlier.

It is very necessary that the soil must be mixed very thoroughly. The mixing of the soil is done by kneeding and with the help of spades. With the help of animals also the mixing could be done. When the brick manufacture is to be done on a large scale, the pugging of the soil could also be done with the help of a machine called pug-mill.

2.10.2 Pug-mill

It consists of a circular iron cylinder 1.25 to 2 m high, diameter at top 1 to 1.25 m and diameter at bottom 0.75 to 1 m. This is partially burried underground to avoid unnecessary lift of the clay to be fed. In the centre of the cylinder there is a vertical iron shaft to which is fixed a number of horizontal arms. To each arm are attached vertical wedge shaped steel knives. The shaft is revolved by a long arm. The machine can be worked with electricity or power.

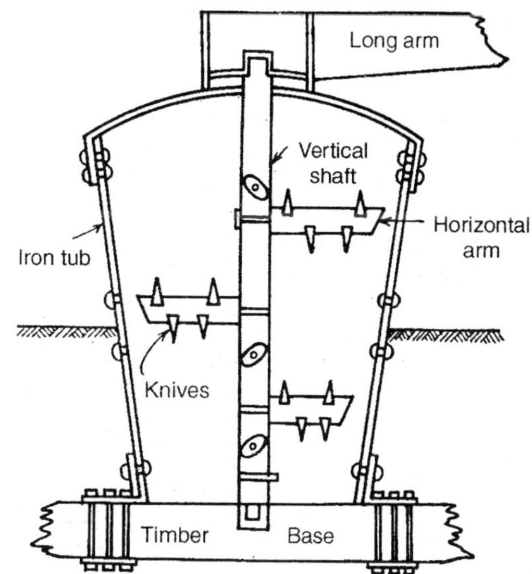

Fig. 2.8. Pug mill

The mill is mounted on a pair of square timbers fixed at right angles to each other. Clay and water are fed from the top of the pug-mill in the required proportions. As the shaft revolves, the blades churn the soil, while the kneeding is done by the knives. The clay, when ready for moulding is forced out through an opening, provided at one side near the bottom of the cylinder (not shown in the sketch).

2.11 MOULDING OF BRICKS

2.11.1 A Brick Mould

In the moulds shown in Figs. 2.9 and 2.10 bricks are cast. Generally the mould is made of wood. Metal moulds can also be used. Mould is a rectangular box without top and bottom. The long arms are projecting a little to act as handles.

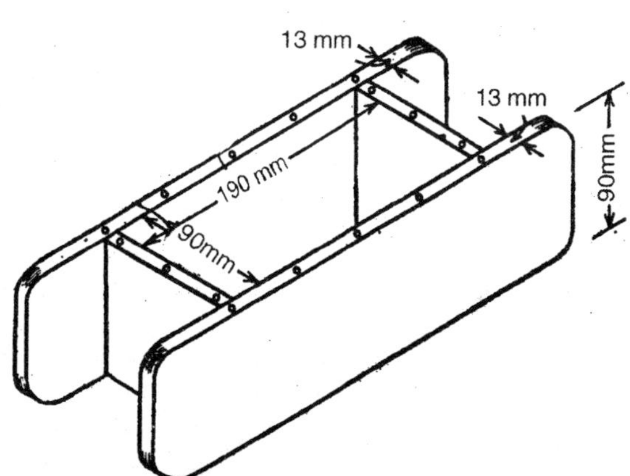

Fig. 2.9. Wooden mould

Seasoned wood should be used to prevent changes in dimensions. The edges should be protected with strips of brass or steel screwed on. The inside of the mould may be lined with steel. Steel moulds, made of plate, 60 mm thick, are used if bricks are to be manufactured on a large scale. They maintain their shape and are economical in the long run due to longer life.

The moulds should be of such dimensions which would give the finished burnt brick of the required size. Normally, shrinkage allowance, varying from 10 to 12 percent is provided but it depends upon the nature of raw material, design of the kiln and the moulding and burning operations. However the exact amount of shrinkage allowance to be provided should be determined experimentally by actually burning bricks.

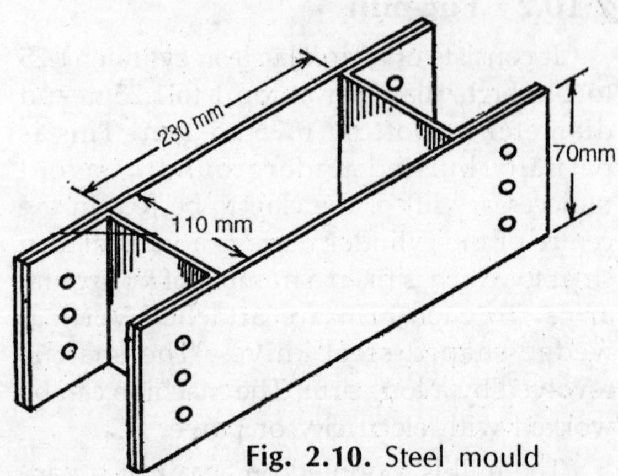

Fig. 2.10. Steel mould

2.11.2 Machines for Moulding

Normally bricks are hand moulded but for moulding of bricks at a faster rate, machines can also be used.

Ellson Brick Making Machine. About 2000 bricks per day can be moulded by this machine. Five men are required to operate it. It has a box like matal mould in which the clay for brick making, which has been thoroughly pugged, is filled. By pressing the handle down and releasing the plunger which is connected with the lever, the bricks are pressed. In this manner a total pressure of about 20 tonnes is applied on the bricks.

Different types of moulds can be used in this machine. The legs and the handle of the machine can be removed and the machine can be transported very easily on the site for brick making. The weight of the machine is about 230 kg.

Stabilized Soil Blocks. The brick making machines can also be used for making stabilized soil blocks. CINVARAM is another such machine which is used for making stabilized soil blocks. The machine is very simple in design and easy to operate.

2.11.3 Methods of Moulding Bricks

There are twc ways of moulding bricks, based on the method employed for moulding, namely :

1. *Hand Moulding* which is done in two ways :

(*i*) Ground moulding

(*ii*) Table moulding

2. *Machine Moulding.* The methods of moulding, depending upon the consistency of the soil and the amount of water added, are classified as :

(*i*) Soft mud process ;

(*ii*) Stiff mud process, and

(*iii*) Dry press process.

In the soft mud process, the clay is tempered with enough water to form a mass of soft consistency. Less water is used in the stiff mud process so that the mixture is much more rigid than in the soft mud process. In the dry process, the clay is pressed to form bricks in a dry state. This is used where the clay cracks very easily on drying.

The various methods are explained below :

2.11.3.1 Hand Moulding

Ground Moulding

The different processes involved are :

(*i*) Before moulding each brick, the mould should be either dipped in water for some time to remove soil sticking to it or should be sprinkled with some dry sand over the inside of the mould to prevent sticking of the soil with the sides of the mould. If water is used, the bricks are said to be *slop-moulded* and when sand is employed, the bricks ate said to be *sand-moulded.*

Sand moulded bricks are superior as they are sharper and cleaner than slop-moulded bricks. A better method would be to dip the mould in water and also sprinkle dry sand. Sand should be sprinkled very well in the inner surface of the mould so that bricks do not stick to the mould.

(*ii*) Take a lump of well mixed soil, the volume of which should be little more than that of the brick. This is done because wet soil shrinks on drying and burning. This lump should be shaped in hand to the size and shape of the brick.

(*iii*) After this it should be rolled in sand and with a jerk the lump is thrown into the mould in such a manner that the mould is completely filled with the soil. As far as possible no extra soil should be required to fill the mould. To fill the mould properly a wet soil or a lump should be pressed with the help of a thumb.

(*iv*) The surplus soil should be scrapped off and the top surface levelled. For scrapping, metal plate 300 × 37.5 × 5.0 mm in size is used. This is called 'strike' and may also be made of wood. Normally a wire fixed in lever or a wooden frame is used for removing the extra soil in the mould.

(*v*) With the help of the handle which is attached to the mould, the mould is lifted up and struck with a hard thing so that by the jerk, the brick in the mould may get detached from it.

(*vi*) On a level ground sprinkle some sand so that bricks do not stick to it. Now the mould is turned over on the ground and is then lifted leaving the brick to dry. Often on a metal plate the mould is turned over to separate it from the brick. The brick is then taken and kept at the place for drying.

Bricks moulded directly on the ground have their underside rough and can have no *frog*. A better method is to mould them on a wooden block called moulding block having a projection 7.5 mm deeper than the thickness of the brick and is placed on the moulding block, fitting closely round the projection. The frog required on the finished brick is obtained by providing corresponding raised portion on the projection itself. After the clay has been forced into the mould and the excess clay projecting above the mould struck, a pallet board which is a thin piece of wood, of a size a little larger than the size of the mould, is placed on the top of the mould. The mould alongwith the brick inside is lifted up and placed upside down. The mould is lifted, leaving the wet brick on the pallet. Another pallet is then placed on the top of the brick and between the two pallets the brick is carried to the drying ground, where it is placed on its edge. The pallets are removed.

Table Moulding

The various processes involved in the table moulding are the same as in hand moulding. The moulder, instead of sitting, stands at the table. On the table is fitted a board, known as *stock-board.* This is of the same size as the mould and has a projection for the frog. The mould is dipped in water or sand is sprinkled and is placed to fit the stock-board. The moulder then dashes the prepared clay into the mould and presses it into the corners with his fingers. Any surplus clay is scrapped off and top is levelled by means of the strike or a wing fixed in a wooden frame. The brick is then removed and placed for drying as explained above.

2.11.3.2 Machine Moulding

Machine moulded bricks are much superior to hand moulded bricks. They are heavier, compact, stronger and uniform. The edges are straight and sharp.

Soft-mud Process. Machines described under article 2.11.2 can be used for moulding bricks by soft-mud process. For this the soil is prepared as in the case of hand moulded bricks and machines are used to mould bricks by applying pressure by hand.

Stiff-mud Process. In the stiff-mud process, the tempered clay from the pug-mill is put into an auger machine where it is further mixed and tempered. The auger is fitted with a die at the front, through which the clay is forced out in the form of a continuous bar. The issuing bar is cut into the required lengths by means of cutting wires fitted on a specially designed cutting table. The wires leave a rough side where they pass through, but it is not objectionable as this is usually on the side of the brick where the mortar joint occurs. More or less laminations are present in the product of auger machines, as the clay bars are given different velocities due to the motion imparted by the auger and friction on the sides of the die. The uniformity in shape and size of stiff-mud brick can be improved, the surfaces made smoother and harder and the density and strength increased by putting these bricks through a press.

Auger brick machines have considerable greater capacities than the machines of vertical press-type. For moulding common bricks, machines equipped with double or triple dies are sometimes used.

Dry-press Method. In this method, vertical press type machines are used. The grounded and screened clay is taken to the machine and placed in the mould in the dry state. Vents are provided to allow the escape of entrapped air. Owing to the heavy pressure being applied, it is possible to obtain bricks of uniform size and with sharp corners. This process also does away with air drying. Face bricks are commonly dry-pressed.

2.12 DRYING OF BRICKS

Moisture is present, in the moulded bricks, in two states namely :

(*a*) Free or uncombined moisture, and

(*b*) Combined moisture or water of crystallisation

The purpose of drying is to remove as much of free moisture as possible and to make the bricks stiff enough for proper handling. The rate at which moisture is lost, when exposed to atmosphere, depends on a number of factors, important of which are temperature, humidity, air movement, percentage of moisture present in the bricks, heating on the exposed surface, etc.

Drying always results in shrinkage of bricks. The amount of shrinkage depends very much on the type of clay or its plasticity. The shrinkage varies from 5 to 10 percent from stiff plastic clays to soft clays. As such the size of the moulded brick is usually kept larger than the required dimensions. The following precautions, if observed during the process of drying of bricks, will yield best results :

(*i*) Drying should be carried at a slow and uniform rate.

(*ii*) It should not be carried out under the direct rays of the sun but preferably under a shade.

(*iii*) The size of the sand particles should be well graded.

(*iv*) There should be uniform and homogeneous distribution of the various constituents.

(*v*) The clay should contain a maximum amount of colloidal material.

(*vi*) The temperature, during drying, should not be more than 120°C.

2.12.1 Methods of Drying Bricks

Drying of bricks can be carried out in two ways:

(*a*) Natural drying

(*b*) Artificial drying.

Natural Drying

(*i*) For drying, the bricks should be arranged in a row. By keeping the bricks at equal distances they receive air and sun for proper drying.

(*ii*) It is necessary to protect bricks from dampness and rain while drying and for this, if required, sheds are also constructed.

(*iii*) The bricks should be allowed to dry till they become leather hard or bone dry (with approximately less then 2 percent moisture content). The bricks may be allowed to dry for about two days in the position as released from the mould. After that they should be turned to stand on edge for another two days or so, they may then be transported and arranged in rows in 'stacks' with spaces of not less than 10 mm between the bricks and allowed to dry in this state, for about a week's time.

(*iv*) After drying, the bricks are taken to the kiln and arranged in it for burning. It is necessary to ensure that the bricks have dried well otherwise the moisture of the damp bricks is driven out when the bricks are burnt and it is harmful to the top surfaces of the dry bricks, thus causing defects in the burnt products.

Artificial Drying

Artificial drier are of two types :

(*a*) Hot floor drier ;

(*b*) Tunnel drier.

The hot floor drier is heated either by a furnace placed at one end of drier or by exhaust steam from the engine used to supply power. Tunnel driers are either periodic which is filled, dried and emptied in rotation or continuous loading being done at one end and unloading at the other. Tunnel driers are heated by underneath steam pipes or by hot air from kilns. These are more economical than hot floor driers. The temperature is usually less than 120°C and it takes 1 to 3 days for drying depending upon the temperature of the drier, the character and the shape of the clay product.

2.13 BURNING OF BRICKS

Bricks are burnt to remove the moisture present in the clay in the mixed state or as water of crystallisation, to impart hardness and strength to bricks and to increase their density so as to make them absorb very little water.

Most of the free or un-combined moisure gets removed during the process of air-drying, but the temperature is not so high as to remove the water of crystallisation or combined water. This water is removed during the process of burning. When a temperature of about 650°C is attained, which is called the temperature of dull red heat, the organic matter contained in the brick is oxidised and also the water of crystallisation is driven away. But heating of bricks is not stopped at this temperature because if the brick is brought back to normal atmospheric conditions, it absorbs moisture from the air and gets rehydrated. Also under favourable conditions, the clay becomes plastic. It is, therefore, essential to burn the bricks at a much higher temperature to bring about reactions between the mineral constituents of clay and giving them new properties such as strength, hardness, less moisture absorption, etc.

When a temperature of about 1100°C is reached, particles of alumina and sand, the two major constituents of brick clay, bind themselves together and increase the strength and density of bricks. Also a small amount of fusible glass is formed which binds the coarse particles of clay together. Further heating is not done and attempt is made to form only a small amount of fusible glass which is just sufficient to impart the required strength and durability. Presence of minerals like lime, magnesium and oxide of iron act like flux.

If the temperature is further raised *i.e.,* above 1100°C a greater amount of fusible glassy mass is formed and the bricks are said to be vitrified. Beyond a certain limit, vitrification results in the general softening of the clay and the bricks begin to loose their shape.

2.13.1 Brick Kilns

(*a*) Temporary brick kiln and open kiln or clamp
(*b*) Intermittent permanent brick kiln
(*c*) Continuous permanent brick kiln

2.13.1.1 Temporary brick kiln or open kiln or clamp

The clamp is formed out of the bricks which are to be burnt in the kiln. Such a clamp can be very quickly made and in one such type of clamps about 1800 bricks can be burnt at one time. By reducing the size of the clamp less number of bricks can also be burnt.

The method of construction of the temporary kiln which is given below can be understood with the help of Figs. 2.11 and 2.12.

First of all separate out a lot of 1800 dry bricks. Now arrange the bricks as shown in Fig. 2.12. In one row set the bricks lengthwise and in the other keep them on their edges (1). This order should be repeated in alternate rows. The bricks should be arranged in such a manner that space for burning fire is left in the middle of the kiln (2). The topmost bricks in the clamp should be kept on their edges very close to each other so that roof of the clamp is formed (3). In the roof so formed leave small spaces (4) for burning of the fire. The outside of the clamp formed with the bricks to be burnt should then be covered with a layer of soil 200 mm thick in order to cover the bricks (5). It should be remembered that the clamp should not be covered with a layer of soil on its roof at this stage. When the clamp is fired then with the heat produced the dampness of the bricks is driven out and the roof of the kiln would sink. Now cover the roof of the clamp with a layer of soil of 25 mm thickness (6). By this the heat of the kiln is conserved and required temperature inside the clamp for burning the bricks is developed. It should be remembered that for burning of the fire, spaces (7) must be left in the roof of the clamp.

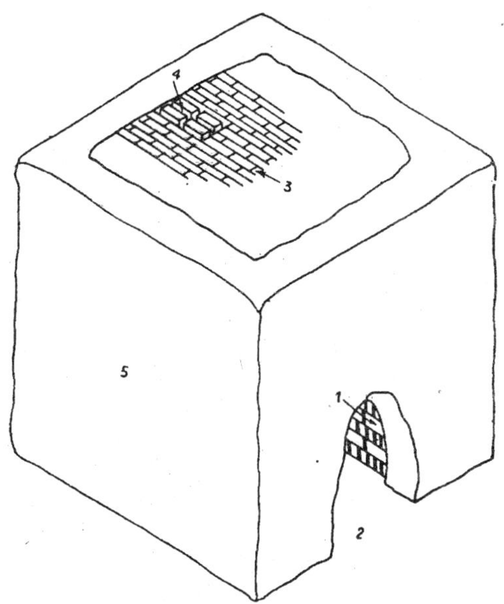

Fig. 2.11. Complete view of temporary kiln

The method of burning. After the construction of the clamp the space which is left in the middle of it should be filled form both sides with dry wooden logs so that it is half-full with fuel. After this the fire should be ignited and should be kept slow and dull so long as the dampness of the bricks is not driven out. After covering the roof of the clamp with a layer of 200 mm of soil, more fuel should be added and the fire should be burnt very briskly.

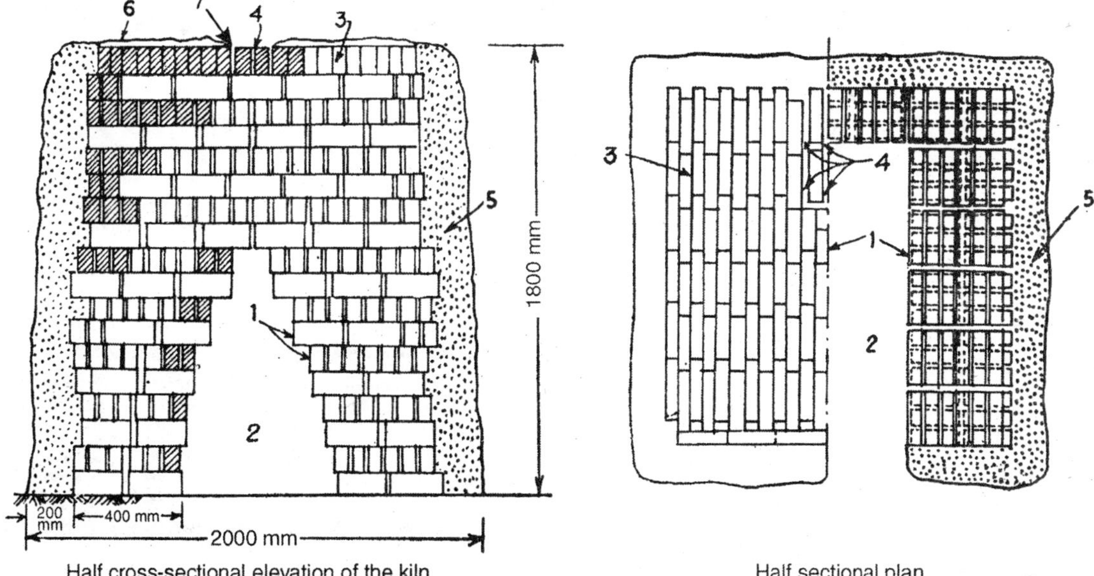

Fig. 2.12. (1) Bricks which are to be burnt. (2) Place of fire. (3) Closely packed bricks to form roof of kiln. (4) Air space between four bricks. (5) Layer of soil over the walls of the kiln. (6) Layer of soil on the roof of the kiln. (7) Space between bricks.

The fire should be kept burning in this manner for about 10 days. After 10 days scratch out the soil covering from the side at some place and see if the bricks have become strong and have burnt properly. If the bricks have not become strong then continue firing them. After proper burning of the bricks allow the clamp to cool. On examining the burnt bricks, it will be found that about 15-20 percent bricks have either broken or are not fit for use.

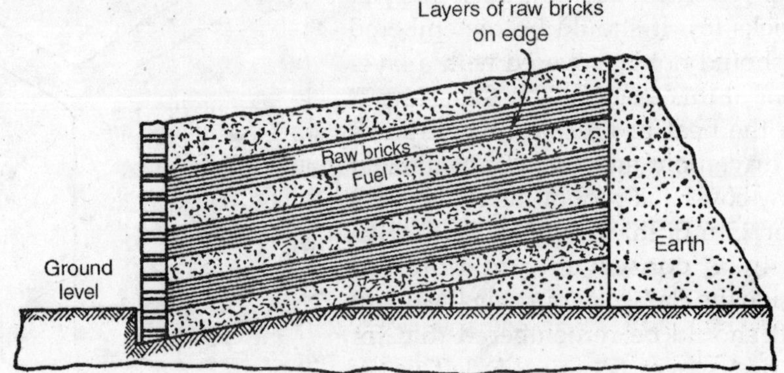

Fig. 2.13. Temporary brick kiln or clamp

Another type of clamp is shown in Fig. 2.13 in which only one wall is permanent. The raw bricks and the fuel are loaded in alternate layers at an angle of about 15° to the horizontal.

2.13.1.2 Intermittent permanent brick kiln

The method of constructing intermittent permanent kilns would be clear from the sketches given in Figs. 2.14 to 2.16. Their advantage is that several loads of bricks can be burnt in the same kiln. The bricks can be set into the kiln very quickly as two of its walls are permanent.

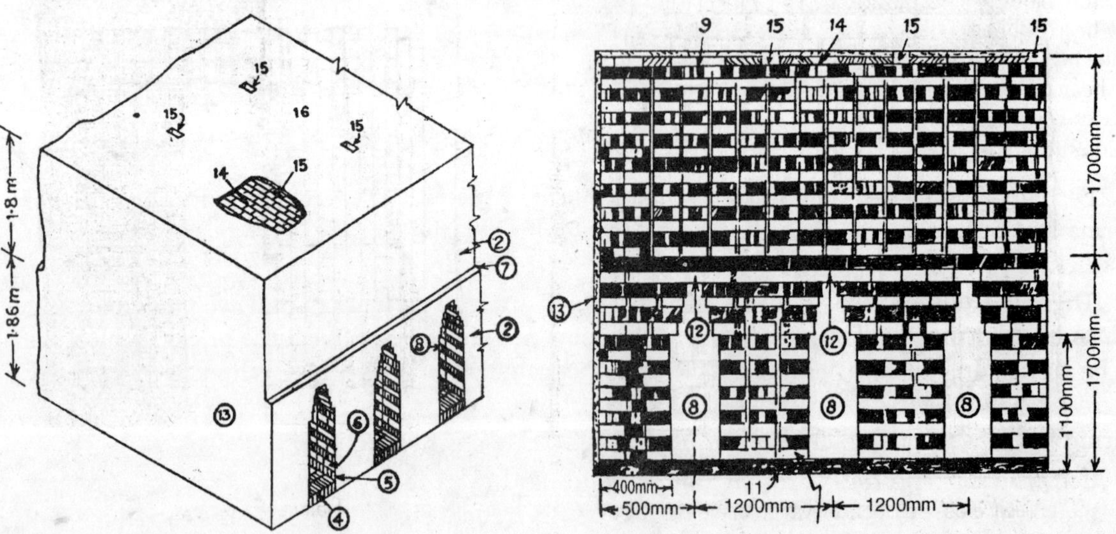

Fig. 2.14. Complete view of the permenent kiln

Fig. 2.15. Sectional front elevation of the kiln

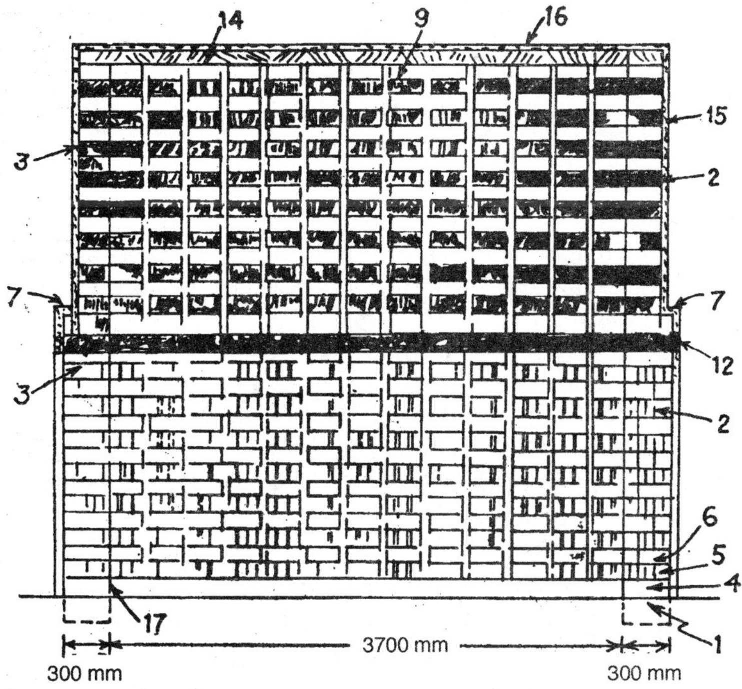

Fig. 2.16. Sectional view of the kiln

(1) Foundation; (2) The front permanent wall which is 300 mm. thick unto height of 1.8 m and rest of which is 200 mm. thick—total height of the wall 3.7 m ; (3) The back permanent wall is constructed in the similar manner as the front wall ; (4) As shown in sketch 'A' in Fig. 2.17 a first layer of the brick in wall; (5) As shown in sketch 'B' in Fig. 2.17 the second layer of bricks in wall ; (6) As shown in sketch 'C' in Fig. 2.17 the third layer of bricks in the wall; (7) From here the thickness of the wall is 200 mm. ; (8) Space for supply of fuel in the front and the back wall ; (9) The heap of bricks which are to be arranged in the kiln for burning; (10) The order of arranging first row of bricks in the heap ; (11) The order of arranging second row of bricks in the heap; (12) The arrangement of bricks to form the heap up to top of the kiln and for forming the roof of the kiln; (13) Cover of mud on the walls ; (14) Roof of the kiln formed by closely set bricks; (15) Holes for flue in the roof of the kilns ; (16) Cover of mud on the roof; (17) The inside surface of the kiln.

Permanent kilns are made only at such places which have been selected on somewhat permanent basis for brick manufacturing. Depending on the size of the kiln definite number of bricks can be burnt at a time.

The two front and rear walls of the kiln which are *only* permanent and are made of burnt bricks in mud mortar. The side walls of the kilns are constructed with the bricks, which are to be burnt. In the kiln, about 16000 bricks can be burnt at a time. When necessary, by decreasing the dimensions of the kiln lesser number of bricks can also be burnt. In any case burning of less than 10 000 bricks in the kilns does not prove to be economical.

Construction. The trench for the kiln should be dug at such a place where the front and the back walls of the kiln can be constructed. This trench should be 3700 mm long and 300 mm deep. It is better if the trenches are dug deep so that hard soil is met with and the foundation of the permanent walls of the kiln could be strong. The front and the back walls of the kilns are made of the same shape and dimension. The thickness of the permanent walls should be 300 mm upto 1.8 m height. The lower portions of the walls should be in the trench and if some space is left around, it should be filled with the soil and rammed properly. The lower portion of the walls which is below the ground level should be made of burnt bricks and the part of the wall above the ground level could be made of unburnt bricks and mud mortar. When the walls had been constructed upto the height of 1.8 m. then the thickness of the rest should be reduced by 100 mm. *i.e.*, the thickness should be reduced from 300 to 200 mm. In this manner the top portion of the walls are constructed thinner than their lower portions. By making the lower portions thicker, the walls are able to withstand the pressure of the bricks and by keeping the upper portions thinner the load on the wall is reduced. The upper portion of the wall which is 200 mm thick is further raised to a height of 1700 mm.

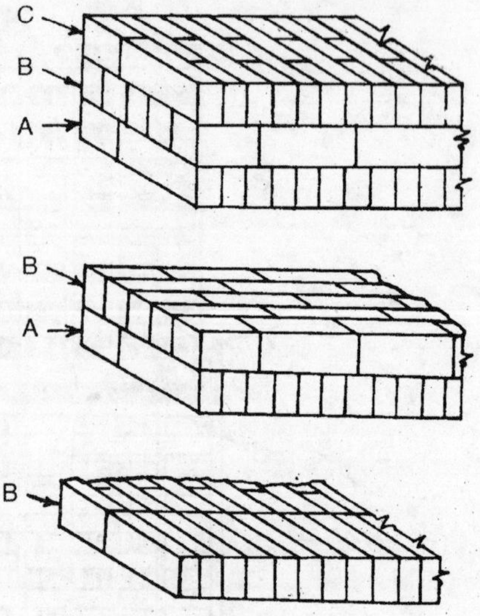

C
B
A
B
A
B

Fig. 2.17. Methods of laying bricks to form the front and back walls of the permanent kiln

The method of construction of this wall is very important. The first layer of the bricks in the wall is laid on its edges. After this apply on it a layer of mud mortar and lay the second layer of bricks. Again apply mud mortar over it and lay the third layer of bricks (See Fig. 2.17). In this way a layer of mud mortar and another of bricks are laid. When the thickness of the wall equals the length of the brick then half bricks are used.

For supply of fuel in these two walls at least three spaces should be left. These spaces should be made in the form of tunnels running through the kiln.

Loading the Kiln. When the front and the back walls of the kiln have been constructed, brick paving is done between these walls to form the floor of the kiln. For making the floor the bricks should be set very closely to each other and mud mortar should be used ; unburnt bricks could also be used. Unburnt bricks are arranged in square heaps in the kiln. The square heaps of the bricks are of three bricks length. These are raised from the floor up to the top of the kiln.

For the first loading of the kiln three bricks are laid on their edges with their faces parallel to the walls. Behind these at a distance of 16 mm another row of bricks is laid in the similar manner. After this three bricks are laid so that their edges are parallel to the walls at a distance of 16 mm to each other. Like this the two rows or bricks are laid at right angles. In this order the bricks are laid to form the heap up to the top of the kiln. In every heap a space of 16 mm should be left so that the bricks could be heated from all sides.

While forming the heaps of bricks in ,the kiln three spaces should be left for the fuel. A space of 450 mm width and 1500 mm height should be left. When the bricks have been arranged in the kiln upto the height of 1500 mm then the roof of the kiln is formed by filling in a solid layer of bricks in between the walls. By forming the roof of the kiln, firstly the temperature of the kiln is reduced a little and bricks are also burnt well. On shrinking of the bricks some cracks are formed in them and the heat coming out through these helps in burning the bricks. The roof of the kiln is covered with wet soil so that the heat of the kiln may not be wasted and bricks are burnt well. Now a layer of 15 to 20 mm thickness of wet soil is pasted on all sides of the kiln. Like this the kiln is prepared which is ready for firing.

Burning the Kiln. First of all through the front and the back wall, fuel is filled in the kiln. In the beginning the fires are kept dull so that the dampness of the unburnt bricks can be extracted in the form of steam. In the day time the dampness of the bricks in the form of steam is net visible, but at night it is clearly visible. For two days and two nights the dampness of the bricks should be allowed to go off in the form of steam. The steam as it comes out dries the bricks which from the roof of the kiln. When the bricks forming the roof of the kiln have dried then the roof should be covered with a paste of wet soil completely, but in the roof four spaces must be kept open for the flue.

Before pasting the soil on the roof, bricks standing on edges should be laid flat so that they do not crack. After covering the roof, the intensity of fire should be increased. With the increase of temperature in the kiln, the bricks shrink and the roof starts sinking a little, it should be kept in mind that the roof should not be allowed to sink from here and there. Where there is more of heat there the roof sinks. For decreasing the temperature inside the kiln around that spot, the hole left for flue in the roof or near to it should be closed by putting a brick over it and covering it with a paste of soil. By doing so the temperature around that spot is reduced and the roof there stops sinking further. Due to loss of heat at some places the roof does not sink at these places. In such circumstances remove one brick from the roof of the kiln to provide the hole for the flue. By doing so that part of the roof gets more of heat and it sinks. In this manner every effort should be made to keep the entire roof at the same level.

Ordinarily bricks are burnt in a period of seven days. As for as possible the temperature of the kiln should be kept constant. For this the quantity of fuel required is increased or decreased as the case may be and an effort is made to keep the roof of the kiln at the same level as mentioned above.

When the bricks have burnt, the kiln is left to cool down and this takes about three to four days. When the kiln is cooled, bricks which have broken or badly deformed or have clogged to each other should be sorted out. The rest of the bricks could be used for construction.

When the burnt bricks have all been removed from the kiln, the permanent wall of the kiln and the floor are left as such. For reusing the kiln a fresh load of unburnt bricks is put into the kiln and the process of burning is repeated.

2.13.1.3. Continuous Permanent Kilns

In the continuous permanent kiln, there is a better control on heat. As such the burning of bricks is done properly and since the heat losses are minimised, the consumption of fuel is also less. Thus the cost of burning bricks is also reduced. In these kilns, the out turn is large as all the processing ; loading, burning, cooling and unloading are done simultaneously. The types of continuous kilns are :

(*i*) Bull's trench kiln

(*ii*) Hoffman's continuous kiln.

Bull's Trench kiln. Bull's kiln that have been commonly adopted in our country are usually oval in plan, but a circular plan can also be followed. Details of oval type of kiln are shown in Fig. 2.18. The dimensions given are for the most suitable size of small kiln, capable of burning 10,000 bricks daily. This type of kiln is made in a trench excavated below the ground level to about 2 to 3 metres depth. If full depth cannot be dug, a portion of it can be made above the ground with banks of earth. It is about 50 to 60 m in length, .and 5 to 7 m wide.

In the trench, the bricks are laid with spaces between them and forming flues at the bottom. At the top, flues are provided with flueholes at 1 m interval, to charge the fuel from time to time during the firing process. Bricks are stacked in sections of about 3-5 m length, each section being separated by a gap of about 150 mm over which two chimneys are placed. In order to avoid the heat loss by radiation, the loaded bricks in the kiln are covered with earth and coal ashes.

The outgoing gases from one section which is being fired, is led to the next section for warming the stacked bricks, before they escape out of the chimney. It takes about one complete day to burn each section. As the burning of one section is complete, the fuel holes of this section are closed and it is allowed to cool gradually. The fresh air is pre-heated by the cooling bricks. The next section is then fired. As the fire advances to next sections, the previous burnt sections get cooled. The cooled section is then unloaded and dry raw bricks can again be stacked. Thus the various processes of loading, warming, burning, cooling, unloading and stacking raw bricks can be done simultaneously, hence the kiln is said to be continuous.

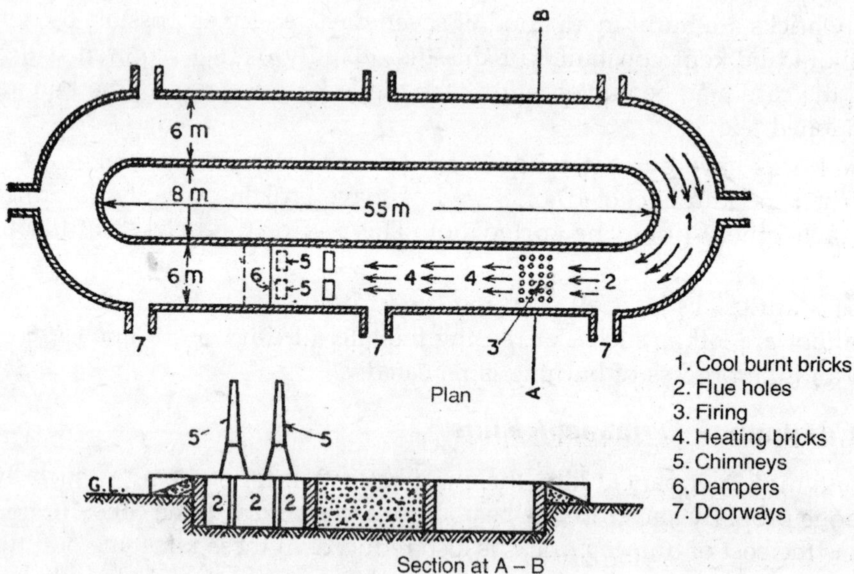

Fig. 2.18. Bull's trench kiln

Since the kiln is not covered at the top, it has to stop working during rainy season. The kiln needs about 5 tonnes of coal to start the fire and then 20 tonnes of coal is required to burn 1 lakh of bricks.

Hoffman's kiln. Hoffman's kiln has a chimney at the centre, around which are 12 chambers in the form of a ring, as shown in Fig. 2.19. Every chamber has a door opening outside which can be closed or opened, when desired with dry brick walls and sand packed in between them. Each chamber has a flue leading to the chimney which may be closed or opened, when required by using the relevant damper. Each chamber has to perform a specified function.

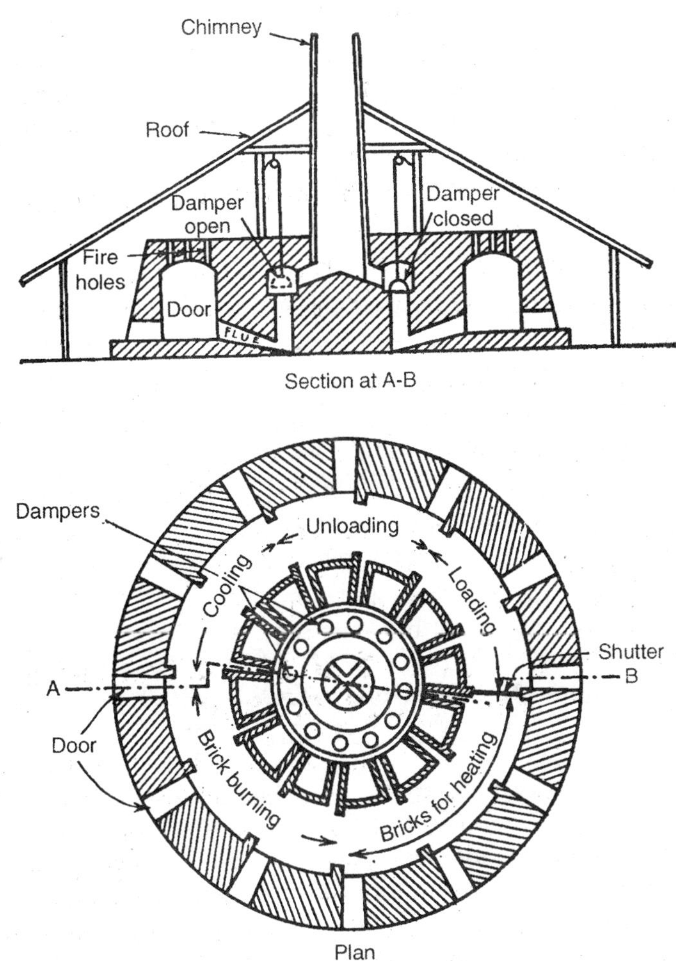

Fig. 2.19. Hoffman's continuous kiln

In the Fig. 2.19, the bricks in the three continuous chambers are being fired, the bricks in next three chambers are getting dried and heated, the bricks in the next two chambers being loaded with raw bricks, two chambers being unloaded and the last two chambers being cooled. The air enters into the chambers through two doors where the bricks are being loaded and moves forward. It gets heated by the cooling bricks. Then it reaches the burning zone, where it helps the burning of the fuel which is coal dust. The fuel is fed through fire holes, provided for the purpose. The hot gases of combustion pass through the raw bricks in next three chambers. These gases are then obstructed by the iron shutter and led through a flue into the chimney from where these escape into the atmosphere.

In a moderately sized kiln of 12 m length, 5 m wide and 2.5 m high, on an average of about 20,000 bricks per day can be manufactured.

Two to three hundred kg. of coal dust is required for burning 1000 bricks against half a ton in the Bull's trench or other intermittent kilns.

The advantages of Hoffman's kiln are :

(*i*) The fuel consumption is low. This is due to the fact that the kiln is never allowed to cool down to a low temperature. Moreover the heat of combustion gases is fully utilised in heating the raw bricks and the cool air is heated in the cooling chambers before being passed to the burning chambers.

(*ii*) The kiln being closed at the top, can be made to work during the entire year. Thus regular supply of bricks is assured.

(*iii*) Drying space and time is saved, as raw bricks are loaded directly into the chambers.

(*iv*) Bricks are burnt uniformly by the hot gases and there is no rapid draught to cause uneven burning.

(*v*) Burning can be regulated, through the fire holes.

(*vi*) The height of the chamber being only 2.5 m there is no danger of the lower layers of bricks being crushed ; thus there is high percentage of good quality bricks.

(*vii*) As the changes of temperature are not sudden, bricks do not get distorted.

(*viii*) Due to the combustion of fuel being perfect, no smoke is produced.

The only disadvantage with the Hoffman's kiln is its high initial cost. This is economical only if large and continuous supply of bricks is required.

METHOD OF MAKING GOOD BRICKS FROM BLACK COTTON SOIL

Black cotton soil is found extensively in the central and western parts of India. Because this soil is very plastic, it is difficult to make good quality bricks from it. This soil shrinks very much and contains lime nodules which expand on burning causing lime bursting in bricks. For these reasons difficulty is experienced in mixing and pugging of the soil. While moulding, the bricks made from such soils stick to the mould, crack on drying, and after that on burning, bursting of lime nodules takes place. Therefore good and strong bricks from black cotton soil as such cannot be produced.

The black cotton soil can be made suitable for brick making in the following two ways :

(*i*) By addition of 40 to 50 percent coal ash with black cotton soil. The coal ash should be finely ground and should not contain unburnt carbon. Used coarse coal ash results in weak and porous brick.

(*ii*) By grinding and burning black cotton soil and then adding in it certain proportions with the unburnt black cotton soil, good bricks can be formed. This is done in the following manner :

Dig the damp black cotton soil from a pit and separate it in a layer of 50 mm thickness over the ground. After about 2 hours, grind this soil in a grinding mill in which about 275 to

325 litres of soil can be ground at a time. After grinding the soil for 40 minutes the soil is taken out from the grinding mill and screened. One labourer is required for driving the bullocks and two more labourers are required for putting the soil in the grinding mill and for removing it from there and for screening it.

About 1100 litres screened soil is taken and water is added to it in required quantities and lumps of soil are formed out of it. For this work two labourers are required. The lumps so formed are kept dry for one night after which they are burnt in an ordinary potter kiln for 18 to 20 hours ; the temperature of the kiln is kept 500 to 600°C. The lumps should be kept in the kiln as long as the soil does not become red and when water is sprinkled on the lumps they become soft. The burnt soil lump should be removed from the kiln and ground in the grinding mill again. The burnt ground soil should then be screened and 25 to 30 percent of this is added to the unburnt black cotton soil and mixed thoroughly. For mixing the burnt and unburnt soil, form two separate heaps of the two types of soil and with the help of spade mix them well with each other. While mixing, it is important to remember that the two soils should be mixed very thoroughly. After this water in sufficient quantity should be added to the mixture and pugging done by feet. Here also it should be remembered that pugging should be done very thoroughly. This mixture is then allowed to remain for one night.

Next day with the help of wooden moulds which have a metal lining inside, the bricks are moulded. Burnt ground soil is used for sprinking in the mould and for dusting the lump of the soil which is thrown into the mould to fill it and to form the brick. The bricks are burnt in kiln. The bricks after burning are arranged in a row and are then dipped into water completely to prevent lime bursting.

For making more than 5000 bricks per day power driven grinding mill is required.

CLASSIFICATION OF BRICKS

I.S. 1077 : 1992 has classified common burnt clay building bricks as in Table 2.1

Table 2.1: Classification as per IS Code 1077

Sr. No.	*Class Designation*	*Average compressive strength not less than N/mm²*
1	35	35.0
2	30	30.0
3	25	25.0
4	20	20.0
5	17.5	17.5
6	15	15.0
7	12.5	12.5
8	10	10.0
9	7.5	7.5
10	5	5.0
11	3.5	3.5

However, bricks, burnt in a kiln, are generally classified as follows :

(*i*) First class bricks

(*ii*) Second class bricks

(*iii*) Third class bricks

The reject of the kiln are

(*i*) Under burnt or pilla bricks

(*ii*) Jhama bricks

(*iii*) Brick-bats, which are broken pieces of well burnt or under burnt bricks.

These are described below :

2.14 FIRST CLASS BRICKS (GRADE A)

The first class bricks should conform to the following requirements :—

(*a*) The size of bricks should be :

Modular	190 × 90 × 90 mm
	190 × 90 × 40 mm
Non-modular	230 × 110 × 70 mm
	230 × 110 × 30 mm

(*b*) They should be made from good brick earth, free from saline deposits and should be hand moulded.

(*c*) They should be thoroughly burnt without being vitrified and should have uniform deep red, cherry or copper colour.

(*d*) They must be regular and uniform in shape and size with sharp and square corners and parallel faces.

(*e*) They must be homogeneous in texture and emit a clear ringing sound on being struck.

(*f*) They should be free from flaws, cracks, chips, stones, nodules of lime or kankar and other blemishes.

(*g*) A first class brick should not absorb water more than 20% of its own weight after 24 hours' immersion in cold water.

(*h*) The first class bricks must have a minimum crushing strength of 15 N/mm^2 (150 kg./sq. cm). Any individual brick shall not fall below the average strength by more than 20 percent.

(*i*) First class bricks should show 'Nil' signs of efforescence either in dry state or subsequent to soaking in water.

2.15 SECOND CLASS BRICKS (GRADE AA)

Second class bricks should have the following characteristics :

(*a*) They should be as well-burnt as first class bricks or slightly over burnt but not vitrified in any part.

(*b*) They must give a clear ringing sound when struck.

(*c*) They may have slight irregularities in size, shape and colour provided these irregularities are not such as to give uneven courses when used for construction.

(*d*) They may have slight chips, flaws or surface cracks but must be free from lime or kankar nodules and be homogeneous in texture.

(*e*) The minimum crushing strength of second class bricks shall 7.5 N/mm^2 (75 kg. per sq. cm.) Any individual value shall not fall below the average strength by more than 20 percent.

(*f*) They shall not show any appreciable sign of efflorescence either in dry state or subsequent to soaking in water.

(*g*) They should not absorb more than 22% of their own dry weight when immersed for 24 hours.

2.16 THIRD CLASS BRICKS

(*a*) These need not be so fully burnt as first or second class bricks. These may be slightly be under-burnt or slightly over-burnt.

(*b*) They may be distorted and have rounded edges and may not be uniform in shape. These defects, however, shall not be such as to cause difficulty in obtaining uniform courses with their use.

(*c*) They shall not absorb water more than 25% of their own dry weight after 24 hours' immersion in cold water.

(*d*) Third class bricks may show moderate signs of efflorescence.

2.17 PILLA BRICKS

Underburnt or *pilla* bricks are those, which remain half-burnt and have a yellowish colour. These bricks are easily breakable and their use is prohibited, except in sun-dried brickwork.

2.18 SPECIAL TYPES OF BRICKS

Special types of bricks are sometimes required for use at specified locations. These are shown in Figs. 2.20 and 2.21.

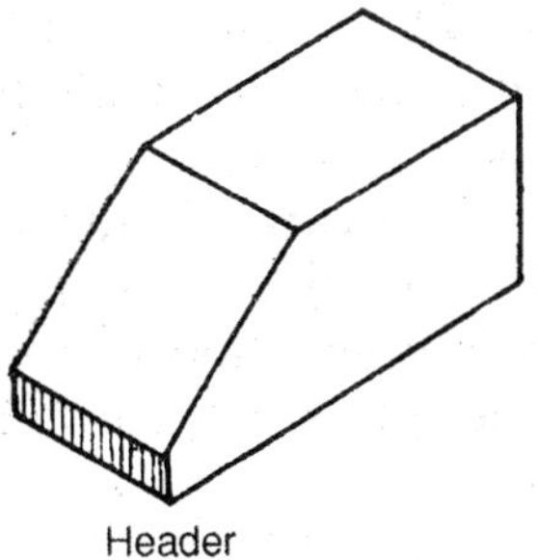

Header

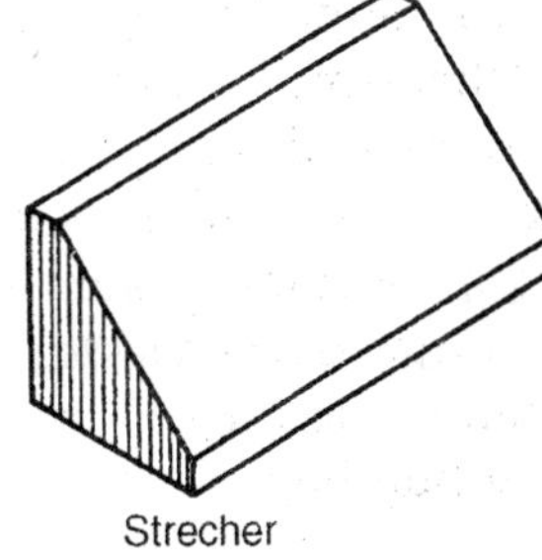

Strecher

Coping Brick

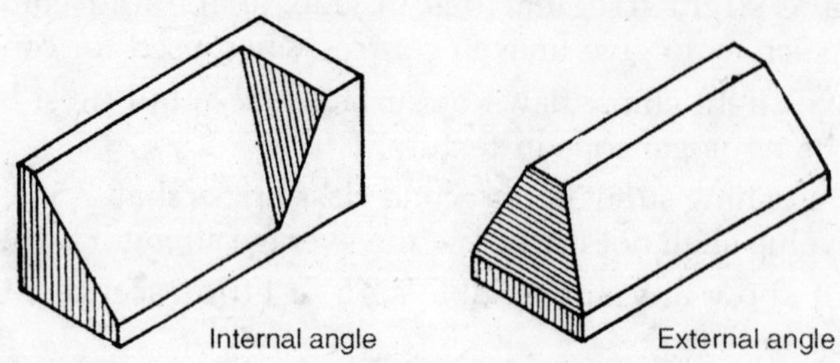

Fig. 2.20. Special bricks

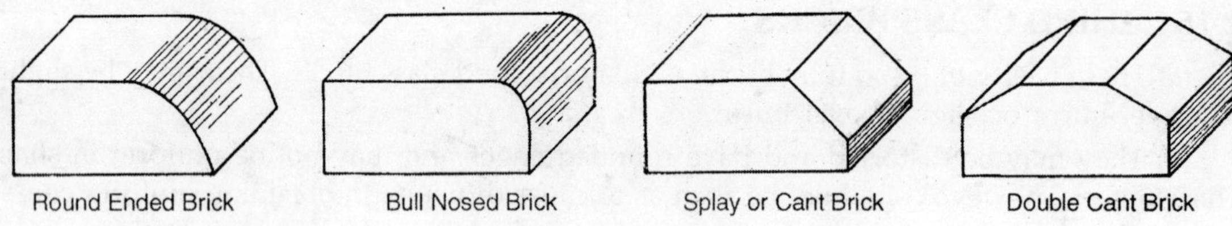

Fig. 21. Special bricks

2.19 JHAMA BRICKS

Jhama bricks are over-burnt bricks, which et vitrified or distorted so as to be useless for exact work. They may be broken up for ballast provided the vitrified mass has not become porous of spongy in the process of being over-burnt. Selected *jhama* bricks may be allowed for use in foundations of temporary buildings or for soling coat of unimportant roads.

SAND LIME BRICKS

Although not a clay product, they are used as a subsitute for common building bricks.

Sand lime bricks are made from a dry mixture of sand and 5 to 10 percent of hydrated lime which is damped slightly and pressed into bricks under a pressure of 105 N/mm^2 (1050 kg./ cm^2.). The bricks are cooked in superheated steam at 0.711 N/mm^2 (7.11 kg./cm.^{2}) pressure for about 8 hours, and at this temperature, some of the lime reacts with the surface of the sand particles to form a hydrated silicate which acts as a bond. When the bricks are removed from the cooker and stacked in the air, the balance of the hydrated lime reacts with the carbon dioxide of the air to form calcium carbonate which increases the strength of the bricks. On account of their smooth surfaces, even shape, freedom from efflorescence, uniform colour and satisfactory strength, sand lime bricks are sometimes preferred to clay bricks for use on frontages of very important building.

2.20 PAVING BRICKS (IS 3583 : 1988)

Special type of bricks for use on road surface are manufactured. These bricks vary in sizes from 190 × 90 × 90 mm to 190 × 90 × 40 mm. The 90 mm depth bricks are commonly used. These bricks have plain or wire cut surfaces with two or more lugs (extension 3 to 6 mm beyond the face) on one side and a vertical bar lug on each end. These bricks are not provided with frogs.

Paving bricks are made from fire clay or shale and their approximate composition is as follows.

Silica (SiO_2)	56	percent.
Alumina (Al_2O_3)	22	"
Lime (CaO)	1	"
Magnesia (MgO)	1	"
Alkalies K_2O and Na_2O	4	"
Organic matter	7	"

These bricks are burnt in kilns at a temperature ranging from 900 to 1200°C and take about 7 to 10 days for proper annealing.

Paving bricks should be free from marked distortion and should have one plane face. It should have closed-grained structure and should produce ringing sound when struck together. It should be uniform in quality so that the pavement may wear evenly. The crushing strength should be greater than 40 N/mm^2. The absorption should not more than 5 percent when immersed in water for 24 hours.

REFRACTORY BRICKS

Refractory bricks are those which are capable of enduring high temperatures and are commonly used in the lining of heat resisting structures such as furnaces, fire places, chimneys, boilers, ovens, kilns, converters, driers etc. Following are the main requirements of a refractory brick :

(*i*) The ability to withstand high temperatures.

(*ii*) The possession of suitable physical and chemical properties to resist the action of hot gases and slag produced during the process.

(*iii*) Power to resist abrasion when hot, and

(*iv*) Low thermal conductivity.

Refractory Clay

Refractory clay is that clay which can stand very high temperatures without becoming soft or melting. In other words, the fluxing materials which are mainly responsible for lowering the melting point, should not be present. Thus refractory clay should be composed of heat resisting materials such as silica, alumina, magnesite, bauxite, chromite, etc. and should contain substances like oxide of iron, lime, alkalies, etc. in as small amounts as possible.

2.21 CLASSES OF REFRACTORY BRICKS

Refiactory bricks are of three types, *i.e.*

(*a*) Acid bricks

(*b*) Basic bricks and

(*c*) Neutral bricks.

These are described below :

2.21.1. Acid Bricks

Bricks commonly used to resist the action of silicious and other acid slags such as fire-bricks, silica bricks come under this category.

Fire Bricks. Fire bricks are made from fire clays or refractory clays which are generally mixed with small percentages of sand to reduce shrinkage during burning. The clay, generally, comprises 50 to 75 percent silica, 20 to 40 percent alumina and less than 10 percent of fluxes. The more refractory bricks are made from clays having low flux contents and high alumina contents. A very high alumina content, however, gives the bricks a basic reaction. The bricks are moulded by either the soft or stiff-mud processes or repressed after partial drying. The first class fire-bricks should not soften at a temperature less than 1700°C and should withstand a load of 0.2 N/mm^2 (2 kg/cm^2). at a temperature of 1350°C.

Silica Bricks. Silica bricks are made from quartzite and sandstone or silica sand which contains more than 95 percent of silica. If the silicious rock contains small percentages of clay, it is sometimes possible to mould the brick without artificial addition of lime. Ordinarily, however, the ground silicious rock or the sand are not sufficiently plastic when tempered with water and are mixed either 2 percent of high calcium lime to act as a binding material or very small percentages of fire clay are added. The silica bricks are moulded by the soft-mud process. They are burnt at a temperature of about 1483°C which is slightly higher than that of fire bricks. Silica bricks have a large coefficient of expansion over a portion of the temperature scale as well as a considerable permanent expansion during firing. As such they are laid with wide joints. The compressive strength of silica brick tested flatwise is often more than 15 N/mm^2 (150 kg./mm^2). The softening temperature ranges from 1700 to 1800°C.

Their chief use has been found in the steel industry and in coke ovens.

2.21.2 Basic Bricks

Basic bricks are manufactured from magnesite, dolomite, bauxite etc. Common types of basic bricks are described below :

Magnesite Bricks. These consist essentially of magnesium oxide with approximately 12 percent of other oxides. The magnesite is first calcined at 800°C and carbon dioxide is driven out. This is usually done at the mines. A large proportion of this calcined magnesia is sinstered at 1800°C. It is then mixed with 1 to 5 percent of calcined magnesia and the mass is tempered with a small amount of water. For bricks which must resist high temperature, the proportion of calcined magnesia must be low. Since such mixtures are of low plasticity, a very little tar or magnesium chloride is sometimes added to act as a binder. The bricks are then hand moulded, partially dried and repressed. After further drying, the bricks are burnt.

These bricks are used for lining basic Bessemer converters, basic open hearth furnaces, blast furnaces, copper furnaces and in various metallurgical processes.

Dolomite Bricks. Although not so satisfactory as magnesia, calcined dolomite, the carbonate of magnesia and lime are often used. The natural rock or a mixture of the rock and clay is ground, tempered with water and moulded. After drying, they are burnt like magnesia bricks. These bricks suffer greater contraction at high temperature than magnesia bricks.

Bauxite Bricks. Bauxite, which consist of 50 percent or more of alumina and iron oxide, silica as impurities, is calcined and crushed to a fine powder. It is then mixed with 15 to 30% of fire clay with water and moulded into bricks. Although properly burnt bauxite bricks are highly refractory yet they have such a high shrinkage at high temperatures that they have not come into general use.

2.21.3 Neutral Bricks

In a furnace and in certain flues where the reaction of the surrounding medium may be either acidic or basic, refractory bricks which are neutral in reaction are used. These bricks are moulded, dried and burnt in the same way as magnesia brick. The various types are :

Chrome Bricks. Contain 30 to 40 percent of chrome oxide and the remainder being mainly oxide of magnesium, aluminium and silicon. They possess great resistance to the corrosive action of slags and gases. They are less refractory and weaker than magnesia bricks.

Chrome–magnesite Bricks. Chrome magnesite bricks are made from mixtures of dead burnt magnesite and chrome ore. They possess good mechanical properties.

Spinel. Spinel is a special refractory produced in an electric furnace consisting of equal parts of magnesia and alumina.

Forsterite Bricks. Forsterite bricks are made from the mineral forsterite ($2MgO\ SiO_2$). They have less shrinkage and possess good mechanical properties at high temperatures.

2.22 STRENGTH OF REFRACTORY BRICKS

Refractory bricks can stand compressive stresses of about 150 to 200 N/mm^2 (1500 to 2000 kg./cm^2). They weigh about 2400 kg./m^3. Water absorption varies from 4 to 10 percent.

2.23 LAYING OF REFRACTORY BRICKS

Refractory bricks are always set in a mortar of fire clay and not in lime or cement mortars.

TERRA-COTTA AND FAIENCE

2.24 TERRA-COTTA

Terra-cotta means burnt clay. It is a kind of earthen-ware which is used as a substitute for stone for the ornamentation of buildings and structures. It is less costly than stone, when a number of pieces of an intricate design of one pattern are frequently repeated in a structure. When properly made, terra-cotta weathers well and is not affected by atmospheric influences and acids. Because of its glazed surface, it can be cleaned more easily than a porous stone. Moreover it is very light, strong and hard. It is also fire proof.

Terra-cotta clay. Terra-cotta is made from a finely ground mixture of refractory brick clay and shale or refractory clay and impure clay which is mixed with ground brick or other burnt clay products to reduce shrinkage. The approximate composition of a good terra-cotta clay is as follows :

Silica	70.0	percent
Alumina	10.0	"
Ferric oxide	6.0	"
Lime	1.0	"
Magnesia	Traces	"
Water	5.5	"
Alkalies	0.5	"
Organic matter	7.0	"

Clay having higher percentage of iron and smaller percentage of lime are preferred for the manufacture of terra-cotta. Usually one single clay does not conform to the desired chemical composition. Thus a mixture is prepared of fire clay, silica sand, crushed old terracotta or pottery, powdered glass etc.

Preparation of Terra-cotta Clay. The various ingredients are first crushed and finely pulverised in a dry state and mixed in the required proportions to obtain the desired composition. The mixture, thus obtained, is thrown into large tanks containing water. It is mixed up thoroughly and lifted through wooden boxes with loose joints through which the water runs off. The strained mixture is pugged a number of times until it becomes uniform.

Moulding. Plain forms of uniform section are extruded through dies and other simple shapes are hand moulded in moulds made of plaster of Paris. Usually an allowance of 5 to 10 percent is given for shrinkage during drying and burning. Soft soap powder is smeared on the moulds to prevent adhesion of the clay to the mould. Intricate patterns are moulded in green clay by sculptors and used for making plaster moulds. Thirty to fifty pieces can be reproduced from a given mould. Since the mould does not last long, patterns are made from the full size drawing and moulds of plaster of Parirs are taken from these patterns. The moulding is done by pressing the clay carefully in the moulds in a manner similar to brick moulding.

Drying and burning. After being moulded, terra-cotta pieces are slowly dried to avoid warping. This is usually done in a room in which temperature is controlled. Then they are sprayed with a "slip". This slip made up of clay, felspar, flint etc., is opaque and imparts the desired colour to the pieces on burning. Burning is done with utmost care to prevent distortion or discoloration by flames and gases. Special kilns in which the terra-cotta does not come in contact with gases are desired. It is usually burnt in tunnel kilns, the dried pieces being loaded on cases which are slowly moved through the kiln. Usually 5 days are required to burn the product. The maximum temperatures attained are about 1100° to 1200°C.

The natural colour of terra-cotta is buff and grey but any desired colour can be given by glazing before burning. Glazed terracotta has a matt glazed surface.

Uses. Hand made special shapes of terra-cotta are used on the ornamental parts of buildings such as cornices, copings, sills, jambs, arches, bases of pillars etc. Blocks of terra-cotta are used for wall facing since it is hard. It can also be used for floors. These blocks are either machine moulded or hand moulded.

Size. Machine made terra-cotta blocks for masonry facing ranges in size from 200 mm by 400 mm to 600 mm by 1200 mm. The blocks may be solid and of 50 mm thickness or they may be formed with hollow cores, the total thickness being 100 mm. Hand made slabs range from 300 mm to 750 mm in width, from 100 to 300 mm thick and are of various lengths. They are usually hollow and have open backs so that they can be bonded to masonry walls. In a solid condition, terra-cotta weighs about 1800 kg/m^3.

Porous terra-cotta. Porous terra cotta member is made by mixing straw or saw dust in the clay. It is fire-proof, can be sawn, can hold nails easily and stick well to plaster. This is very rarely used in India.

2.25 FAIENCE

It is similar to terra-cotta but the pieces are burnt twice. They are first fired like terra-cotta when it has reached a biscuit colour, commonly known as biscuit stage. They are then removed from the kiln, allowed to cool, coated with glazed compound and fired again. This is stronger and has a greater colour range than terra-cotta. This is used in industrial areas since it is ordinarily unaffected by adverse atmospheric conditions and for ornamental purposes.

EARTHEN-WARE AND STONE-WARE

2.26 EARTHEN-WARE

Earthen ware is made from ordinary clays, similar to brick or tile clays. They are burnt at a comparatively low temperature. Since the body of the ware is not completely fused, it is more or less weak and porous and is liable to be damaged by frost. It approximates to tarra-cotta in character.

It is used in making ordinary drain pipes, electrical cable conduits, partition and patent floor blocks.

2.27 STONE-WARE

It is a name given to articles made from refractory clays to which powdered stone-ware, sand etc. have been added and burnt at high temperatures like fire bricks. The articles are impervious to moisture, hard, and closed grained and durable.

The plastic clays, from which it is made, contains about 76 percent silica, 24 percent alumina with a very small percentage of other ingredients. The clay contains very little infusible matter and is generally mixed with a certain proportion of powdered stoneware, ground decomposed granite or sand to prevent excessive shrinkage. A certain percentage of oxide of iron is added to give it the desired colour and for proper fusion.

The clay is prepared carefully and moulding is done accurately to the required shape. It is then dried and the material burnt in a domed kiln, raising the heat gradually to a high temperature and maintaining this for 1 to 3 days depending on the size of the kiln and the articles in it. The stoneware, during burning, acquires a glaze due to the ingredients added in the clay. The burnt stoneware is allowed to cool down gradually. If stoneware is to be used for resisting the action of acids, salt glazing is done.

Sewer pipes, sanitary fittings, etc. are made of stoneware material.

2.28 CLAY PIPES

Sewer Pipe. In the construction of sewers, clay pipe has been used with advantage. These should have high strength, and be impervious and hard-burnt. These should be capable of resisting action of acids and gases in the sewage. They should be straight, smooth on the inside and free from cracks, etc.

Sewer pipe is made from stoneware clays. It is moulded in presses from a stiff-mud or dry-mud. Sewer pipe is burnt at temperatures between 1050 to 1300°C. Salt glazing is done to ensure a smooth and impervious surface.

The pipe is made in lengths of about 9 m and in diameters upto 100 mm. Sewer pipes are commonly provided with a belt on one end into which the small end of the adjacent pipe gets fitted. The absorption should be less than 7 percent after boiling for 5 hours in distilled water.

Drain Pipe. Drain pipe is a type of earthen-ware. This is moulded in machine or sewer pipe presses. They are burnt in kilns at a temperature lower than those used in burning sewer pipe. Salt glazing is sometimes employed. These are usually cylindrical in form and laid with uncemented butt joint and are used for drainage of seepage water. The absorption should not be more than 11 to 13 percent.

Conduit. In large cities, conduits made of clay are usually used for carrying underground cables and wires. Conduits are manufactured from fire clay and are given salt glazing to render them impervious to water.

HOLLOW OR PERFORATED CLAY BRICKS

In many countries solid bricks are being replaced by hollow bricks which are lighter and bigger in size. A new type of brick, called perforated brick is of particular advantage in the construction of multi-storeyed buildings to reduce the dead load of the structure and load on the foundation. The use of such bricks also provides better thermal insulation. Proportion of perforations in perforated bricks that are manufactured range from 25 to 50% while the size of perforation varies from country to country. According to the German and U.S.A. standards, bricks with perforations upto 25% are classified as solid bricks. The following are the manufacturing advantages of perforated bricks. For manufacturing hollow and perforated bricks, it is necessary to resort to mechanised means.

2.29 MANUFACTURING ADVANTAGES

(*i*) The cost of raw material (clay) constitutes approximately 25 percent of the cost of production of solid bricks in a mechanized plant. Introduction of perforations proportionately decreases the amount of clay required to make a brick thereby affecting substantial saving in the production cost.

(*ii*) Perforations make it easy for the clay bricks to be extruded out. Therefore, the rate of production is increased and the power required is less.

(*iii*) Because of perforations the circulation and penetration of heat is quicker thereby facilitating the oxidation of iron compounds, carbon and sulphur in the clay and uniform burning of the clay from inside to the outer surface. Thus it speeds up the burning rate and it has been claimed that by replacing solid bricks by perforated bricks, as much as 50% increase in the kiln output can be achieved.

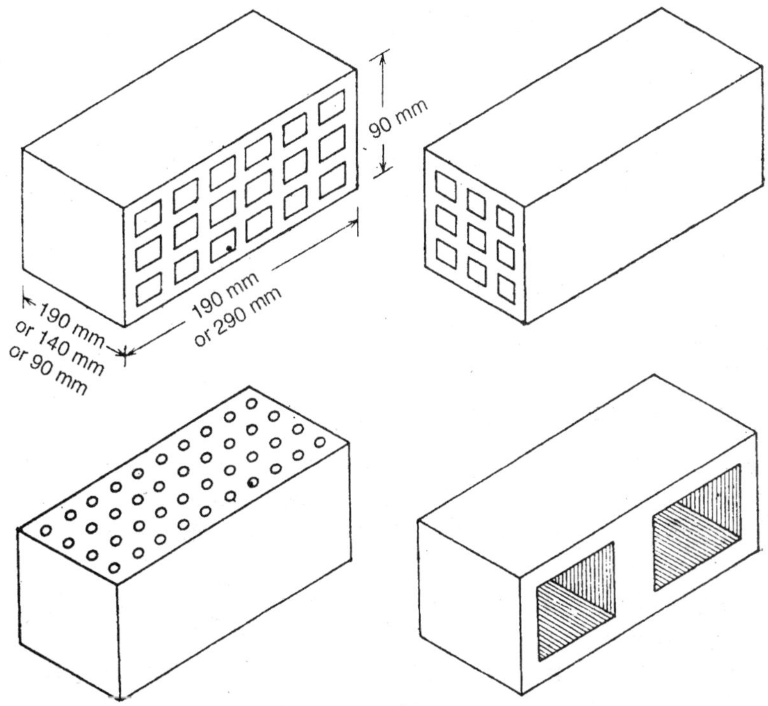

Fig. 22. Various types of perforated bricks

With normally adopted pattern of perforations there is no thickness generally more than 20 mm which requires drying. As such the perforated brick dries in much shorter time in the sun than the solid brick thus saving yard space required for drying and reducing lead for sending the bricks in the kiln. This results in increased production with less cost.

(*iv*) In addition to permitting increase in output, perforations, in the bricks also lead to economy in fuel used for burning. It acts in two ways to produce this result ; there is smaller weight of clay to be heated in burning a given volume of bricks and also the faster fire travel makes it possible to reduce the time for which each chamber or section is under fire. Thus the amount of heat lost, both from the kiln structure as well as out going gases, is reduced. It is claimed that as much as 33% fuel economy can be achieved.

(*v*) Due to less weight of the perforated bricks larger number of bricks can be transported for a given tonnage rating of the carrier, thus affecting saving in the transport cost of the bricks.

2.30 PROPERTIES OF PERFORATED BRICKS

Strength. Hand moulded bricks which cannot be compressed and burnt properly, made from the alluvial soils of Indo-Gangatic plain, are known to have crushing strength values ranging from 8 to 12 N/mm^2. It is claimed that a perforated brick made by extruding with de-airing and burning at the optimum temperature under controlled firing conditions will have crushing strength greater than 12 N/mm^2. It may, therefore, be generalized that perforated bricks may by mechanized means are expected to have higher crushing strength than those made from the same clay be hand moulding methods.

Durability. Since the quantity of raw material used for making the perforated bricks is less, proportionately the quantity of soluble salts will be less. In addition, the perforations interrupt the passage of water in the brick and therefore there is liklihood of less efflorescence occuring. As perforated bricks made with mechanized means will be better burnt and less porous, there will be still less tendency to efflorescence in the perforated brick.

Resistance to rain penetration. Hand moulded bricks as produced around Delhi are known to possess 22-26% of porosity. This high porosity is known to absorb appreciable quantities of rain water (if there is no outside rendering) and transmit moisture to the inner side of the wall. Perforated bricks made by mechanized means and burnt properly are expected to have porosity of not more than 20 per cent. This low porosity, it is claimed, will reduce penetration of water through the brick and any penetration therefore will only be through the joints. In addition, perforations will be beneficial in interrupting the direct path to the rain water from outside to inside of the wall over a significant part of its area.

Thermal insulation. Air space in the perforated brick will substantially reduce the passage of heat through it. By experimental means it has been established that a perforated brick with 50% voids will permit only half the quantity of heat which passes through a solid brick with 20% voids. In addition, decrease in solid matter will proportionately decrease the conductance of heat through a wall built with perforated bricks. The heat capacity of the wall is however reduced proportionately.

2.31 APPLICATION ADVANTAGES OF HOLLOW OR PERFORATED BRICKS

The following are the advantages in the use of perforated bricks in construction :

(*i*) *Rapid construction.* Bigger sizes of perforated bricks for the same convenient weight, from handling point of view, can be manufactured ; thereby resulting in rapid construction and saving in the unit labour cost of construction.

(*ii*) *Less mortar consumption.* Use of bigger units in construction will also decrease the number of joints ; thereby reducing the quantity of mortar required to lay a specific volume of wall. This feature of application of perforated brick will also decrease the cost of construction. It may be mentioned here that the view held by a section of engineers that a quantity of mortar will be consumed in filling the perforations has been proved to be erroneous by practices. The entrapped air in the perforations resists its filling by upward thrust. In addition the drag of the wire-cut partially closes the mouth of the perforations also.

(*iii*) A perforated brick made, as it is, by mechanized means and properly burnt will have low porosity compared to the present hand moulded bricks. In addition to the advantages of better resistance to rain penetration and thermal insulation, the uniformity in sizes and texture will enable the use of these bricks in exposed exterior brickwork thus economising in the cost and the materials required for external rendering necessary in the case of the present type of bricks.

CLAY ROOFING TILES

Clay products that resemble bricks but are thinner and called bricks tiles have been used for structural work in the construction of walls, for flooring and for making the flat-roof waterproof. Still thinner clay tiles of particular shape and size, as may facilitate throwing off rain-water quickly, have been used for covering the roofs and are called roofing tiles. The use of such tiles has been extensively practiced, specially in the rural areas where sloping roofings are commonly built. Also in the urban areas, particularly in the regions of heavy rain-fall, clay roofing tiles have been employed for the construction of sloping roofs.

2.32 ADVANTAGES OF CLAY ROOFING TILES

Clay roofing tiles are locally manufactured and are cheap. Generally, good soil for making the tiles is found in almost every part of our country. The method of tile making is simple and it requires semi-skilled labour. On account of these advantages, the use of clay tiles for roofing is suitable for building low cost houses. Besides the above, the clay tiled roof offers many advantages, which are as follows :

(*i*) Clay tile roof performs admirably in the tropical climate as is ours as clay tiles possess good heat insulating properties and therefore prevent penetration of solar heat into the rooms from the roof top. Thus during summer months the house remains cool, whereas in winter months it does not become excessively cool.

(*ii*) Clay being inorganic material, clay tiled roof is more durable and offers greater resistance to fire, as compared to the roof made of grass-thatch or palmyra leaves.

(*iii*) Clay tiles being thin and light in weight require a light roof structure to support them and as such the cost of roofing is low.

(*iv*) Clay-tile roof, if constructed properly, does not involve much of maintenance job as parts of roof where the tiles might have been broken can be replaced by new tiles.

(*v*) A clay-tile roof can be built on self-help basis as it is easy and quick to construct.

2.33 DEFECTS OF CLAY ROOFING TILES

Despite the advantages, as given above, the use of clay tiles for roofing is becoming less popular and some reasons for this are as follows :

(*i*) Often it is difficult to make the clay-tile roof leak proof. Clay tiles being porous absorb rain-water and due to this the roof remains damp and the tiles also become weak and break. As such the roof may start leaking.

(*ii*) Tiles being light and if not properly fixed with the roof structure, may be dislodged or thrown away by strong winds thus uncovering the roof at some places.

(*iii*) The tiles may also slide away due to sharp slope to which the roof is laid, thus opening out the roof.

(*iv*) If the tiles are not properly laid with sufficient over-lap and to the required slope, it becomes difficult to stop the roof from leaking. This work requires great care in laying the tiles over the roof as well as it is an important consideration in the design of the shape of tile.

2.34 IMPROVEMENTS OF CLAY ROOFING TILES

From the type of defects as mentioned above it would be observed that if care is taken in design and to produce tiles of suitable design and if sufficient care is exercised in laying the tiles, the above mentioned defects of the clay-tiled roof could be avoided. To achieve this the following improvements are required :

(*i*) To ensure draining of the rain-water quickly so as to prevent leakage from the roof, proper shape of tiles should be evolved.

(*ii*) In the design of tiles, proper overlapping at the joints should be ensured to prevent leakage through the joints.

(*iii*) The tile should be designed in such a manner that they permit their being fixed on the roof structure securely to prevent displacement.

(*iv*) The shape of the tiles should be such that to form the roof a light structure is required which can be built at a low cost.

(*v*) The shape of tiles and the particular shape to which these should be laid to form the roof, should be evolved considering the climatic conditions particularly the quantity of rain-fall which occurs in a particular area.

(*vi*) The tiles should be strong and dense so that they do not break easily and absorb little water.

2.35 COMMON TYPES OF CLAY ROOFING TILES

The common types of clay roofing tiles which are used in our country have been briefly described below and need further improvement both in design and quality.

2.35.1 Country Tiles

Country tiles are made by the village potters and have been extensively used in the rural areas. The usual shape is half round tile or flat tile and are classified as class AA and class A tiles with characteristics given in Table 2.2. The dimensions of the tiles is given in Table 2.3.

Table 2.2: Classification of Clay Country Roofing Tiles

Sl. No.	*Characteristic*	*Requirement*			
		Half Round Tiles		*Flat Tiles*	
		Class AA	*Class A*	*Class AA*	*Class A*
(1)	(2)	(3)	(4)	(5)	(6)
(*i*)	Water absorption percent, Max	19	24	19	24
(*ii*)	Breaking load, kN, Min				
	(*a*) Average	0.40 (40 kg)	0.30 (30 kg)	0.35 (35 kg)	0.25 (25 kg)
	(*b*) Individual	0.35 (35 kg)	0.25 (25 kg)	0.30 (30 kg)	0.20 (20 kg)

Table 2.3 Dimensions of Tiles

	Dimensions (mm)					
	Overall length		*Overall width*		*Overall height*	
			Wide end	*Narrow end*	*Wide end*	*Narrow end*
Half Round Tiles						
	Size I	250	105	85	50	40
	Size II	250	120	95	60	45
Flat Tile						
	Size I	200	175	145	20	20
	Size II	250	200	160	25	25

A country tile is shown in Fig. 2.23. The method of making it is very simple. The village potter prepares a hollow cone from the lump of clay on his wheel. This is cut into two longitudinally. The tiles after drying are burnt in the common potter's clamp.

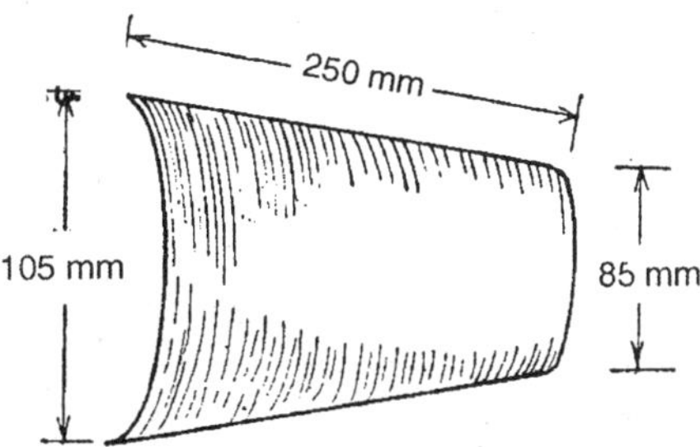

Fig. 2.23. Country Tile

Though country tiles can be made easily, the product generally is porous and not quite strong. As a result, the tiles often break and become weak due to absorption of water. The country tile roof therefore requires repairs and it is difficult to make it leak-proof. The quality of such tiles can be improved by selecting good soil and by proper burning of the tiles. The suitability of the soil can be evaluated by conducting the four simple field tests described under article 2.7.

2.35.2 Guna Tile

When the hollow cone of clay prepared by the potter on his wheel is not cut longitudinally, it is called Guna tile.

The common shape and size of the tile is shown in Fig. 2.24. Guna tiles can be used for the construction of sloping as well as flat roofs. The tiles should be carefully moulded so that the smaller end of one tile can fit into the bigger end of the other.

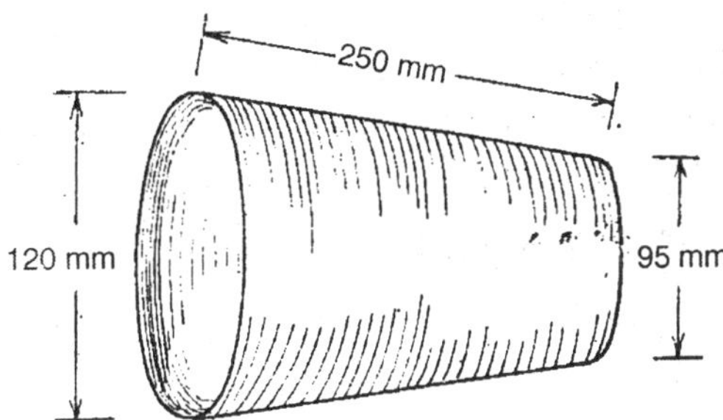

Fig. 2.24. Guna Tile

2.35.3 Allahabad Tiles

Allahabad tiles consist of two units called undertile and overtile. The undertile is flat and has a level depression with up-turn flanges along with the two edges (Fig. 2.25). Fig 2.26 shows a flat ridge tile. The overtile is half-run in section and is tapering as shown in Fig. 2.27. Another type is semi-hexagonal (Fig. 2.28.)

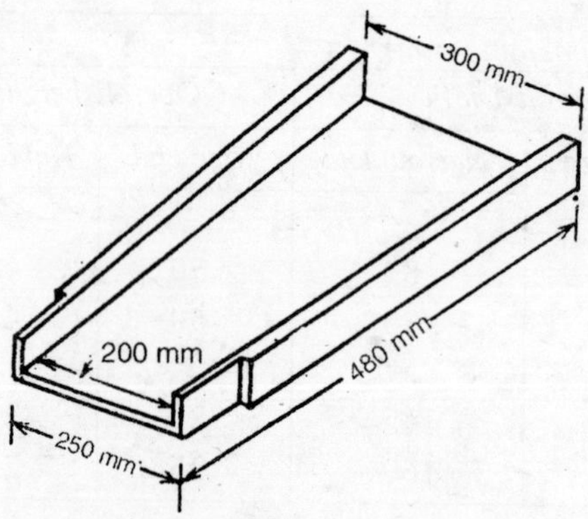

Fig. 2.25. Allahabad undertile-Flat type

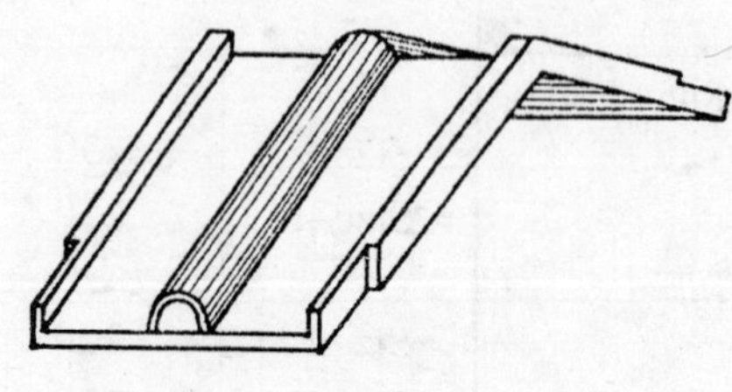

Fig. 2.26. Allahabad flat ridge tile

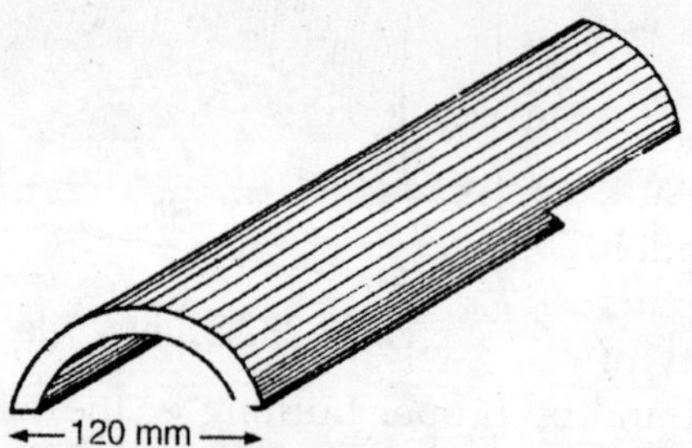

Fig. 2.27. Allahabad overtile—Half round type

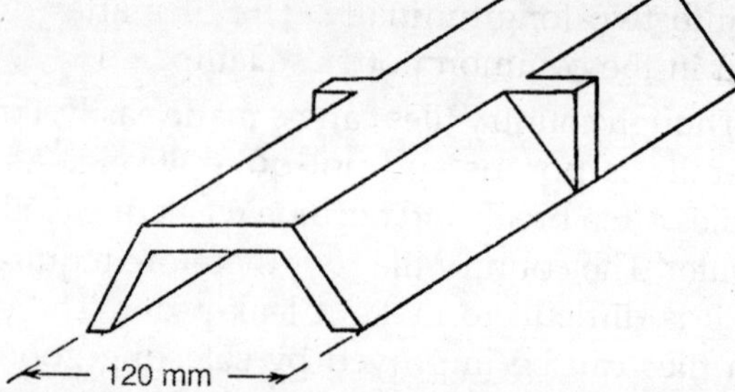

Fig. 2.28. Allahabad overtile—Semi-hexagonal type

The tile is cast on a wooden mould by putting wet plastic clay of required thickness on the mould. The mould is then removed by sliding and the wet tile is left to dry. It is then burnt in a clamp. Allahabad tiles have been commonly used for making good and pleasing roofs.

2.35.4 Raniganj Tiles and Other Local Tiles

Moulded tiles of different shapes and sizes are produced in different parts of the country. Raniganj tiles is one such variety. Pan tile, which is also commonly used, is another common type of tile.

2.35.5 Mangalore Tile

Machine moulded clay roofing tiles, called Mangalore tiles have been manufactured and used extensively, especially in South India. Being machine pressed, these are very strong and dense. These are of definite shape and size as shown in Fig. 2.29 and have such arrangements which provide for better fixity with the supporting roof structure. The tiles are machine moulded and are burnt in kilns of better design.

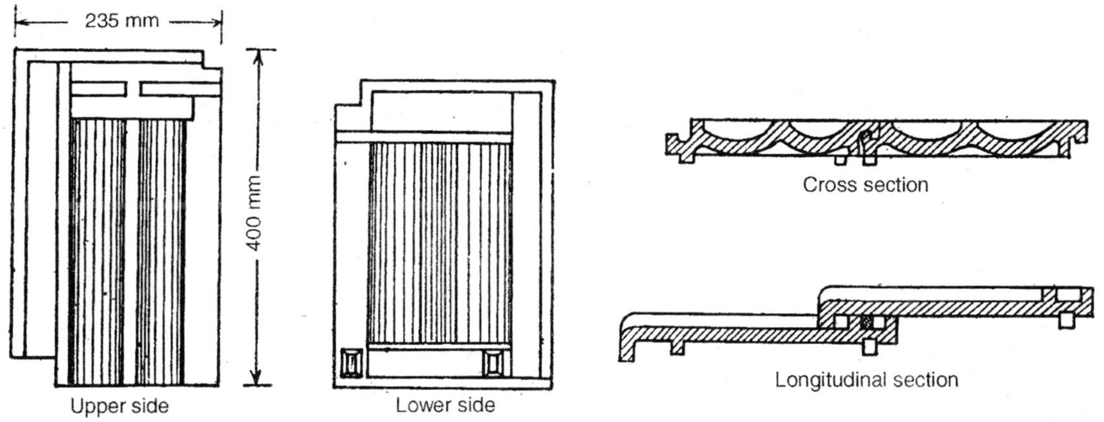

Fig. 2.29. Details of Mangalore tile

Thel Bureau of Indian Standards (I.S. 654 : 1992) specifies that the Mangalore tiles are of two classes, Class AA and Class A with the characteristics as specified in the Table 2.4:

Table 2.4: Classification of Roofing Tiles

Sr.No.	*Characteristic*	*Requirement*	
		Class AA	*Class A*
(*i*)	Water absorption percent, Max	18	20
(*ii*)	Breaking load, kN, Min		
	(*a*) Average	1.0 (for 410 × 235 mm) 1.10 (for 420 × 250 mm and 425 × 260 mm)	0.80 (for 410 × 235 mm) 0.90 (for 420 × 250 mm and 425 × 260 mm)
	(*b*) Individual	0.90 (for 410 × 235 mm) 1.00 (for 420 × 250 mm and 425 × 260 mm)	0.68 (for 410 × 235 mm) 0.78 (for 420 × 250 mm and 425 × 260 mm)

Three dimension of the tiles specified are :

Overall length mm	*Overall width mm*
410	235
420	250
425	260

Minimum overlap should be 60 mm lengthwise and 25 mm width wise. Average weight of 6 tiles should not be less than 2 kg and not more than 3 k]g.

It has been found that as much as 40% of the production of Manglore tiles is wasted on account of warping during burning. The Central Building Research Institute, Roorkee, has found that if aluminium chloride is added to the soil in certain specified proportion, 15% wastage in the production can be avoided.

2.36 BURNING OF TILES

Tiles have to be burnt more gradually than bricks to prevent warping. In case the same kiln is being used for burning both bricks and tiles, the tiles are set in the top portion of the kiln. This is because of the fact that tiles being thinner require less heat.

For better control, tiles are burnt in special types of kilns One such type is shown in Fig. 2.30. Flooring is made by a course of bricks laid flat. The tiles are set on this floor, edge wise and close to each other. The fuel used is wood. In the first instance, the fire is kept low till the gases are driven out. The temperature is then raised gradually till the inner portions of the flues become red hot. The fire is then slackened for about 6 hours, raised again to white heat and kept in this condition for three hours. The process of slackening, raising and keeping constant the temperature is repeated again. Then the flues are filled with fuel and their openings closed with bricks in mud mortar. The kiln is allowed to cool down slowly for a period of about three days and the tiles gear uniformly burnt.

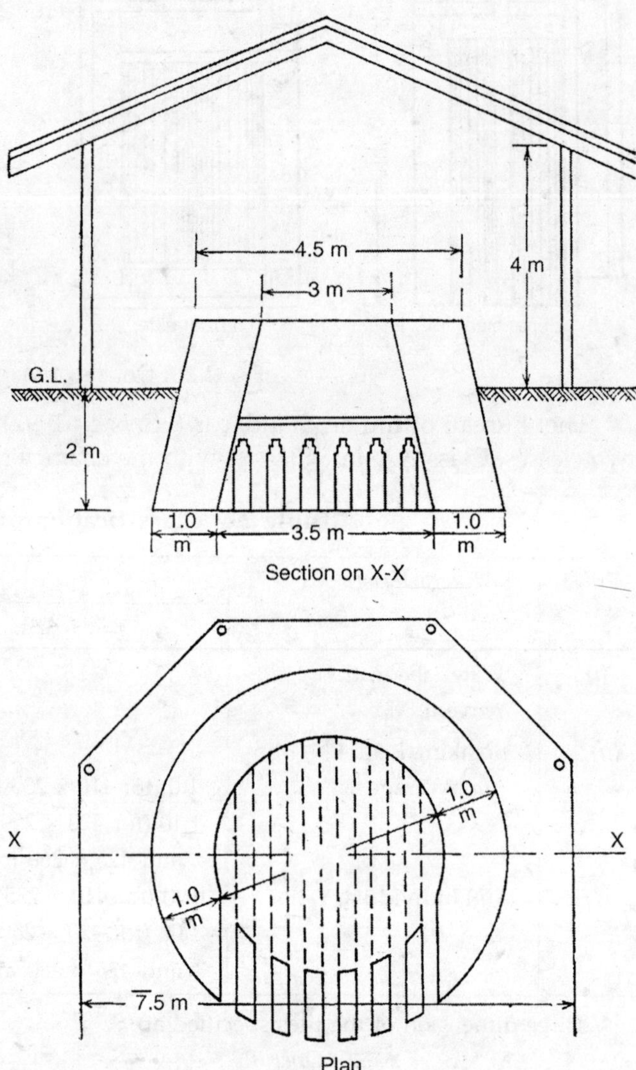

Fig. 2.30. Sialkot pattern kiln for burning tiles

FLOORING AND WALL TILES

2.37 FLOORING TILES

Clay tiles of various shapes, sizes, thicknesses, colours and surface finishes are manufactured for use as surface covering for floor. The method of manufacture is the same as for superior types of roofing tiles. If need be, to the available clay may be added ground glass, sand etc., to still improve the non-shrinkage quality. Water is mixed in the required proportion and the slurry thus formed is sieved through a fine wire mesh. The mixture so obtained is termed as 'slip'. Slip is taken to tanks, where it is allowed to stand for a sufficient time. The water from the tanks is removed and the settled material is allowed to dry. Further drying is carried out in ovens and the clay thus obtained is ground to a fine powder. Iron, if suspected, is removed by electromagnets.

Floor tiles are classified into two types depending upon the method of moulding. One is the designed face floor tile, also called architectural tile and the other plain tile. Plain tiles are usually made in the form of mosaics, either vitreous or semi-vitreous and are composed of the same clay throughout. Plain face tiles are made by covering the die, forming the base of the mould, with a thin layer of specially prepared clay, the rest filled with ordinary clay and compacted with hand press. In making designed face floor tiles, a cellular frame is placed on the bottom of the mould and clays which have been properly coloured are screened into the different cells in accordance with the pattern scheme. The frame is then removed, the backing inserted and pressure applied to compact the tile. The clay is compressed under heavy pressures using a 150 ton press to form tiles. Drying of the tiles is carried out very slowly, preferably under a shade, to safeguard against warping. They are carefully stacked in the kiln and burnt properly. The size of the floor tile recommended is 290 mm × 140 mm × 50 mm. Smaller sizes 140 mm × 140 mm × 100 mm or 100 mm × 100 mm × 10 mm are also made. Special tiles are moulded for use at corners and curved places.

Floor tiles should show little absorption, high transverse strength and a high resistance to abrasion. They should be uniform is texture so that the wear is not uneven.

2.38 WALL TILES

Tiles, for use on wall surfaces, are similar to the floor tiles, except for their design and degree of burning. They are burnt at a comparatively low temperatures, glazed and fired again at a still lower temperature. Different designs and colours are available. They are used on face work and to some extent in arches and ceilings. Recommended size for the face tile is 190 mm × 90 mm × 40 mm.

STRUCTURAL HOLLOW CLAY TILES

Structural tile is light in weight, fire proof, rat proof, termite proof and is free from decay caused by the contact with chemicals. The use of these tiles reduces the maintenance and constructional cost. Structural clay tile is a term referred to the various sized and kinds of hollow and practically solid structural units which are moulded from various types of clays. These tile units are used to build foundations, walls, partitions, floors and other structural members.

It should be noted that these units can be used while being laid on end or on side and are designed accordingly. Load bearing tiles are made in various grades *e.g.*, load bearing (L.B.) and extra load bearing (L.B.X.). Specifications have been framed in different countries to account for the various physical properties of the tiles. Generally absorption is not greater than 16 percent for good quality tiles. The compressive strength based on gross area in N/mm^2 should not be less than 20 and for individual tiles not less than 15. For non-load bearing walls these limits are relaxed.

The dimensions specified in IS 3951 (Part 2) : 1975 are :

Length	—	290 and 390 mm
Width	—	90 to 190 mm in stages of 50 mm
Height	—	125 to 200 mm is stages of 25 mm

The shell of a tile constitutes the four sides which surround the hollow interior and the webs are the partitions between the shells. The average overall thickness of the shells should not be less than 12 mm and of the web not less than 10 mm for end construction tiles. The shell for side construction tile should not be less than 12.5 mm. The width of any shell in the side construction tile measured in the direction of wall thickness, should not exceed 4.5 times the average overall thickness of either the upper or lower bearing shells. All tiles must be free from laminations, cracks, blisters, surface roughness or other defects which would interfere in the construction.

If plaster is to be applied directly to the sides of a tile the surfaces should be grooved to provide keys for holding the plaster in place. Common and face tile units may have grooves on one or two faces and both ends grooved. Groove should not be less than 10 mm and not more than 5 mm in depth. They should not be more than 25 mm in width. The area covered by grooves should not exceed 50% of the area of the face. For walls and partitions where no plastering in needed, smooth tiles may be used.

Glazed surfaces can be obtained for the tiles in various colours. Other finishes resembling rocks, bricks etc. can also be obtained.

EFFLORESCENCE IN BRICKWORK AND REMIDIAL MEASURES

Efflorescence is a white deposit which may occur on the surface of any masonry wall and is caused by the crystallisation of soluble salts brought to the surface by moisture in the masonry. These salts may have been in any or all of the masonry materials in the wall, in facing, backing, or mortar.

Staining on brickwork differs from efflorescence in that it is permanent and does not, like efflorescence, dissolve and disappear when the brickwork is wetted. Staining, moreover, has a different origin and mode of formation.

2.39 MECHANISM OF EFFLORESCENCE

Most of our walling materials have a porous structure, and when rain is driven against the outer face of the wall, water is absorbed and soluble salts present in the body of the wall are thus taken into solution. When drying conditions prevail during spells of drying winds following rains, evaporation takes place at the outer face of the wall and the water previously absorbed is drawn to that face and, with it, the salts in solution. As evaporation proceeds, the concentration of the salts in solution increases until the point is reached where they crystallize out. The exact position where crystallization takes place depends largely on the capillary properties of the materials. It may be on the surface or in a layer just behind the surface. If it is on the surface, it provides a serious disfigurement, while, if it is behind the surface, it may lead to disruption and spalling of the surface layer.

The salts most commonly encountered in efflorescence are the sulphates of magnesium, sodium and calcium and calcium carbonate.

2.40 EFFECTS OF EFFLORESCENCE

According to their nature and quantity, the soluble salts in bricks may produce effects of four main kinds. They may:

(*i*) cause temporary, but unsighly, white efflorescence,

(*ii*) cause efflorescence that gives more or less permanent coloured stains,

(*iii*) crystallise in or under and disrupt either bricks, plaster or decorations, and

(*iv*) react in solution with certain compounds in Portland cement or hydraulic lime mortars, causing the mortar to expand and ultimately to soften. Sulphate attack, as this last effect is called, may show itself in serious cracking and distortion of the brickwork, and in cracking and displacement of renderings. Efflorescence may be caused by salts from the mortar or from external sources, as well as by salts from bricks, and it is seldom possible to be sure where the slightly temporary efflorescences come from. Both bricks and cement contain more than enough soluble salts to supply the tiny quantity required to form even a prominent efflorescence and the position of the efflorescence is not a clear indication of its origin. The more serious effects are usually due to salts derived from the bricks.

2.41 PREVENTION AND PRECAUTIONS FOR EFFLORESCENCE

Prevention of efflorescence must begin with the design of the building. It should follow through the selection of materials and the method of work during the construction period, and continued during maintenance of the structure after its completion. The neglect of certain simple precautions may bring very unsatisfactory results.

The most important precaution is to prevent moisture from entering the walls, either during construction or later. Some of the precautions which will help in preventing water from entering the wall are :

(*i*) use of impervious coping,

(*ii*) pointing of joints carefully with water-repellent mortar,

(*iii*) flash parapet walls, carrying the flashing completely through the wall one or two courses above the roof level, and waterproof the entire inner side of the wall,

(*iv*) provide adequate drips on all copings, cornices and window sills making the grooves at least 1.5 mm wide and 10 mm deep and provide projection of 5 mm or more so as to keep the drip away from the wall,

(*v*) where practicable, avoid all joints in the sills. If brick sills are used, flash under them with metal flashing provided with drips,

(*vi*) protect walls under construction to prevent rain from entering,

(*vii*) keep gutters and down spouts in good repair,

(*viii*) point up cracked joints promptly,

(*ix*) coat the backs of retaining walls with an impervious layer,

(*x*) provide adequate and effective damp-proof courses.

The presence of soluble salts in clay products can be practically eliminated by the manufacturer by certain procedures in the drying and burning of the ware, together with the judicious use of barium compounds in the clay. The resultant increase of cost renders this procedure impracticable in the manufacture of the less expensive building products, though it has been generally used in the face brick industry. From the standpoint of efflorescence it is therefore very essential that the backing up materials be protected from excessive water penetration.

One of the important sources of efflorescence is free lime, which occurs in the lime mortar and also in cement. The development of efflorescence from the mortar may be largely prevented by the use of water repellent materials, especially calcium or ammonium stearate (2% by weight of the cement and lime used in the mortar).

Mortars so treated are effective in preventing the passage of moisture where only capillary forces abstain. Calcium stearate is a powder which is mixed dry with the cement before adding sand and water, while ammonium stearate is a paste which is usually added to the mixing water. On an ordinary building job the powder has proved more practical and accurate to use with the equipment commonly available. Although the best practice calls for the use of stearate in mortar for backing as well as facing materials, it is especially important that it be employed in laying up the facing where the joints will be subjected to unusual exposure, such as rain or water running down the surface of the wall.

Efflorescence can also be reduced by mixing barium carbonate with mortar. This has proved effective in preventing the temporary efflorescence of calcium and magnesium sulphates. Barium carbonate will not prevent the scum (layer of dirt) from lime hydroxide suspensions from Portland cement or other cement products, or from uncarbonated lumps of lime in lime mortars. It will also not prevent the scum from salt water.

Addition of diatomaceous earth to the cement is said to prevent efflorescence completely. These substances probably combine with the hydrated lime to produce an insoluble silicate.

Ammonium carbonate added either to the cement or gauging water is also stated to prevent efflorescence, and is moreover said to exert a favourable influence upon the setting of the cement.

Joints of copings, window sills and parapet walls need special care, as well as repointing of joints that have cracked.

The measures necessary to prevent defects resulting from the crystallisation of salts are :

(*a*) Correct design of constructional details, such as providing damp-proof courses, flashings of roof drainage to prevent entry of water into the body of the brickwork.

(*b*) The use of mortar of suitable texture and containing little or no soluble salts.

(*c*) Avoidance of the use of bricks with a high soluble salt content in very exposed positions. Use of a hard fired brick for facing work is recommended. The conjoint adoption of the hard fired brick with a suitable water repellent mortar will operate to eliminate efflorescence not only because both components of the brickwork contain a minimum amount of salts, but also because the relative pore characteristics of the brick and mortar will be such that evaporation will take preferentially from the joints, so that such salts as are in the brickwork will collect therein rather than appear as an efflioresence on the bricks.

(*d*) Avoidance of the use of soft porous bricks in contact with limestone or cast stone, string courses and the like unless experience has shown that the bricks are resistant to damage from crystallisation.

(*e*) Protection of the brickwork against contamination with salt bearing materials during building operations, (*f*) Protection of the brickwork from being thoroughly soaked during construction. The proper covering of the work during rain in order to exclude the entry of superfluous water, which will not only delay drying out, but will also facilitate solution of salts and hence possible efflorescence, is recommended.

TESTING OF BRICKS, TILES, ETC.

The samples of bricks, tiles etc. shall be taken so that they form a fairly good representative of the entire number of bricks, etc., which are required to be tested. A sample of 50 shall be taken from every consignment of 50,000 or part thereof. The samples can be taken by either of the two methods :

(*a*) *Sampling in motion*

In this method, samples can be taken when the bricks are in motion, *i.e.*, while they are being loaded or unloaded, effort being made to collect the samples at regular intervals so as to get a representative sample of the whole quantity.

(*a*) *Sampling from a stack.*

In this method, the bricks are taken out at random from a stack of bricks. The number of bricks required shall be taken from across the top of the stack, the sides accessible and from the interior of the stack by opening trenches from the top.

The samples taken by either of the two methods shall be stored in a dry place until these are required for the tests. Whenever, tests are to carried out, bricks, etc. be taken at random from the sample.

2.42 TEST FOR DIMENSIONS OF BRICKS, ETC.

Twenty whole bricks should be selected at random from the sample selected as described above. All blisters, loose particles of clay and small projections should be removed. They are arranged upon a level surface in contact with each other and in a straight line. The overall length of the assembled bricks is measured with a steel tape or other suitable inextensible measure sufficiently long to measure the whole row in one stretch. Measurement by repeated application of short rule or measure is not permitted. If for any reason it is found impracticable to measure 20 bricks in one row, the sample may be divided into two rows of 10 bricks which should be measured separately to the nearest millimetres. All these dimensions should be added together.

The dimensions of 20 bricks when tested in accordance with the procedure shall be within the following limits

Length = 3720 to 3800 mm
Width = 1760 to 1840 mm
Height = 1760 to 1840 mm

2.43 TEST FOR DETERMINATION OF WATER ABSORPTION OF BRICKS

(*a*) *Laboratory Tests (IS. 3495 (Part 2) 1992)*

The test specimens consist of five whole bricks, tiles etc. selected at random from the sample obtained as already described above.

The apparatus consist of a balance sensitive to within 0.1 percent of the weight of the specimen.

The test specimen is dried to a constant weight in a ventilated oven at 110° to 115°C. If the specimen is known to be relatively dry, this may normally he accomplished in 48 hours but if the specimen is wet, several additional hours may be required to attain constant weight. The specimen should then be cooled approximately to room temperature and weighed. In a ventilated room, bricks etc. properly separated require four hours for cooling unless an electric fan passes air over them continuously, in which case two hours may suffice. Specimens noticeable warm to the touch should not be used for the absorption test. The dry specimen shall be completely immersed without preliminary partial immersion, in clean water at 15.5° to 30°C for 24 hours. Each specimen should then be removed, the surface water wiped off with a damp cloth and the specimen weighed. Weighing any one specimen should be completed within three minutes after removing the specimen from the water.

The percentage of water absorption by weight should be calculated as :

$$\text{Water absorption, percentage by weight} = \frac{W_2 - W_1}{W_1} \times 100,$$

where W_1 = weight of dry specimen

and W_2 = weight after soaking in water.

The average value of the five specimens should be taken as the water absorption of the lot.

The water absorption should not be more than 20 percent by weight upto class 12.5 and 15 percent by weight for higher classes.

(b) Field Test.

The test specimen consists of five whole dry bricks, tiles etc., and should be selected at random from the sample obtained.

The apparatus consists of a balance sensitive of within 0.2 to 0.3 percent of the weight of the specimen.

Test specimens are weighed and then completely immersed in clean water at room temperature and allowed to remain in this state for a period of 24 hours. The specimen is then taken out, wiped with a damp cloth and then weighed immediately.

The percentage of water absorption by weight is calculated as follows :

Absorption, percent by weight after

$$\text{24 hours water immersion} = \frac{100(b-a)}{a}$$

Where

a = weight of the dry specimen, and

b = weight of the specimen after 24 hours' immersion in cold water.

2.44 TEST FOR DETERMINATION OF EFFLORESCENCE OF BRICKS, TILES, ETC.

(*a*) *Laboratory Test (IS 3495, Part I, 1992)*

Not less than five dry bricks are selected at random from the sample of bricks obtained.

Each brick is placed on end in a shallow flat bottom dish containing distilled water, the depth of immersion of the brick being not less than 25 mm. Thy whole arrangement is allowed to stand till the water in the dish gets evaporated. When the water has been absorbed and the bricks appear to be dry, a similar quantity of distilled water is again placed in the dish and the same allowed to evaporate as before. At the end of this period the bricks shall be examined for efflorescence.

The liability to efflorescence is reported as nil, slight, moderate, heavy or serious in accordance with the following definition.

(*a*) *nil*—when there is no perceptible deposit of efflorescence.

(*b*) *slight*—when not more than 10 percent of area of the brick is covered with a thin deposit of salts.

(*c*) *moderate*—when there is a heavier deposit than under 'slight' and covering upto 50 percent of the area of the brick surface but unaccompanied by powdering or flaking of the surface.

(*f*) *heavy*—when there is a heavy deposit of salts covering 50 percent or more of the brick surface but unaccompanied by powdering or flaking of the surface.

(*e*) *Serious*—When there is a heavy deposit of the salts accompanied by powdering and or flaking of the surfaces and tending to increase with repeated wettings of the specimen.

The rating of efflorescence should not be more than 'moderate' upto class 12.5 and 'slight' for higher classes.

(*b*) *Field Test of Efflorescence*

Five bricks are selected at random from the sample of bricks. Each brick is placed on end in a shallow dish containing clean, potable water. The quantity of water in the dish should be such that the brick is immersed to a depth of not less than 25 mm. The brick it allowed to stand in this position for a few days under atmospheric conditions and room temperature until all the water in the dish is evaporated. When the water has been absorbed and the bricks appear to be dry, a similar quantity of clean potable water is placed in the dish and the same allowed to evaporate as before. At the end of this period, the bricks are examined for efflorescence, as above.

2.45 DETERMINATION OF COMPRESSIVE STRENGTH OF BRICKS (IS 3495 PART I, 1992)

Five whole bricks are selected at random from the sample of bricks. The bricks are immersed in water at 25° to 29°C for 24 hours. They are removed and allowed to drain at room temperature for about five minutes and wiped free from surplus moisture. Their frogs are filled with mortar composed of one part of Portland cement and one and a half part of clean, coarse sand graded to 3 mm and down. The bricks are then stored under damp sacks for 24 hours. After the expiry of this period, they are immersed in water for seven days.

At the end of seven days, the sample of bricks are taken out, wiped dry and placed with the flat surfaces horizontal and the mortar filled face upwards between 2 three-plywood sheets each approximately 3 mm thick and carefully centred between the plates of the compression testing machine. The compression plate of the testing machine should have a ball-seating in the form of a portion of a sphere, the centre of which coincides with the centre of the face cf the plate. The load is applied axially at a uniform rate of approximately 14 N/mm^2 (140 kg per cm^2) per minute until failure occurs.

The maximum load at failure divided by the area of bricks is taken as the compressive strength. The arithmetic mean of the compressive strength of the five bricks tested is taken as the compressive strength of the lot.

The compressive strength of any individual brick tested should not at all be below the minimum compressive strength specified for the corresponding class of brick in Table 2.1. The lot should then be checked for the next lower class.

QUESTIONS

1. To what uses can the clay products like bricks, tiles, etc. can be put in building and road construction.
2. List the constituents of brick making soil and briefly discuss the functions performed by them.
3. What are the harmful constituents of brick making soil? Explain briefly their harmful effects.
4. Give the physical, mechanical and chemical composition of brick making soil.
5. Describe the field test performed to determine the suiability of brick earth for :
   (*a*) Consistency
   (*b*) Moulding
   (*c*) Shrinkage and deformation after burning of bricks
   (*d*) Strength and quality.
6. What points should be kept in mind while preparing the soil for tests in order to achieve better results?
7. List the steps involved in the manufacture of bricks. What considerations govern the selection of site for brick making?
8. How will you prepare the site and soil for brick making?
9. With the help of neat sketches, explain the methods of hand moulding of bricks. What advantage are gained by machine moulding?
10. What purpose is served by drying of bricks? List the precautions to be observed while drying bricks. Explain briefly the method of drying bricks.
11. Describe the manufacturing process of clamp burnt bricks.
    [*Mumbai Univ. Nov 2013*] [*RGTU Bhopal ; June 2013*]
12. Explain the method of burning bricks in intermittent permanent brick kiln.
13. With the help of diagram, explain the manufacture of bricks in either of the following continuous permanent kilns:

(*a*) Bull's trench kiln
(*b*) Hoffman's continuous kiln

14. Explain the method of making good bricks from black cotton soil.

15. Give the classification of common burnt clay bricks as per Indian Standard Code. List the requirements for first class bricks (Grade A) an second class bricks (Grade AA).

16. Distinguish between modular and non-modular bricks. Give their dimensions. With the help of sketches. Explain the following terms :

(*a*) Header and stretcher [*VTU Karnataka June 2012*]
(*b*) King and queen closer
(*c*) Coping brick

17. Write Short notes on :

(*a*) Sand lime bricks
(*b*) Paving bricks
(*c*) Jhama brick

18. Where are refractory bricks commonly used? What are their requirements? Name the classes of refractory bricks and discuss the types of acid bricks.

19. Give details of the types of following refractory bricks :

(*a*) Basic bricks ; and
(*b*) neutral bricks

20. What do you understand by the term terra-cotta? Give the approximate composition of good terra-cotta clay.

21. Distriguish between terra-cotta and faience. How are these manufactured and what are their uses?

22. Write short notes on :

(*a*) Earthen-ware
(*b*) Stone-ware
(*c*) Clay pipes

23. List the manufacturing advantages of hollow or perforated clay bricks. Discuss their properties and application advantages.

24. Enumerate the advantages of clay roofing tiles. Why is the use of clay roofing tiles becoming less popular? Suggest methods for their improvements.

25. List the common types of clay roofing tiles. Give details of Mangalore tiles and their classifications and characteristics.

26. Write briefly on :

(*a*) Burning of clay roofing tiles
(*b*) Flooring tiles
(*c*) Wall tiles
(*d*) Structural hollow clay tiles

27. What is the mechanism of efflloresence in brickwork? What are its effects? How can you prevent it?
28. What tests are performed to check the quality of bricks? Give details of the test for checking the dimensions of bricks. *[JNTU-Hyderabad ; Dec. 2013]*
29. Give details of the following tests on bricks and tiles :
    (*a*) Efflloresence test
    (*b*) Compressive strength test

OBJECTIVE TYPE QUESTIONS

1. The principal constituent of all kinds of clay is
   (*a*) Lime (*b*) Magnesia
   (*c*) Kaolin (*d*) Iron
2. Alkalies in suitable brick soil should not be more than
   (*a*) 3% (*b*) 4%
   (*c*) 5% (*d*) 6%
3. Silica (SiO_2) in good brick soil should not be less than
   (*a*) 45% (*b*) 50%
   (*c*) 55% (*d*) 60%
4. Silt is an intimate mixture of :
   (*a*) Clay and fine sand (*b*) Clay and medium sand
   (*c*) Clay and coarse sand (*d*) None of these
5. The loss on ignition for brick soil should not be more than
   (*a*) 6% (*b*) 7%
   (*c*) 8% (*d*) 9%
6. The percentage of clay and silt in good brick soil should vary from :
   (*a*) 15 to 30 (*b*) 25 to 35
   (*c*) 25 to 40 (*d*) 30 to 45
7. Hand moulding of bricks is done by :
   (*a*) Dry press process (*b*) Soft mud process
   (*c*) Table moulding process (*d*) Stiff mud process
8. Shrinkage, during drying of building bricks, varies from :
   (*a*) 5 to 10% (*b*) 8 to 12%
   (*c*) 10 to 15% (*d*) 15 to 20%
9. The modular brick size is :
   (*a*) 180 × 80 × 80 mm (*b*) 190 × 90 × 90 mm
   (*c*) 200 × 100 × 10 mm (*d*) 210 × 110 × 110 mm
10. The non-modular size of the brick is :
    (*a*) 220 × 100 × 80 mm (*b*) 210 × 90 × 90 mm
    (*c*) 200 × 110 × 70 mm (*d*) 230 × 110 × 70 mm

11. Water absorption after 24 hrs immersion in cold water for bricks with compressive strength upto 12.5 N/mm^2 should not be more than :
(*a*) 10% (*b*) 15%
(*c*) 20% (*d*) 25%

12. Water absorption after 24 hrs immersion in cold water for bricks with compressive strength greater than 12.5 N/mm^2 should not exceed :
(*a*) 10% (*b*) 15%
(*c*) 20% (*d*) 25%

13. The bricks get vitrified in the kiln, when burnt above :
(*a*) 1100° C (*b*) 1200° C
(*c*) 1300° C (*d*) 1400° C

14. The average compressive strength for burnt clay facing bricks should not be less than
(*a*) 3.5 N/mm^2 (*b*) 7.5 N/mm^2
(*c*) 10 N/mm^2 (*d*) 15 N/mm^2

15. The average compressive strength for paving bricks should not be less than
(*a*) 35 N/mm^2 (*b*) 40 N/mm^2
(*c*) 45 N/mm^2 (*d*) 50 N/mm^2

16. Jhama bricks are :
(*a*) Under burnt (*b*) Properly burnt
(*c*) Over burnt (*d*) None of these

17. Frogs are not provided in :
(*a*) Wall bricks (*b*) Paving bricks
(*c*) Facing bricks (*d*) None of these

18. Efflloresence is caused due to the presence of ________ in brick soil
(*a*) Silica (*b*) Magnesia
(*c*) Lime (*d*) Iron oxide

19. The average out turn of suitable bricks in clamp kiln is :
(*a*) 70% (*b*) 75%
(*c*) 80% (*d*) 85%

20. It takes about ________ days for bricks to burn properly in clamp kiln :
(*a*) 5 (*b*) 10
(*c*) 15 (*d*) 20

21. Which kiln for burning bricks has low fuel consumption?
(*a*) Clamp kiln (*b*) Intermittent permanent brick kiln
(*c*) Bull's trench kiln (*d*) Hoffman's continuous kiln

22. Black cotton soil possesses the following property :
(*a*) It is very plastic (*b*) It shrinks very much
(*c*) It contains lime nodules (*d*) All the above

23. The minimum crushing strength of second class bricks should be :
(a) 7.5 N/mm^2 (b) 10 N/mm^2
(c) 12.5 N/mm^2 (d) 15 N/mm^2

24. First class bricks are of grade
(a) A (b) AA
(c) B (d) None of these

25. Sand lime bricks contain the following percentage of lime :
(a) 3 to 6 (b) 5 to 10
(c) 8 to 12 (d) 10 to 15

26. Which of the following types of bricks possess low thermal conductivity :
(a) Sand lime bricks (b) Paving bricks
(c) Refractory bricks (d) Common burnt bricks

27. Which of the following are the refractory bricks :
(a) Acid bricks (b) Basic bricks
(c) Neutral bricks (d) All of these

28. Which one of the following clay products are burnt twice :
(a) Terra-cotta (b) Faience
(c) Refractory bricks (d) Stone-ware

29. The maximum percentage of water absorption for class AA Mangalore tiles should be :
(a) 15% (b) 16%
(c) 17% (d) 18%

30. For testing of bricks and tiles, a sample of ________ is taken out of a lot of 5000 bricks
(a) 50 (b) 75
(c) 100 (d) 125

31. The breaking load in kN for class A Mangolare tiles (410 × 235 mm) should be :
(a) 0.6 (b) 0.7
(c) 0.8 (d) 0.9

32. The minimum breaking loa in kN for class AA individual Mangalore tile of size 410 × 235 mm should be
(a) 0.8 (b) 0.9
(c) 1.0 (d) 1.1

33. The efflorescence is termed as 'slight', when not more than ________ percent of area of the brick is covered with a thin layer of salts
(a) 5 (b) 10
(c) 15 (d) 20

34. The bricks should be burnt at a temperature ranging from
(a) 1000 to 1200° C (b) 1200 to 1400° C
(c) 1400 to 1600° C (d) 1600 to 1800° C

35. The length of a curved tile varies from :

(*a*) 80 to 100 mm (*b*) 100 to 120 mm

(*c*) 120 to 150 mm (*d*) 150 to 200 mm

Answers

1. (*c*)	**2.** (*b*)	**3.** (*d*)	**4.** (*a*)	**5.** (*b*)	**6.** (*b*)
7. (*c*)	**8.** (*a*)	**9.** (*b*)	**10.** (*d*)	**11.** (*c*)	**12.** (*b*)
13. (*a*)	**14.** (*c*)	**15.** (*b*)	**16.** (*c*)	**17.** (*b*)	**18.** (*c*)
19. (*c*)	**20.** (*b*)	**21.** (*d*)	**22.** (*d*)	**23.** (*a*)	**24.** (*a*)
25. (*b*)	**26.** (*c*)	**27.** (*d*)	**28.** (*b*)	**29.** (*d*)	**30.** (*a*)
31. (*c*)	**32.** (*b*)	**33.** (*b*)	**34.** (*a*)	**35.** (*d*)	

FURTHER REFERENCES

1. S.K. Sharma "A Text-book of Building construction" S. Chand, New Delhi.
2. Indian Standards (IS) 654, 1077, 3495 (Part I & II), 3583 and 3951 (Part II)

LIME

3.1 INTRODUCTION

In our country, upto the thirties of 19th century, lime had been the principal cementing material for all structures. Magnificent palaces, massive forts, dams, bridges and other structures stand today testifying as much to the glory that was India's as to the skill and art with which lime was used as a cementing material. These structures have stood for centuries and are still sound and intact.

The lime industry lost ground and popularity in recent times with the advent of cement. Cement began to replace lime as it is available in ready-for-use form and the technology of its use is greatly developed. But lime possesses some unique properties which result in better quality construction and for this reason in the industrially advanced countries lime is being used in plasters and mortar to a very great extent. The use of cement in conjunction with lime in composite mortar (cement-lime-sand mortar) further extends the usefulness of lime and gives the best results.

Besides these technical advantages in the use of lime, the manufacture and use of lime are cheaper.

Lime should therefore be used in building construction on its merits and not merely as a substitute for cement. In order, therefore, to use lime, firstly, it is necessary that the lime industry should be developed and modernised to produce required quantities of reliable quality of lime ; secondly, to facilitate its use, marketing of lime should be specialised so as to offer it in easily consumable form ; thirdly, the correct techniques of preparation and application of lime mortars and plasters should be publicised so that the building work is of the required quality and without any defects. The neglect of some of these factors has been responsible for many shortcomings in construction with lime and this, in fact, has led to reluctance in the use of lime.

3.2 GOOD QUALITIES OF LIME

Lime possesses the following unique properties which result in better quality construction :

(*i*) better workability and plasticity,

(*ii*) early stiffening,

(*iii*) low shrinking on drying which makes for durability,

(*iv*) good strength,

(*v*) resistance to moisture,
(*vi*) freedom from major cracking,
(*vii*) excellent adherence to masonry units,
(*viii*) adjustment to stresses in the masonry in course of time due to slow setting, and
(*ix*) Improvement in strength continuously with time.

MANUFACTURE OF LIME

3.3 RAW MATERIALS

Lime is made from limestone, $CaCO_3$. Generally two kinds of limestones are found in India, kankar limestone and massive limestone. Besides these two as may be expected in a sub-continent of varied geology as ours, there are many more forms of calcium carbonate shells along the coastal areas, coral at the southern tip near Tuticorin, tufa limestone in many old stream and river beds, marble and calcite. But none of these raw materials contribute to any substantial production of lime on a country-wide basis.

Kankar limestone is a major source of limestone for burning lime in India. It is extensively used in Andhra Pradesh, Tamil Nadu, Mysore and to some extent in Uttar Pradesh, Rajasthan and Bihar. In these states it is available in most localities near the surface and is collected by individuals and burnt to obtain lime. Kankar limestone is usually found in pockets and as such it does not lend itself to organised large scale working.

Organised lime production will always have to be based on massive limestone, deposits of which India has extensive reserves. Brief details of the deposits in each state are given below :

Uttar-Pradesh (U.P.). The limestone in Dehradun is in a wide band running from Hatipoor Estate in east about 16 km east of Mussoorie, to past Rishikesh in the west. The overburden varies, but in many places the rock is exposed. In most places workings are not more than 3 m deep : but it is estimated that the thickness is 300 m in certain places. The belt runs along the top of the mountains as over 900 m above sea-level. The stone is also found in the riverbeds of the Doon valley as it is washed down by monsoon rains from the hills.

The quality of stone varies considerably from predominantly high calcium limestone ($CaCO_3$—98% and over) in the east to predominantly high magnesium limestone (MgO—8% to 15%) in the west of Rishikesh. It has very little silica impurity and iron or alumina. There is a high percentage of sulphur and if the stone is struck it gives a sulphur smell. An analysis range of the stone quality is given below :

Range of impurities in Dehradun Limestone

SiO_2	0.1 to 2.6 percent
Fe_2O_3	0.05 to 2.16 percent
P_2O_5	0.01 to 0.41 percent
S	trace to 0.61 percent
CaO	By inference and making allowance for MgO ; CaO can be 99% and if MgO is high CaO is 85 to 87%.

The colour of the stone is grey, but there is also white and black marble in certain area. The top layers of the stone are often weathered giving a product which is brownish in appearance. In general the stone is brittle and shatters easily. River-bed stone is a mixed quality of calcium limestone, dolomite and magnesium limestone. As river bed stone is hand picked the bulk average quality is dependant on the skill of picking. In general it is a varying mixture of calcium and magnesium limestone.

Andhra Pradesh. Andhra Pradesh has many good deposits of limestone. The State is also very rich in Kankar limestone and limestones suitable for building slabs. High calcium limestone is found in Anantpur district in the south of the State, and is being used for manufacturing chemical lime. Except in Anantpur district most limestones have high silica and low magnesia. The western part of the state has a band of limestone 40 km in length running through Cuddapa and Kurnool districts. In the east there are vast deposits in Guntur, Bezwada and also in Godavari districts.

Tamil Nadu. The state has an abundant well-distributed supply of kankar limestone. This stone is used extensively for lime burning. The quality is not known and it is presumed that it is a hydraulic limestone. The state has also considerable reserves of massive limestone. Tamil Nadu has also large deposits of coral limestone off the coast near Tuticorin.

Kerala. The state has neither large deposits of massive limestone nor kankar limestone. The only lime producing raw material which is available is shell.

Mysore. Limestone in Mysore is found in Shimoga district. There is limestone in Chitaldrug district. Now the Shahabad district is also in Mysore which is a large limestone area-an extension of the Cuddapah band running up through Kurnool from Andhra Pradesh.

Madhya Pradesh. Limestone deposits in Madhya Pradesh are extensive. There are two general locations. The Vindhyan limestones are found in a band running from Jabalpur to Satna and the Cuddapah limestones in the Raipur-Bilaspur area. The band is many km wide in certain areas, and is over 30 km long. The depth of working is generally over 7.5 to 12 m. The quality of this band varies from a high calcium limestone in the direction of Jabalpur to a high silica (8 to 15%) stone in the area around Satna.

The reserves of limestone in the Raipur-Bilaspur area are even more extensive. They belong to the Guddapah age. There is a band running between Raipur and Durg which has been estimated at about 40 million tons. Similar limestones exist in the Bilaspur area and the reserves here too are considerably estimated at over 10 million tons. Those limestones are high silica 7% and over. Magnesium is present but to a small extent.

Odisha. There is a long band of limestone in the State running very near and parallel to the main Nagpur — Kharagpur railway line. Together with the limestone there is a band of dolomite and also calcitic marble. The deposits of calcitic marble are broken and haphazard. The deposit of limestone is extensive. The resources have been estimated at about 27 million tons of less than 6% silica stone, good for metallurgical purposes.

Rajaslhan. The deposit of limestone at Sojat Road is 3 km from the railway station and accessible by all-weather road. The deposit is extensive and has under 600 to 1200 mm of earth. The workings show limestone beds of 1.2 to 2.4 m. The stone at Sojat Road is a fat-limestone. Rajasthan has abundant limestone in all areas of the State.

The chief limestone areas are situated on two bands of the Vindhyan system running north to south through Gotan in Jodhpur, Sojat Road in Jodhpur to Chitor in Udaipur and west to east from Abu Road through Kotah, Lakheri and Sewai Madhopur. Besides this fat-magnesium and semi-hydraulic limestone deposits of varying quality exist in every locality, some with very high silica. In the area between Rewari and Jaipur limestone is abundant, the quality is mixed, contains high silica and high magnesium. Limestone is worked at Maonda for export and for production of lime, and in Jaipur for local burning. The limestones of this area are high in sulphur. The sulphur anhydride at Sojat Road analysis is 0.34%. Some of the limestones like those at Gotan have traces of mineral oil.

Maharashtra and Gujarat. The southern part of Kathiawar peninsula is chiefly limestone. The two general geological formations in the area are Deccan trap in the north and limestone in the south. Limestone of various qualities is available everywhere, including areas near Jamnagarh, the south of the coast is limestone.

In two areas near Porbandar and Verbal, the quality is suitable for chemical lime. In other areas the quality varies from a high silica to a high alumina. Kankar is also found.

The deposit at Rajpur is a band of Vindhyan limestone running north-west to south-east from Wani to beyond Sonapur. The band is 21 to 24 m deep and has 1 to 2 m of overburden. In one area there is an out-crop which runs above the surface level for 150 m.

3.4 ACCESSIBILITY TO THE DEPOSITS

The second and a very important consideration is access to the limestone areas. The two factors of importance in this respect are : the distance of the deposits from rail-head and the conditions of extraction. In India where the density of railway track per square km is low and where road transport for many reasons is prohibitive many of the proven reserves are unavailable on economic considerations only. Under the present conditions for a lime to be competitive, 9 to 11 km on a first class road is the maximum permissible distance from rail-head. Further physical conditions such as overburden, height from ground level and excessive limestone bed thickness 30 m is the reasonable exploitable bed, the rest substantial proven reserves of limestone are in practice unavailable.

There are many areas having limestones which are unsuitable for any purpose. The quality required for the building lime industry should generally have silica with iron and alumina (10 to 15%) for hydraulic lime, (5 to 10%) for semi-hydraulic lime and (0 to 5%) for fat-lime. Besides this the chemical lime industry requires as nearly as 100% calcium carbonate stone. When a stone meets the chemical specifications set for a building lime it also needs to have certain other physical properties such as crushing strength, friability, and density to suit it to particular processes and purposes.

3.5 QUARRYING

In general, most of the limestone deposited in India lie beneath the surface but in a few places like Birmitrapur Japla, Porbandar, Dehra Dun and Benjari they are in the form of hills, and are much easier to work. Limestone also exists at considerable depths but deep limestone quarries are not economical and are usually outside the purview of present methods of exploitation. Normal procedure is to consider 30 m bed as a workable limit.

The best form of the limestone bed that can be obtained is a solid massive deposit of 7.5 m and above. The earthwork, if it is less than 1.5 m is of considerable assistance. Such deposits are usually found on the Vindhyan and Aravali regions from Japa and Bisra in the east through Katni and Bilaspur to Gotan and Sojat Road in the west. In the centre of India the area comprises of south Bihar, eastern U.P., M.P., Odisha, Andhra Pradesh and south Rajasthan for such deposits. The stone is usually black and hard. The earth overburden can vary considerably, the stone is normally being quarried in areas where the earthwork is upto 12 m. There is another large belt in the Deccan section of the country running from Rayalcheruvu in the south through Kurnool, Morchela, Wadi and Shahbad.

Of primary importance in any lime production operation is the quantity and quality of the available limestone. A close estimate of the available material can be made by boring, probing, and digging. Some estimate of the thickness of the deposit can be made by examining exposed out croppings, but drilling is the surest method, although expensive for small operations.

If there is a choice of locations, the quarry should be near the centre of the area where the lime is to be applied ; and it is desirable to have a good firm road with easy grades leading from it. A good road out to a paved highway is almost a necessity for a neighbourhood quarry. Limestone for lime burning is in general extracted by manual methods. At present only a few manufacturers have to some extent mechanised their quarry operations, this too because a major portion of the stone production goes for steel production.

At the present time nearly all the mines do their earthwork and also removal of stone by head-loads. Breaking is done by hammer and crowbar and blasting holes by crowbar. The explosive is usually gun powder. However in advanced countries, quarrying is done by mechanical means. The procedure adopted is described below :

Stripping. The removal of overburden is usually done with a drag scraper pulled by a team of tractors. Such work can be done with labour and equipment, with little cash expenditure. For large operations power shovels are used.

Drilling. In many cases the upper strata of the formation are weathered to such a degree that quarrying can be done by hand. Where the strata of stone are not clearly defined, are thick, and only slightly weathered, drilling and blasting will be necessary. Drilling may be done with hand or power drills. Choice of the two methods will depend on such factors as the thickness of the ledge, availability and cost of labour the quality of the stone and the capacity of the pulverizing machinery.

Where quarry operations are on a large scale and the limestone formation is horizontal or nearly so, well drills are used. Where operations do not warrant so large an investment, jack hammer drills worked by compressed air, electricity, or steam may be used. If the grinding machinery is operated it is usually most convenient to use the steam jack-hammer. Disanvantage of steam are the high heat loss by radiation and condensation, and the possibility of damage to steam pipes when the blasts are fired. Electricity has the advantage of being readily available at all times and the equipment is easy to move. Compressed air has the advantage of portability, not only in the quarry but from one quarry to another.

Blasting. After the holes have been drilled they should be tested before being loaded, to be certain that they are open to permit the free passage of the explosive. The type of explosive to be used is determined by the conditions in each quarry. Where the drill holes are filled or partially filled with water, or where the stone is to be well shattered, it is advisable to use gelatin dynamite. Where the holes are merely damp a moisture resistant explosive should be used. In a few cases it will be advisable to use an explosive less than 40 percent strength.

The amount of explosive necessary to blast down the rock will vary in the different formations. Since the breakage per kg of explosive varies from 6 to 13 tons, a common method is to assume a breakage of 6 or 8 tons per kg and load at that rate. After a number of trials, the rate of loading for a particular formation may be determined. These trials will also help to determine the correct spacing for drill holes.

This is done by loading to a point below the seam, priming with an electric cap, and filling with well packed earth to a point above the seam after which the loading of the hole is completed and primed with an electric cap. The use of electric caps permits simultaneous explosion of the separated parts of the charge caps and the fuse cannot be used for this kind of shot since two charges primed with cap and fuse cannot be exploded simultaneously.

Secondary Blasting. Occasionally masses of rock loosened first by the primary blast are found to be too large to handle. They may be broken down with explosives by either block holding or mud capping method known as secondary blasting.

Block holding consists in drilling a hole in the rock and loading it with a small charge of explosives. The amount of shattering will determine the accuracy of the load, and the succeeding loads should be decreased or increased accordingly.

In mud capping the explosive is held against the boulder to be broken by a 150 to 200 mm layer of well packed wet earth. As the explosion of dynamite has the same effect as the blow of a hammer, common rule is to place the charge at the weakest point. Concave surfaces are favourable places for loading because they help to confine the explosive gases whereas convex or rounded surfaces have a tendency to dissipate them. The explosive may be removed from the wrapper or used in the stick form. If removed from the wrapper it should be thoroughly packed and the blasting cap embedded in it before the mud is applied.

For block holding and for mud capping either electric or common fuse caps may be used as detonators, but the electric method of firing avoids hang fires which are always dangerous and the consequent loss of time, waiting for the charge to explode.

3.6 TRANSPORT

The lime industry being based on mined limestone employs various kinds of transport, carts, tramline, track, rail, waterways etc., to get the stone from the quarry to the manufacturing site. Once the lime is calcined, it again has to be transported to the place of use and at this stage the chief forms of transport are road and rail. Where transport facilities have been available, lime industry has developed considerably and in other places where the demand for lime and the limestone were both present, the industry did not develop on account of inadequate transport facilities.

3.7 FUEL

The fuel may be of various kinds ; traditionally wood and coal have been used. Recently coke, oil, natural gas, producer gas and waste materials are also being employed. In India wood was extensively used for producing an excellent quality lime ; but with the rise in prices of firewood, wood burnt lime has priced itself out of the market except in certain areas like Kerala. Coal is now the chief fuel for the production of good lime. Cinder from the railways though extensively used, gives a very unsatisfactory product.

Coal. Coal is used at all the major lime producing centres in India. The whole pattern of development of the lime industry has been dependent on the coal supply position. On account of this, the limestone areas near the coal-fields where coal is easily available have grown into major lime centres. Good limestone areas which have appeared away from the coal-fields, in spite of other advantages, have found their development handicapped by shortages of coal. At this time when coal supplies are becoming increasingly difficult due to inadequate wagon availability and as supplies will continue to be difficult due to shortfalls in the target of the Plan, more than ever, the state of the lime industry will reflect the attention given and the resourcefulness employed in solving the problem of supply of coal.

Direct Fired Gas Kilns. Some of the kilns in sugar and chemical works use coke. However, it is felt that the availability of coke generally for lime burning is remote. As the coal being allotted for lime burning is normally of inferior quality, high in volatiles and ash, some considerations must be given to the matter of production of gas from low quality coal for lime burning. East European countries faced with this problem have developed their kiln design for producer gas.

Cinder. This is an important fuel for lime burning in India. The chief sources for cinder is the railways. They get a good price for their waste product from lime burners who cannot get coal.

There are numerous lime kilns in the south using Kankar, all of them use cinder now.

The chief disadvantage in burning with cinder is that it tends to produce a pseudo-lime not calcined at a sufficiently high or uniform temperature to give a satisfactory product and there is little that can be done to rectify this. A second disadvantage is that cinder being a weak fuel introduces a much larger percentage of ash in the mixed-feed kiln product. This disadvantage is real in terms of absolute CaO availability ; but if a mortar is designed keeping in mind the ash content of the product (lime) it can be compensated for. Some attempts to reduce the silica in the lime product can be made by separating the lumps from the dust, but this increases the cost as the dust lime is wasted and handling cost increased. Besides this, the quality of cinder burnt lime is usually doubtful. It gives the appearance of having been completely burnt ; however, reports on cinder-burnt lime are that it is incompletely burnt chiefly because the cinder being a weak fuel fails to produce sufficiently high temperature.

Oil. Burning with oil is a new process in India and steps must be taken to pioneer its use. The availability of crude oil, with the coming into production of three refineries, has increased considerably. It is used extensively in other countries and in the U.S.A. and is one of the major fuels for lime burning. Oil has a special significance in the south where the possibility of delivering coal is remote due to the lack of adequate movement capacity. It will also be of value to burners in the west coast of Maharashtra State and Gujarat and Rajasthan where the landed cost of coal is high.

3.8 BURNING OF LIMESTONE

The amount of heat required to burn limestone depends on its character and varies from 760° to 1205°C. Greater heat is usually required for calcium limestone than for magnesium limestone. Where such impurities as silica, alumina and iron are present, the temperature must be kept below their melting points to prevent the formation of slag and the making of "dead burnt" or "overburnt" lime. About 2 percent of silica will not affect the burning of the stone, but the alumina and iron content should not exceed 0.5 to 1 percent. Less, heat is required to burn a dense, fine-grain stone than one which is coarsely crystalline and porous. In burning, a high temperature may be used for a short time or a lower temperature for a longer time. The more nearly the amount of heat used approaches the minimum required, the better the quality of the lime.

3.8.1 Precautions in Burning

In burning limestone, the following points should be kept in view :

(*i*) The heating should be gradual. Sudden heating would release moisture and carbon dioxide quickly. This will blow off the limestone, kankar or shell into pieces.

(*ii*) When the colour of the raw material in the kiln is changed to bright red, it shows completion of the burning process.

(*iii*) Denser and clayey limestone requires more fuel than pure limestone kankar or shell.

(*iv*) There should be neither under-burning nor over-burning of limestone. If it is so, it is called 'dead-burnt', lime which cannot be slaked properly.

(*v*) There should be as little loss of heat during burning as possible in kilns for burning limestone.

3.8.2 Kilns

There are two main types of kilns which are common in India :

(*a*) Country batch type kilns, and

(*b*) Vertical Intermittent kilns.

All the kilns in the country at lime centres and wayside stations are modifications of these two types. The chief technical problems in lime burning are transfer of heat to the stones, complete combustion and maximum utilisation of the fuel, handling of larger quantities of stone and lime. The kilns in India have been constructed to meet these prerequisites within the limitation of know-now available to the burners.

In other developed countries, continuous kilns and rotary kilns are used.

3.8.2.1 Country Batch Type Kilns (Bhattis)

These kilns may be seen in every part of the country. They vary in size and shape from area to area. Generally they are rectangular and less than 2 m in height ; sometimes these are also cylindrical. The sizes vary in the rectangular type from height 1 m by 2 × 3 m to height 3 m by 3 × 6 m and the cylindrical type from 1.25 m diameter and 2 m height to 3 m diameter and 4.5 m height. The chief common characteristic of all these klins is that they are operated batchwise. The kilns are filled with limestone and fuel in alternate layers and fired. The fire rises through the bed of fuel and stone slowly, calcining the limestone to lime. The only control on the fire are the air holes provided at the bottom.

The layers of stone and fuel vary from area to area. Normally 200 to 300 mm layer of stone is taken and the layer when coal is used is $\frac{3}{2}$ to $\frac{1}{2}$ of the layer of stone. When cinder is used the layer of cinder is generally equal to the layer of stone. These kilns are heat-wasteful as they do not embody any heat exchange principle to utilise the heat available below 900°C, which goes waste. This is the inherent drawback in the design of such kilns which cannot be eliminated by improvements of the present structures.

The kilns are constructed of cheap local materials using third class brick and mud walls or stone and mud walls of 300 to 450 mm thickness. In a few cases part of the kiln is below the ground level to economise the cost of construction.

A major part of the lime in India is produced in such kilns. They have become popular as lime burners have found that due to inadequate coal receipts they have to depend on cinder. Cinder cannot be burnt in the continuous tall vertical kilns and our present experience of the urban semi-organised burner is that we must not rely on coal. It is known that the village lime burner has also to depend on cinder. As the country batch kiln is so widely dispersed and responsible for such a large percentage of Indian lime production, the problem of overcoming these handicaps has to be carefully considered. It is felt that in the interest of quality lime production and conservation of fuel and the best utilization of raw-materials it is necessary to establish a policy to eliminate such kilns slowly. If coal were to be supplied to all country batch-kilns, the problem of under burning and high ash contamination would be eliminated but another more serious problem of waste of coal resources and transport capacity would be created. The country kilns have low heat efficiency. The lime to fuel ratio is around one. In the better kilns ratio of 2 : 1 to 3 : 1 and higher may be expected. This means for the same coal, two to three times the lime of superior quality could be produced in better kilns.

3.8.2.2 Vertical Intermittent kilns

Kilns of this design exist in many centres of production in the country (Fig. 3.1). The size varies from 2.5 to 4 m in diameter and from 5.5 to 12 m in height. These kilns are generally mixed feed type and as a rule burn steam coal. Coke can be used with advantage but it is not available for lime burning in India. The kilns of better design are suitable for continuous operation but are generally not run in this manner for various reasons, the chief being the inabiltiy to feed the kilns sufficiently by the manual methods which are being used at present. This may soon be eliminated as many of the burners using tall kilns are actually considering and working on designs for less labour and intensive mechanized methods of feeding these kilns.

Many of the vertical semi-continuous kilns in the lime industry need considerable improvement. Some of the kilns in Satna (Madhya Pradesh) and Birmitrapur (Odisha) are quite satisfactory and would seem to point the way to further development in the design of these kilns is that they rely to a great extent on local materials.

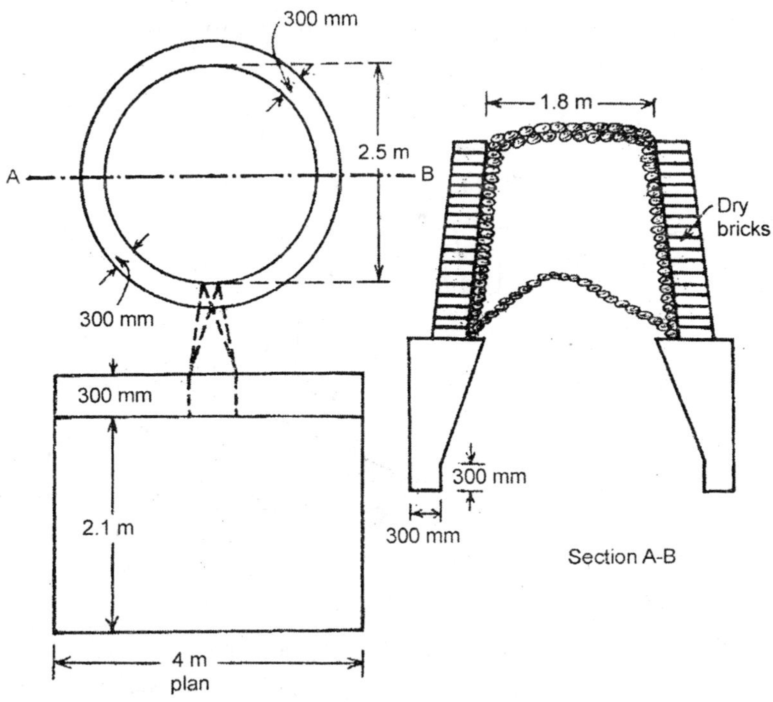

Fig. 31. Vertical intermittent kiln

Suggestions for Improvement of Vertical kilns.

(*a*) *Raising equipment design.* Design of indigenous equipment for raising stone to feed the kiln and a distribution system for stone and fuel, when the kiln is mechanically fed should be evolved.

(*b*) *Lining of kilns with fire bricks.* The manufacturing capacity of fire bricks in the country is far behind requirements and cost are high. Both these factors handicap kiln construction. Alternatively, natural volcanic rocks can be used to replace fire bricks when heat resisting properties are required. The insulation type of fire bricks in the unit masonry construction of Indian vertical lime kilns, are not required.

(*c*) *External Gas firing.* In this country the quality of coal available for mixed feed vertical lime kilns will always be low grade and not coke as in Europe and America and so to get lime of standard quality external gas firing will have to be developed along East European lines. Steam coal in mixed feed kilns is heat wastful as up to 30% of the heat value of coal is lost when the volatiles are burnt in preheating zone. This could be conserved in gas fired kilns.

3.8.2.3 Continuous kilns

These are not much in use in India but are very common in. advanced countries. These are also of the vertical type, explained above and consist of cylinderical steel shells lined with refractory bricks. The kiln is either mixed feed type or separate feed type. In the mixed feed type (Fig. 3.2) coal and limestone are fed from the top of the kiln in alternate layers, the lime being removed at the bottom.

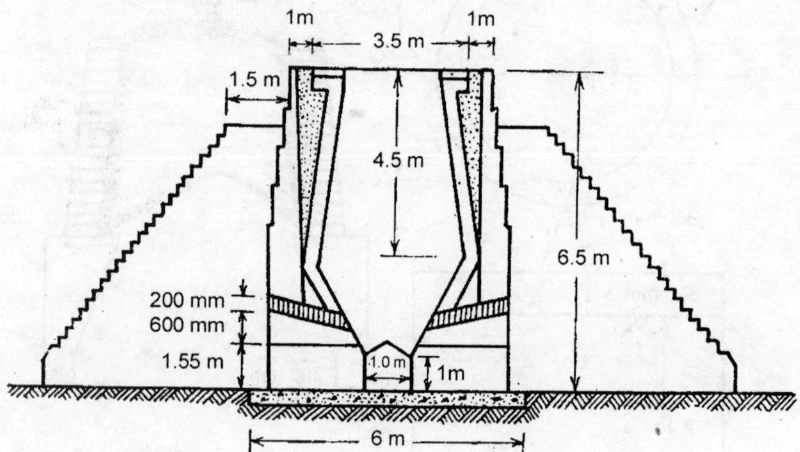

Fig. 32. Continuous kiln

The vertical kiln with separate feed are so designed that the fuel and limestone do not come in contact during the burning process. The fuel is burnt in separate fire-places which is set either in the wall of the kiln or outside the kiln shell. This is so arranged that the heat produced will ascend into the stack. Thus the limestone comes in contact only with the hot gases of combustion. The fuel may be coal or wood or producer gas. Although the fuel consumption in the separate feed kilns is more, yet this is off set by the high grade quality.

Rotary kilns. Rotary kilns about 45 m long, similar to those used in burning Portland cement are extensively used in U.S.A. for burning limestone. They have the disadvantage that the limestone is required to be finely crushed prior to calcination and the product is also finely divided. Rotary kilns make possible improved control of feed and temperature.

DEFINITIONS

Quicklime. A calcined material, the major part of which is calcium oxide in natural association with a relatively smaller amount of magnesium oxide, capable of slaking with water.

Fat-lime. The lime which has high calcium oxide content and is dependent for setting and hardening solely on the absorption of carbon dioxide from the atmosphere.

Hydraulic Lime. Lime containing small quantities of silica and alumina and or iron oxide which are in chemical combination with some of the calcium oxide content, giving a putty or mortar which has the property of setting and hardening under water.

Hydrated lime. A dry powder resulting from treatment of quicklime with water enough to satisfy its chemical affinity for water under the conditions of its hydration.

Lump lime. Quicklime as it comes from kilns.

TYPES OF LIME

Based on the chemical composition there are three types of lime :

(*a*) High calcium lime (*b*) Magnesium lime, and (*c*) Hydraulic lime.

3.9 HIGH CALCIUM LIME

Pure limestone with little impurities yields high calcium limes which are most commonly drawn as fat limes. Such limes can be readily slaked, and mortars prepared from these possess excellent working qualities, but they take a very long time to develop adequate strength.

3.10 MAGNESIUM LIME

Lime from limestone having a high proportion of the magnesium oxide is called magnesium lime. According to Indian Standards, if limestones has more than 50% of magnesium oxide, the lime produced from it is called magnesism lime. But there are, in fact, a few varieties of magnesium limes which contain less than about 35%. Such lime is generally produced from dolomitic limestones. Magnesium oxide is slaked with difficulty and therefore, the slaking of magnesium limes needs more attention. The properties of magnesium lime, when properly slaked, resemble those of high calcium lime. That is to say they develop strength slowly. It is to be noted that they have no hydraulic properties but they are sometimes wrongly called hydraulic limes.

3.11 HYDRAULIC LIME

Some limestones contain clay more or less uniformly distributed. Limes obtained from such limestones are self hardening *i.e.* having hydraulic property and are, therefore, called hydraulic limes. Hydraulic limes may be described as either eminently hydraulic or semi-hydraulic depending on their hydraulic properties. This distinction, however, is an arbitrary one. Mortars made from hydraulic lime develop early strength.

3.12 OTHER TYPES OF LIME

Limes are generally called by different names depending on the source and form of limestone. Stone lime is produced from limestone and kankar lime from kankar. These two types of limes have been extensively produced all over the country. Shell lime is obtained from sea shell along the coastal areas and is a very fat lime; coral lime from coral reefs at the southern tip near Titicorin ; Sweet lime or Gypsum lime from gypsum. These types of limes are produced in certain localities only.

CLASSIFICATION OF LIME

Bureau of Indian Standards has classified lime in six categories (I.S. 712—1984).

Class A—Eminently hydraulic lime used for structural purposes.

Class B—Semi-hydraulic lime used for masomy mortars, lime concrete and plaster undercoat.

Class C—Fat lime used for finishing coat in plastering, white washing, composite mortars, etc. and with addition of pozzolanic materials for masonry mortar.

Class D—Magnesium/dolomitic lime used for finishing coat in plastering, white washing, etc.

Class E—*Kankar* lime used for masonry mortars.

Class F—Siliceous dolomitic lime used for undercoat and finishing coat of plaster.

Note 1: Lime shall be available either in hydrated or quick form, except that of Classes A and E which shall be supplied in hydrated form.

Chemical and physical requirements of these classes of lime are given in Tables 3.1 and 3.2.

Table 3.1: Chemical Requirements

Sl. No.	*Characteristics*	*Class*									
		A	*B*		*C*		*D*		*E*	*F*	
		Hydrated	*Quick*	*Hydrated*	*Quick*	*Hydrated*	*Quick*	*Hydrated*	*Hydrated*	*Quick*	*Hydrated*
(1)	*(2)*	*(3)*	*(4)*	*(5)*	*(6)*	*(7)*	*(8)*	*(9)*	*(10)*	*(11)*	*(12)*
(i)	Calcium and magnesium oxide percent, *Min* (or ignited basis)	60	70	70	85	85	85	85	50	70	70
(ii)	Magnesium oxide percent (on ignited basis) *Max*	6	6	6	6	6	—	—	6	—	—
	Min	—	—	—	—	—	6	6	—	6	6
(iii)	Silica, alumina and ferric oxide (percent Min. (on ignited basis)	20	10	10	—	—	—	—	20	10	10
(iv)	Unhydrated magnesium oxide (Percent) on ignited basis	—	—	—	—	—	8	8	—	8	8
(v)	Insoluble residue in dilute acid and alkali Percent Max. (on ignited basis)	15	10	10	2	2	2	2	25	10	10
(vi)	Carbon dioxide percent Max. (on oven dry basis)	5	5	5	5	5	5	5	5	5	5

(*vii*)	Free moisture content percent, Max.	2	—	2	—	2	—	2	2	—	2
(*viii*)	Available lime as CaO percent Min. (dry basis)	—	—	—	75 (Dry basis)	75 (Ignited basis)	—	—	—	—	—

Table 3.2: Physical Requirements

Sl. No.	Characteristics	Class A	B		C		D		E	F	
		Hydrated	Quick	Hydrated	Quick	Hydrated	Quick	Hydrated	Hydrated	Quick	Hydrated
(1)	(2)	(3)	(4)	(5)	(6)	(7)	(8)	(9)	(10)	(11)	(12)
(*i*)	Fineness										
	(*a*) Residue on 2.36 mm IS Sieve, percent, *Max*	Nil	—	Nil	—	Nil	—	Nil	Nil	—	Nil
	(*b*) Residue on 300 micron IS Sieve, percent, *Max.*	5	—	5	—	Nil	—	Nil	5	—	Nil
	(*c*) Residue on 212 micron IS Sieve, percent, *Max*	—	—	—	—	10	—	10	—	—	Nil
(*ii*)	Residue on slaking										
	(*a*) Residue on 850 micron IS Sieve, percent, *Max*	—	10	—	5	—	5	—	—	10	—
	(*b*) Residueon on 300 micron IS Sieve, percent, *Max*	—	—	—	5	—	5	—	—	—	—
(*iii*)	Setting time										
	(*a*) Initial set, Min, h	2	—	—	—	—	—	—	2	—	—
	(*b*) Final set. max, h	48	—	—	—	—	—	—	48	—	—
(*iv*)	Compressive strength *Min*, N/mm^2										
	(*a*) at 14 days	1.75	1.25	1.25	—	—	—	—	1.0	1.25	1.25
	(*b*) at 28 days	2.8	1.75	1.75	—	—	—	—	1.75	1.75	1.75
(*v*)	Transverse strength at 28 days, N/mm^2, *Min*	1.0	0.7	0.7	—	—	—	—	0.7	0.7	0.7

(*vi*)	Workability bumps, *Max*	—	—	—	12	10	12	10	—	—	—
(*vii*)	Volume yield ml/g *Min*	—	—	—	1.7	—	1.4	—	—	—	—
(*viii*)	Soundness, Le Chatelier expansion, mm, *Max*	5	—	5	—	—	—	—	10	—	10
(*ix*)	Popping and pitting	Free from pop and pits	—	Free from pop and pits	—	Free from pop and pits	—	Free from pop and pits	—	—	Free from pop and pits

STORAGE OF LIME

Quicklime should be run to putty as soon as possible after delivery as it deteriorates quickly when stored. It attracts moisture and carbon dioxide from the atmosphere and ultimately reverts to its original composition, calcium carbonate. In doing so, lump quicklime falls to a powder, becomes what is known as "air-slaked" and thereby loses a great deal of its useful properties.

The reaction between quicklime and water liberates a large amount of heat accompanied by expansion. This can be observed by dipping a piece of quicklime into water. Under normal storage conditions, the small amount of heat which may be generated is liberated so gradually that it is dissipated without causing a big increase in the temperature of the mass. If, however, considerable quantities of water are allowed to come into contact with the stored quicklime, say, through a leaky roof, the mass may become sufficiently hot to set fire any combustible material with which it may be in contact and by expansion (200% to 300%) damage a container or structure. For this reason it is clear that during transport and whenever it is necessary to store lime, particular care should be taken to ensure that quicklime is kept dry. These precautions apply equally to lump quicklime and ground quick lime.

If unslaked lime is to be stored for a short time it should be piled up and covered with a blanket of lime powder to exclude moist air.

Hydrated lime on the other hand can safely be stored for considerable periods without deterioration with the simple precaution, that the bag should be kept in a dry place ; presence of moisture forms lime cakes.

MARKETING OF LIME

Standard quality lime is superior to cement in workability and elasticity which go to make an excellent mortar giving an extremely durable bond in masonry construction. The proper use of lime would assist considerably in raising the standard quality of masonry construction in the country which has come down partly due to indiscriminating use of cement. The labour involved in the use of cement is less and it is, therefore, preferred by labour saving craftsmen and builders who are interested in completing the job quickly regardless of quality of work. It is to be granted to the advocates of cement that the ease in use is a desirable quality, but the extra labour in using lime is well worth the returns through better quality work. It is felt that no engineer who carefully examines the problem of lack of water-tightness in present day construction will grudge the extra labour involved in the use of lime which will give this quality.

Lime is available either as (*a*) Hydrated lime or (*b*) Quicklime.

3.13 HYDRATED LIME

Hydrated lime is used widely in the U.S.A. It is produced by chemical combination of water with calcium or dolomitic lime under specific conditions and it is sometimes further air blown and sized. This product is stable and is easily standardized. Hydrated lime has the special advantages of cement, it is ready for use and is a standard product.

There are some drawbacks to the introduction of hydrated lime in India, the main being that in the process of hydration, the weight is increased by some 25% on account of water and the increase in volume is about 200 to 300% due to expansion of lime while slaking. These two increase the cost of handling, transportation and packing. In a country where freight rates are high and packing material is costly, these are very important considerations. A start has however been made in India and to limited extent manufacture of hydrated lime is being done.

The hydrated lime should be supplied in suitable containers, such as jute bags lined with polythene or high density polythene woven bags lined with polythene or craft paper bags, preferably containing 50 kg of lime. The use of liner can be dispensed with if the lime is used within 30 days.

3.14 QUICKLIME

Quicklime is essentially calcium oxide. In India it is produced in lumps and is marketed as ground quicklime.

Quicklime is supplied in containers like metal container filled with 50 kg lime. It is suggested that in India we should follow the practice of marketing crushed unslaked lime packed in paper paper bags, as is the practice in European countries. These paper bags being air-tight prevent lime from slaking and expanding which usually leads to bursting of jute bags. The lime will be crushed to pea size before it is packed in bags. Bagged crushed lime will eliminate transportation of the water which is added to make the hydrated lime. The weight to volume ratio being much higher than in hydrated lime, the bagging cost will also be lower than that for the hydrated lime. Bagging will ensure quality and instructions for slaking, mixing and use can be printed on the bags. The important factor to consider, while deciding in favour of crushed lime as against hydrated lime, is that crushed lime will save considerable transport capacity upto 30% of the total lime requirements in the country at any given time.

USES OF LIME

Use of lime will be dealt with under two heads ; namely,. (1) Use of lime as a building material ; and (2) Use of lime other than as a building material.

Lime is being put to a number of uses for construction purposes, important of which are :

3.15 WHITEWASH

Lime is most commonly employed for whitewashing the outside and inside of rooms and buildings as it renders an attractive 'chalky' white surface which gives an impression of freshness and cleanliness. Quicklime is supplied in containers like metal container, containing 50 kg. of lime.

Whitewashing disinfects the surface and is the easiest and the cheapest way to cover the surface. With the addition of pigments, colouring of wall surfaces can be done in a variety of decorative colours.

Whitewash when used as an external rendering for walls reflects away the sun rays and thus helps in lowering the room temperature. This is an important consideration. Whitewash and cold water paints, containing lime are not as weather resistant as other types, but there is economy in use which will stimulate their future use for insulating purpose. Whitewashing does not make a structure fire-proof ; it is also used to reduce odour and promote better sanitary condition.

In general, any type of quicklime or hydrated lime may be used in the preparation of lime slurry for whitewashing. However, hydrated lime is preferable. But when hydrated lime is used, best results are obtained from the more highly refined types of hydrates which contain virtually no coarse particles or lumps. If unslaked lime is used it is better to strain the lime slurry through a fine cloth in order to remove the coarse particles.

3.16 COLOUR WASH

Colour washes can be prepared by adding certain pigments to the whitewash. The pigmtnts are available in the market as dried powder. Blue shades are obtained by using copper sulphate (*neela thotha*), green by chromium oxide, chrome green and yellow by chrome yellow, zinc-chrome and yellow ochre.

3.17 FLY-ASH LIME BRICKS

Pulverised fuel ash lime bricks (IS. 12894 : 2002) are use in buildings. The size of the brick is 190 to 230 mm in length, 90 to 110 mm in width and 90, 70 and 30 mm in height. For proper bond and modular dimensions in brickwork, bricks of sizes 70 × 110 × 70 ($\frac{1}{2}$ width brick) are also used. They have a compressive strength upto 30 N/mm^2, water absorption of 20% upto class 12.5 and 15% for higher classes is specified.

3.18 LIME CONCRETE

In building foundations and for fillings, lime concrete has been used. For small works like external surface drains, soak pits, tank, etc. lime concrete is extremely suitable and is cheaper than cement concrete. As a matter of fact, where speed of setting and maximum hardness are not important, the foundations can be cast in lime concrete. The strength of lime concrete can be greatly increased by the addition of Portland cement, and the cost is cheaper than cement concrete.

Mass lime concrete had been employed in India for construction of major irrigation works, dams etc. very successfully and pioneering work was done in designing the lime concrete mixes. Use of some pozzolanic material like surkhi has also been made to improve the strength and durability of lime concrete.

3.19 LIME CONCRETE TERRACING

Lime concrete has been used for roof terracing. Adoption of lime terracing gives a water-tight flat roof whose surface being a good reflector improves the thermal comforts of the dwelling. The terrace surface is polished by rubbing.

3.20 LIME CONCRETE FLOORING

Lime concrete has been employed for laying floors. The floor is given a shining and smooth surface by rubbing. With the addition of a variety of pigments and casting decorative designs, the flooring is made attractive to suit various preferences.

3.21 LIME FOR STABILIZATION

It has been established that addition of lime to many soils improves their strength and durability and, as such, lime-stabilized soils have been used for the construction of houses. The addition of lime to the soils helps to reduce the water requirement and excessive plasticity and sensitivity to changes of moisture content.

Generally quantities varying from 2.5 to 5% of lime are added for soil stabilization depending, of course, on the nature of soil. After adding lime the soil has to be worked to promote the lime reaction.

3.22 USE OF LIME OTHER THAN AS A BUILDING MATERIAL

Whereas lime is generally regarded as a building material, actually more of it is used for chemical purposes. The more important uses of lime are in the following ; glass manufacture, steel production, nonferrous metal wares, sewage and waste treatment, insecticides and fungicides, leather tanning, petroleum refining, coke and gas purification, sugar refining, calcium carbonate (precipitated), gelatin and glue, wire drawing, paints and varnish.

The common uses are : — (*i*) When no sterilizing compounds of chlorine are available contaminated water supply can be made safe to drink if excess of lime is thoroughly mixed with water and it is allowed to stand for one or two days before being consumed.

(*ii*) Lime is also used for clarifying muddy water and for treating water which is to be used in the boilers.

(*iii*) Lime is used effectively to reduce odour and promote more sanitary condition in foul smelling and damp, filthy places such as cesspools, drains, garbage cans, poultry houses etc.

(*iv*) During the past several years, lime has attained considerable prominence in road construction. The method of soil stabilization is a wholly new one whose application is expanding rapidly in countries such as the United States and Germany, not only for country roads but also for subgrade preparation of Express Highways and Airport runways as it reduces cost by eliminating the need for heavy subgrades.

Lime pozzolana concrete blocks of size 300 × 300 × 100 mm are used in paving. In view of low abrasive resistance of lime pozzolana concrete, these blocks are provided with a thin wearing course of cement sand mortar of 10 mm, cast integrally with lime pozzolana concrete.

(*v*) Lime is extensively used in agriculture for conditioning certain types of soils. Slaked lime which has become valueless as a building material because of recarbonation, is still of full value for soil correction provided that it has not become bumpy.

SHRINKAGE OF LIME

Fat lime has a great tendency to shrink and crack on drying. To prevent this, a large quantity of sand is required to be mixed with it to prepare mortar. Usually 2 to 3 times sand is added. This prevents shrinkage.

Hydraulic lime has a much less tendency to shrink and crack. Therefore the quantity of sand required is less. Usually equal or $1\frac{1}{2}$ times sand is added. If sand is mixed in excess quantity, the mortar becomes weak.

ARTIFICIAL HYDRAULIC LIME

When natural hydraulic lime is not available or the available hydraulic lime does not possess adequate hydraulicity, artificial hydraulic lime may be prepared from fat lime or quick lime. This can be done by adding to the fat lime necessary ingredients which it lacks in respect of hydraulicity *i.e.* alumina, silica and oxide of iron. There are two methods of making artificial hydraulic lime.

(*a*) *Fat-lime and clay.* Fat-lime is mixed with clay in proper proportions so as to make the composition similar to the natural hydraulic lime. These are intimately mixed and water is added to form a stiff paste. The mixture is thoroughly kneeded, converted into small balls, 25 to 40 mm in diameter with hand and dried. These balls are then burnt in the kiln to the required temperature.

(*b*) *Fat lime, and Surkhi.* Fat-lime is mixed with surkhi, *i.e.*, finely pulverised brick bats in the required proportions. Sufficient quantity of water is added and the mixture ground well in a mortar mill.

LIME SLAKING

Lime as obtained from the kiln is required to be slaked before use. The main object of slaking is to obtain a completely hydrated product with a maximum volume yield. Properly slaked lime consists of fine particles. Slaking of lime is an important process and great care should be exercised to slake the lime completely.

Chemically speaking, the process of slaking means conversion of calcium oxide (kiln lime) into calcium hydroxide (slaked lime) by addition of water. It is important that excess or deficiency of water should be avoided to safeguard against drowning or burning of lime ; both the conditions result in producing an inferior quality of lime with low volume yield. The use of very cold water or uneven distribution of water is also undesirable. For best results lime should be freshly drawn from the kiln for slaking, preferably within seven days of its calcination. All extraneous kiln ash must be removed from the burnt lime before slaking.

During slaking lime expands considerably and therefore if any portion of lime comes into the mortar unslaked, it will slake subsequently thereby expanding and causing damage to the masonry work. It is necessary to adopt different procedures of slaking for various classes of lime because of differences in their chemical composition and in the chemistry of their setting. For instance, fat lime which contains a high percentage of calcium oxide slakes very vigorously with evolution of great heat, while hydraulic lime slakes rather slowly and with less evolution of heat.

3.23 METHODS OF SLAKING LIME

There are several methods that can be adopted for slaking lime. The chief ones used in India are :

(*a*) Tank slaking and

(*b*) Platform slaking.

Slaking by the pressure hydration method which is particularly suitable for dolomitic lime is not employed in India as dolomitic lime is not produced in our country. Mechanical hydration is not much employed as ready made hydrated lime is not yet used extensively, slaking by immersion and air slaking are not desirable as complete slaking is not obtained by these processes.

3.23.1 Tank Slaking

In tank slaking two brick lined tanks are required. The first tank is made at a higher level, generally above the ground level and the second tank is made at a lower level which may be below the ground level. The first tank is about 450 mm deep while the second tank is about 600 to 700 mm deep. The levels of the two tanks are so adjusted that contents from the higher tank are able to flow into the lower tank by gravity.

The higher tank is first filled with water to a depth of about 300 mm and quicklime is gradually added to it, so as to cover the entire bottom of the tank to about half the depth of water. While quicklime is being added water is constantly stirred and hoed. No part of the lime is allowed to get exposed above the water. It is very important that in slaking, lime should be added to water and not water to lime.

As the lime slakes with evolution of heat, water begins to boil and more lime and water may be added till the requisite quantity of lime has been slaked. After the apparent slaking is over, stirring and hoeing should be continued for some time further to make sure that the whole of the lime has been fully slaked. Milk of lime thus formed is allowed to flow through an I.S. sieve No. 30 into the lower tank where it is allowed to mature by standing undisturbed so as to form what is called lime putty. Milk of lime formed by this process loses moisture by evaporation and absorption and thus thickens. For maturing of putty at least 2 or 3 days should be allowed in case of fat lime, but not more than 1 or 2 days should be allowed in the case of semi-hydraulic lime. This ensures complete slaking and at the same time improves the workability. Putty should not be allowed to stiffen or dry till it is used.

To obtain continuous supply of putty sometimes two lower tanks instead of one are used alternatively so that when putty is being used from one tank, fresh putty could be made in the other.

Before starting the second batch slaking in the upper tank, all unslaked stones of lime should be removed and discarded. When quicklime is received from the kiln in very large lumps sometimes it is found expedient to sprinkle on the heap of quicklime a small quantity of water before putting this quicklime into the tank for slaking so as to break the bigger lumps of lime into smaller ones for convenience of handling.

3.23.1 Platform Slaking

In this process quicklime is spread in a 150 mm layer on a water tight masonry platform and water is sprinkled over it in small quantities with a hose pipe until lime disintegrates to a fine powder. As water is added, the heap of lime is turned over and over. Care should be taken that minimum amount of water is added as is required for complete slaking. Slaking should be allowed to continue further for a period of about 24 hours or so. It should then be screened through I.S. sieve No. 30. Slaked lime should be stored in a dry place under cover, well protected from rain.

Slaked lime deteriorates in quality gradually by absorbing carbon dioxide from the atmosphere. Lime should, therefore, be used as soon as possible after it is slaked and should not be kept in storage for more than two weeks. Deterioration of slaked lime, however, can be reduced to some extent by covering the heap with a layer of bricks. This also prevents the lime from being blown away by air, when lime is stored in a shed or in the open.

3.23.3 Tank Slaking Versus Platform Slaking

Tank slaking is suitable for fat lime as well as for hydraulic lime. The putty thus formed gives much better workability than powdered slaked lime. Putty of fat lime actually improves by keeping, provided it is not allowed to stiffen or dry. Fati lime run into putty form can, therefore, be kept without any deterioration for many weeks ; in case of semi-hydraulic lime, however, it loses hydraulic set properties if kept in the putty form too long and it should therefore be utilised within one or two days.

To obtain good workability it is desirable to use semi-hydraulic lime in the form of putty rather than powder. Eminently hydraulic lime should not be run into putty from as it would lose most of its hydraulic setting properties.

SETTING OF LIME

3.24 FAT LIME

Fat lime is hydrated calcium oxide, $Ca(OH)_2$. It absorbs carbon dioxide from the atmosphere to form calcium carbonate *i.e.,*

$$Ca(OH)_2 + CO_2 = CaCO_3 + H_2O$$

Thus it depends, for its setting entirely upon the extent to which it comes in contact with air, from which it derives carbon dioxide. The reaction is usually slow and is confined to the surfaces. For this very reason, fat-lime is mostly used for plastering.

When sand is added to lime in the preparation of mortar, the mixture becomes porous. This enables carbon dioxide, from the atmosphere to reach the interior and make the mortar hard. However, if used in thick walls, the interior often remains in a soft condition. The water gets evaporated, leaving the mortar in a powdered state and the setting of the mortar does not take place, as carbon dioxide cannot reach there.

Fat lime is not suitable for exterior plaster as it does not weather well. For interior plaster work fat lime gives good results and weathers well. Fat lime gives excellent results when used for white washing.

3.25 HYDRAULIC LIME

Hydraulic lime consists of calcium oxide, magnesium oxide, calcium silicates and aluminium silicates. On hydration, it forms-hydroxides of calcium and magnesium. When sand is added to lime for the preparation of mortars, complicated chemical reactions take place. The chief reaction is the decomposition of complex aluminium-calcium silicate into hydrated calcium silicate and hydrated calcium aluminates. Also calcium hydrate is formed due to hydration of free lime. To some extent calcium oxide remains unslaked. All these reactions take place in the presence of water. Calcium silicates and calcium aluminates crystallize and form a hard substance which has great strength. The calcium hydrate formed is highly soluble in water, rises to the surface in the form of solution, gets free supply of carbon dioxide and hardens in the same way as fat lime. It crystallizes to form calcium carbonate.

Thus the setting of hydraulic lime depends upon water because reactions take place only when water is added. It is due to this reason that hydraulic lime is most suitable for use in structures under water. Hardening of lime does not depend on air and therefore, should not be slaked immediately after burning. Water should be added just before use.

TESTING AND SPECIFICATIONS OF LIME

To determine the suitability of lime, the following tests are carried out :

1. Visual examination for physical characteristics.
2. Chemical analysis.
3. Hydrochloric acid test.
4. Soundness test.
5. Workability test.
6. Transverse strength tests.

A brief description of these tests and the limits specified by Bureau of Indian Standards are given below :

3.26 VISUAL EXAMINATION FOR PHYSICAL CHARACTERISTICS

This test gives only a general idea about the type of lime. The lime specimen is examined for colour, *i.e.* dirty white, white or pure white and for the state of aggregation, *i.e.* lumpy, powdery, soft, hard, etc.

Fat lime has pure white colour. Lumpy form may indicate quicklime or unburnt limestone, but the former may be differentiated by its porous structure.

3.27 CHEMICAL ANALYSIS

Chemical analysis of lime is done to know the hydraulic and cementation values of the samples. Table 3.1 gives the limits for the various elements.

In addition, cementation value given by the following formula is also calculated.

$$\text{Cementation value} = \frac{2.8A + 1.1B + 0.7C}{D + 1.4E}$$

Where A is silica oxide (SiO_2)content.
B is aluminium oxide (Al_2O_3) content.
C is ferric oxide (Fe_2O_3) content.
D is calcium oxide (CaO) content.
and E is magnesium oxide (MgO) content.

3.28 HYDROCHLORIC ACID TEST

This test is carried out to determine roughly the class of lime and whether the carbonate content in the lime is excessive or not.

This is done by placing a tea spoonful of powdered lime in a test tube, tapping it gently to get a neat surface on the top. Note the initial level in the test tube. Now add 10 ml of 50 percent hydrochloric acid by volume. Stir with a glass rod, taking care that no inert material is left at the bottom of the tube. The test tube with its contents are kept standing for about 24 hours. The following inferences may be checked :

(1) If the effervescence indicating the liberation of carbon dioxide is abundant, it may be inferred that the lime has a substantial proportion of calcium carbonate.

(2) The volume of insoluble residue at the bottom of the test-tube compared with the original volume of lime will indicate the proportion of inert material and give an idea if it is excessive or not.

(3) In the case of hydraulic lime, a good thick gel will be formed and below it inert material will be deposited. If the gel is so thick as not even to flow when the test-tube is turned upside down, the lime is eminently hydraulic. If the gel formed is not quite thick and tends to flow on being tilted, the lime is feebly hydraulic. If there is no gel formation, the lime is non-hydraulic or fat.

3.29 SOUNDNESS TEST

This test checks the extent of atmospheric deterioration of hydrated lime and whether the field slaking of quick lime is adequate and complete or not.

For hydraulic lime, the test is carried out in the Le Chatelier apparatus, shown in Fig. 3.3. It consists of cylinder, 30 mm internal diameter and 30 mm high and made of brass 0.5 mm thickness. It has a split and indicators 165 mm in length are attached on either side of the split, so that their ends are 5 mm apart. The mould is placed on a glass plate and filled with mortar prepared by mixing cement, hydrated lime and sand in the proportions 1 : 3 : 12 with 12 percent by weight of water. The mould is covered at the top with a glass plate and left undisturbed for one hour. At the end of this period, the distance separating the indicator points is measured. The mould is then placed in a damp air cupboard for 48 hours, transferred to steam boiler and subjected to the continuous action of saturated steam for 3 hours (without immersing in boiling water). The mould is then allowed to cool and the distance between the indicator points is again measured. The increase, if any, over the former measurement should not exceed 10 mm after the deduction of 1 mm (to allow for the expansion of the added cement).

For fat lime, popping and pitting test is performed (Refer to I.S. 1624–1960). The pats prepared by mixing 70 gm of hydrated lime with 70 ml of water and 10 gm of plaster of Paris are subjected to steam action and examined in a good light for disintegration, popping or pitting ; any such defect indicates unsoundness in the lime.

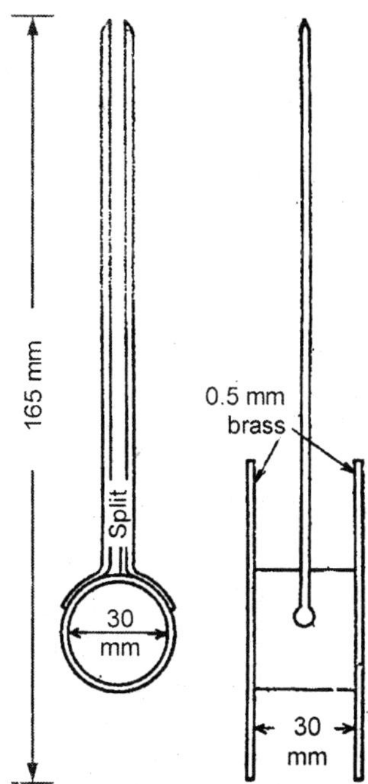

Fig. 3.3 Apparatus for conducting the "Le Chatelier" test.

3.30 WORKABILITY TEST

This test is used to determine whether or not the proportioning and mixing of the various ingredients in lime mortars, namely lime, sand and water with or without the addition of surkhi, cement etc. facilitate easy handling and application in the work. A handful of mortar is thrown on the surface on which it to be used. The extent of area covered and the amount of mortar picked up gives an idea of the workability.

3.31 TRANSVERSE STRENGTH TEST

This test checks whether the hydraulicity as well as the purity of the lime specimen is sufficient or not to develop the expected strength in a mortar of standard proportions and to find the strength of lime mortars being used in the work.

Lime sand mortar of 1 : 3 proportion is prepared by mixing water. Test specimens each 25 × 25 mm in cross-section and 100 mm length are prepared and cured for 28 days in an atmosphere of at least 90% relative humidity and a temperature varying from 24° to 32°C. These are immersed in water for half an hour and tested immediately for transverse strength. The specimen rests symmetrically with their sides on two parallel metal rollers 40 mm in diameter and spaced 80 mm apart. The load is applied through a third parallel roller (See Fig. 3.4) of the same size at a point midway between the other two. The load is applied steadily and uniformly at a rate of about 15 Kg per minute till failure occurs.

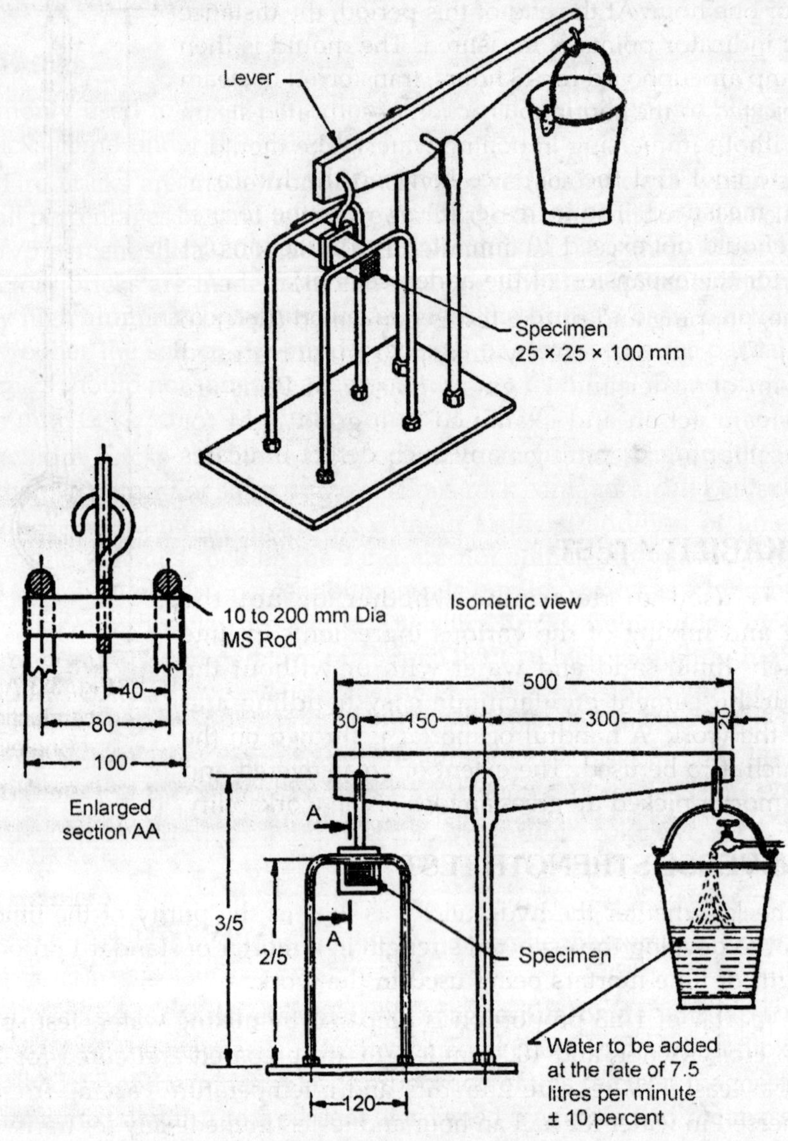

Fig. 3.4. Apparatus for transverse strength test

The modulus of rupture of the test specimen is calculated as follows :

$$m = (3WS/2bd^2) = 0.768\ W$$

where m = modulus of rupture in N/mm^2

W = breaking load in N

S = Span = 80 mm

b = breadth of specimen = 25 mm

d = depth of specimen = 25 mm

The value of m should not be less than 1.05 and 0.7 N/mm^2 for Class *A* and Class *B* limes respectively.

QUESTIONS

1. List the good qualities of lime as a construction material.
2. In which states of the country, the raw materials for lime are available. Discuss briefly their suitability or otherwise.
3. Describe briefly the various operations involved in quarrying of limestone.
4. What precautions will you observe in burning of limestone? Give details of vertical intermittent kiln.
5. Suggest some improvements in the working of vertical kilns. Draw a neat sketch of a continuous kiln and explain its working.
6. How is lime classified based on chemical composition? Give the physical and chemical requirements of different classes of lime as per Indian Standards.
7. Explain the following terms :
   (*a*) Hydrated lime
   (*b*) Quick lime
8. Enumerate the uses of lime as a building material. What are fly-ash lime bricks and state their specifications as per Indian Code.
9. Write short notes on:
   (*a*) Uses of lime other than as a building material
   (*b*) Shrinkage of lime
   (*c*) Artificial hydraulic lime
10. What is meant by lime slaking and give details of two methods chiefly adopted in India. Discuss their relative advantages and disadvantages.
11. Discuss the setting of lime in the following two cases
   (*a*) Fat lime
   (*b*) Hydraulic lime
12. What tests are carried out to check the suitability of lime as a building material? Explain 'Le Chatelier' test to determine its soundness.
13. Write short notes on the following:
   (*a*) Physical and chemical characteristics of lime
   (*b*) Hydrochloric acid test
   (*c*) Transverse strength test

OBJECTIVE TYPE QUESTIONS

1. Lime possesses workability and plasticity
   (*a*) Poor (*b*) Good
   (*c*) Better (*d*) Best

2. Lime gains strength
   (a) Slowly (b) Quickly
   (c) Rapidly (d) None of these
3. The major source of limestone in India is :
   (a) Massive limestone (b) Kankar limestone
   (c) Shells (d) Coral
4. Eminently hydraulic lime should have a minimum of ________ percent of silica, alumina and ferric oxide :
   (a) 10% (b) 15%
   (c) 20% (d) 25%
5. Fat lime, for building lime industry, should contain silica, alumina and ferric oxide
   (a) 15 to 20% (b) 10 to 15%
   (c) 5 to 10% (d) 0 to 5%
6. Semi-hydraulic lime, should have ________ percent of silica, lime and iron :
   (a) 10% (b) 15%
   (c) 20% (d) 25%
7. Chemical lime industry requires as nearly as ________ percent calcium carbonate stone
   (a) 80% (b) 85%
   (c) 90% (d) 100%
8. The fuel used for burning limestone in kilns is :
   (a) Coal (b) Cinder
   (c) Oil (d) All of these
9. Lime is used for
   (a) Masorry work (b) Plastering
   (c) Stabilization of soil (d) All of them
10. Lime consisting entirely of calcium oxide is termed :
    (a) Hydraulic lime (b) Poor lime
    (c) Rich lime (d) Slacked lime
11. The alumina and iron content in limestone should not exceed :
    (a) 0 to 0.5% (b) 0.5 to 1%
    (c) 1 to 1.5% (d) 1.5 to 2%
12. Greater heat is required to burn :
    (a) Calcium limestone (b) Magnesium limestone
    (c) Hydraulic limestone (d) Kankar
13. The temperature required to burn limestone in a kiln ranges from
    (a) 500 to 700° C (b) 700 to 1200° C
    (c) 1200 to 1500° C (d) Above 1500° C

14. The lime used for structural purposes is
(*a*) Fat lime (*b*) Kankar lime
(*c*) Eminently hydraulic lime (*d*) Semi-hydraulic lime

15. Kankar lime is used for
(*a*) Structural work (*b*) White washing
(*c*) Masonry mortars (*d*) Plastering

16. Which of the following types of lime is not recommended for finishing coat of plaster :
(*a*) Siliceous dolomitic lime (*b*) Semi-hydraulic lime
(*c*) Fat lime (*d*) Magnesium lime

17. Maximum free moisture content in lime should not exceed :
(*a*) 2% (*b*) 4%
(*c*) 4% (*d*) 5%

18. The minimum percentage of available lime as CaO in Fat lime should be
(*a*) 60 (*b*) 65
(*c*) 70 (*d*) 75

19. The maximum percentage of carbon dioxide, on oven dry basis, in all classes of lime should be
(*a*) 5% (*b*) 6%
(*c*) 7% (*d*) 8%

20. The minimum initial setting time in hours for eminently hydraulic lime is :
(*a*) 0.5 (*b*) 1
(*c*) 1.5 (*d*) 2

21. The maximum setting time for Kankar lime is
(*a*) 12 h (*b*) 24 h
(*c*) 36 h (*d*) 48 h

22. The compressive strength after 28 days of semi-hydraulic lime should not be less than
(*a*) 1.5 N/mm^2 (*b*) 1.75 N/mm^2
(*c*) 2.0 N/mm^2 (*d*) 2.25 N/mm^2

23. The minimum transverse strength for eminently hydraulic lime at 28 days is specified as :
(*a*) 1 N/mm^2 (*b*) 1.2 N/mm^2
(*c*) 1.4 N/mm^2 (*d*) 1.6 N/mm^2

24. Quick lime should be packed in :
(*a*) Jute bags (*b*) Craft paper bags
(*c*) Metal containers (*d*) Polythene bags

25. Fat lime is :
(*a*) Magnesium lime (*b*) High calcium lime
(*c*) Hydraulic lime (*d*) None of these

Answers

1. (*c*)	**2.** (*a*)	**3.** (*b*)	**4.** (*c*)	**5.** (*d*)	**6.** (*a*)
7. (*d*)	**8.** (*d*)	**9.** (*d*)	**10.** (*c*)	**11.** (*b*)	**12.** (*a*)
13. (*b*)	**14.** (*c*)	**15.** (*c*)	**16.** (*b*)	**17.** (*a*)	**18.** (*d*)
19. (*a*)	**20.** (*d*)	**21.** (*d*)	**22.** (*b*)	**23.** (*a*)	**24.** (*c*)
25. (*b*)					

FURTHER REFERENCES

Bureau of Indian Standards, New Delhi
I.S. Codes 712,1624, 12894

CEMENT

Portland cement is a binding material. This is named Portland because it resembles in colour after setting, to a grey stone found at a place Portland in U.K. Portland cement may be defined as follows :

"Portland cement is the product obtained by finely pulverizing clinker produced by calcining to incipient fusion a well proportioned mixture of argillaceous and calcareous materials, with no additions subsequent to calcination excepting calcined or uncalcined gypsum".

COMPOSITION OF PORTLAND CEMENT

Portland cement is a mixture of lime, silica, alumina, iron oxide, some magnesia and a small percentage of sulphur trioxide. The chemical analysis for a Portland cement is given in table 4.1. In this table minor constituents or impurities such as sodium oxide, potassium oxide, sulphur largely derived from the coal used in the burning operation, phosphorus pentaoxide and manganese oxide are not given. Sodium oxide and potassium oxide are commonly referred to as "the alkalies of Portland cement", although other alkalies are also present.

Tables 4.1 Composition of Portland Cement

Cement Ingredient	*Percentage*
Lime (CaO)	60—67
Silica (SiO_2)	17—25
Alumina (Al_2O_3)	3—8
Iron Oxide (Fe_2O_3)	0.5—6
Magnesia (MgO)	0.1—4

4.1 FUNCTIONS OF CEMENT INGREDIENTS

4.1.1 Lime (CaO)

Oxide of lime forms the major portion of cement and constitutes two-third of it. As such lime has an important bearing on its properties. Quantity of lime should be enough to form the silicate and aluminates of calcium. Over-liming and under liming are both harmful. If present in excess quantity, it results in the presence of uncombined lime in the finished product. Uncombined or 'free lime' hydrates slowly and causes cement to be unsound *i.e.* results in excessive expansion of the hardened cement paste. Under liming may cause improper mix control and is undesirable since the resulting product has low strength producing properties. In actual practice quantity of lime is kept a little below the required amount to avoid cracks.

4.1.2 Silica (SiO_2)

It is present to the extent of one-fifth of cement. It should be present in sufficient quantity to form the di calcium and tricalcium silicates which impart strength to the cement. However, if present in excess quantity, the setting time is prolonged, though the strength increases.

4.1.3 Alumina (Al_2O_3)

This is present in amounts 3 to 8 percent and acts as a flux. Thus it lowers the clinkering temperature. For the formation of a suitable type of cement, high temperature is essential. Thus alumina should not be present in excess amounts as it weakens the cement. However alumina reacts readily with water and helps the cement to set quickly.

4.1.4 Magnesia (MgO)

The proportion of magnesia is limited to 4 percent, but it is generally present in quantity varying from 0.1 to 3 percent. There appears to be no combination between magnesia and other oxides comprising the major part of the raw material. A high magnesia content in cement is undesirable as it produces unsoundness.

4.1.5 Calcium Sulphate ($CaSO_4$)

This is added in the form of gypsum and helps to retard the setting of cement. The initial setting time is increased. Gypsum is added upto 3 to 5 percent during the process of grinding clinker.

4.1.6 Oxide of Iron

Iron oxide imparts colour, combines with lime and silica and helps in the fusion of the material during manufacture. The main role of alumina and iron oxide in cement seems to be one of reducing the calcination temperature to make the process more economical.

4.1.7 Alkalies

They are present in the form of soda and potash. Most of the alkalies present in the raw material is driven out with the flue gases during burning. Though the total content of sodium and potassium in Portland cement is small and their water soluble alkali content is still less, they cause a number of troubles, like alkali-aggregate reaction, efflorescence, and staining when used in concrete, brickwork or masonry mortar. The efflorescence on brickwork or masonry is usually that of alkali or calcium salts derived both from the brick and cement mortar.

The most important damage caused by the alkalies is due to the alkali-aggregate reaction. The reaction is due to the interaction of the alkali hydroxides (sodium and potassium hydroxides) released from the cement and a reactive form of silica.

4.1.8 Sulphuric Anhydride (SO_3)

This is present upto 2 percent. Cement produces slow expansion *i.e.,* becomes unsound, if sulphur is present in excess amounts. A high content of sulphur trioxide also, prolongs the setting time considerably.

4.2 CONSTITUENTS OF PORTLAND CEMENT

When raw material for the manufacture of cement are correctly proportioned, ground and burnt to an incipient fusion, chemical combination takes place in the solid state, resulting in a product called clinker. This clinker when ground to a suitable fineness along with desired quantity of gypsum is Portland cement. The constituents of Portland cement clinker has been a matter of controversy. The important constituents are given in table 4.2.

Table 4.2: Constituents of Portland Cement

Constituents	*Nomenclature*	*Chemical composition*	*Approximate percentage composition*
Di-calcium silicate	C_2S	$2CaO.SiO_2$	27
Tri-calcium silicate	C_3S	$3CaO.SiO_2$	48
Tri-calcium aluminate	C_3A	$3CaO.Al_2O_3$	12
Tetra-calcium aluminoferrite	C_4AF	$4CaO.Al_20_3Fe_20_3$	8

The tri-calcium silicate is the major contributor to strength at all ages. This a particularly true at early ages upto 28 days of curing. At later ages, di-calcium silicate plays an increasingly important part, possibly being largely responsible for strength gains at periods of 1 year and later. Tri-calcium aluminate also contributes to strength. Its major contribution is at one day period, which fact is sometimes taken advantage of in the manufacture of high early strength cement. Tetra-calcium alumino-ferrite contributes little or nothing to strength and is responsible for the colour of the cement.

MANUFACTURE OF PORTLAND CEMENT

4.3 DISTRIBUTION OF PORTLAND CEMENT PRODUCTION

The location of Portland cement factories depends largely on the economic availability of the raw materials. In this respect, it is interesting to note that the distribution of Portland cement manufacturing factories is quite general in India, since every state has atleast one Portland cement factory with the exception of Himachal Pradesh.

4.4 METHODS OF MANUFACTURE OF CEMENT

There are two methods, commonly used for the manufacture of cement :

1. Dry process, and
2. Wet process.

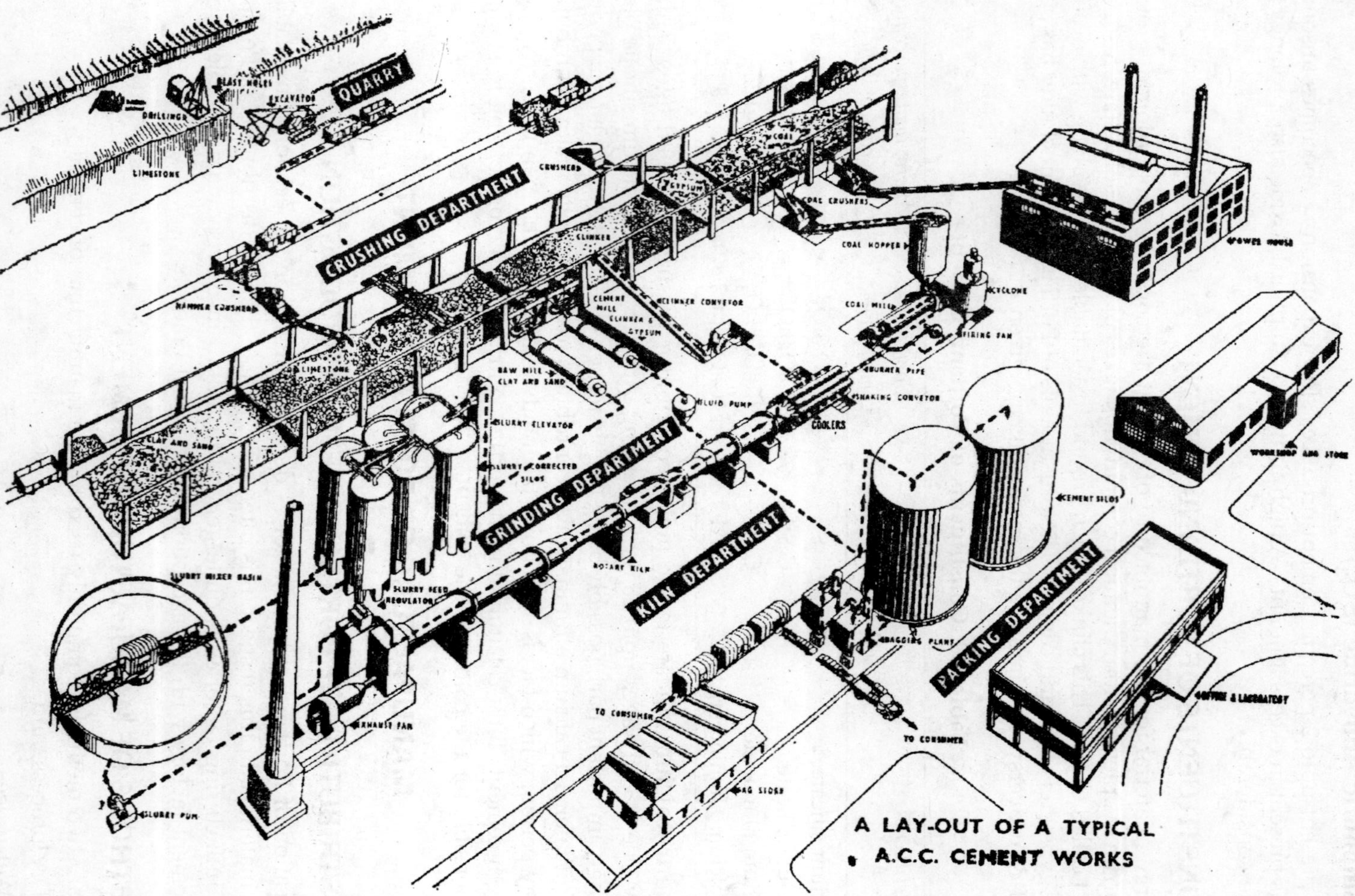

Fig. 4.1. Typical cement factory
(By Courtesy Cement Marketing Co. of India Private Ltd., Bombay)

The principal factor governing the selection of one or the other process is the moisture content in the raw materials. If the raw materials are dry, the dry process is preferred. In case too much water is present in the raw materials, the wet process is favoured. In the dry process the raw materials are mixed in the dry state where as in the wet process mixing is done by the addition of water.

4.5 STAGES OF PORTLAND CEMENT MANUFACTURE

The various stages involved in the manufacture of cement are :

1. Selection of a site where raw materials are available.
2. Preparation of the raw materials.
3. Burning of prepared raw materials in a kiln to obtain clinker.
4. Grinding of clinker along with gypsum.
5. Storage and packing of cement.

An overall outline of the manufacture of Portland cement, from quarry to finished product stage, is shown in Fig. 4.1.

4.6 SELECTION OF SITE FOR RAW MATERIALS

The two principal raw materials are :

1. Calcareous materials ; and
2. Argillaceous materials.

The calcareous material, commonly used, is limestone. The other materials that can also be used include marl, marine shells, and alkaline wastes. These are rarely used in India.

The argillaceous materials are clay, shale, blast furnace slag, slate and ashes.

The raw materials must be economically available and should be mixed in such proportions so as to provide proper chemical composition of the cement.

Limestone, Clay or Shale. Most of the cement factories in India, use a mixture of limestone, clay or shale to manufacture Portland cement. Usually the plant is located at the source of the limestone deposit with the clay or shale deposits located nearby.

It may, however, be pointed out that not all the limestone deposits are suitable as Portland cement raw materials. Deposits of calcareous materials which contain a high percentage of $MgCO_3$ are usually unsuitable. Also the presence of seams of pyrite or gypsum may result in a product having too high a sulphate content. Chalk deposits may be found which have a high percentage of pebbles, thus limiting their use.

The high lime content limestone, to be used as a component of Portland cement raw mix, must be blended with argillaceous materials of a chemical composition which will assure proper proportioning to the desired chemical composition in the clinker.

Marl and Clay or Shale. Marl, which is an earthy, crumbling deposit of limestone or chalk with varying proportions of clay are used to some extent as raw materials for Portland cement manufacture in India.

The nature of the deposit usually determines the methods to be used for their mixing, *i.e.*, whether dry process or wet process is to be employed. Usually marl contains 20 to 35 percent moisture and as. such wet process is economical. They are mixed with clay or shale in water before grinding.

Blast Furnace. Slag. Blast furnace slag, a by-product in the manufacture of steel by blast furnace, comprises essentially of the same oxides as are present in Portland cement, except that it has a low percentage of lime. The limestone and slag are mixed in the desired proportions to manufacture Portland cement.

Chemical Wastes. Alkali wastes, containing precipitated lime or calcium carbonates, as a by-product of the alkali and synthetic ammonium sulphate industries provides a substitute for limestone in. Portland cement manufacture.

4.7 PREPARATION OF RAW MATERIALS

The preparation of raw materials involve the following stages :

1. Removal of overburden.
2. Drilling or blasting.
3. Transportation.
4. Proportioning raw materials.
5. Pulverization.
6. Wet mixing.

4.7.1 Removal of Overburden

The overburden must be removed to permit excavation by means of quarrying or open-pit process. The depth of overburden varies from quarry to quarry. The nature of the overburden will determine the type of equipment that can be economically used. The bull-dozer .can handle shallow overburden to only a few m in depth, whereas larger scale operation will require power shovels or scrapers or slakline cable-way. The cost of overburden removal is, to most cement factories, a complete loss.

4.7.2.1 Drilling and Blasting

After the removal of the overburden, the deposit must be drilled to enable the placement of explosives in such a manner that they are most effective. The nature of the deposit and its location will guide the method to be adopted for drilling.

Blasting of the materials is nothing but a primary crushing-operation, *i.e.,* to break up larger masses of rock, earth, etc. to facilitate economic handling of the material. Generally, primary and; secondary blasting are employed. In the primary method, holes are drilled into the bed of rock, charge is placed and fired to break up-the rock mass. When the size of the blasted rock is such that it cannot be handled conveniently, it must be broken up further by additional drilling and secondary blasting.

4.7.2.2 Digging and Dredging

Materials like clay, shale, sand, etc. do not require blasting and offer little resistance to digging. They arc usually loaded by means of shovels or manual labour.

Other materials like marl, sand and gravel may have to be recovered by dredges.

4.7.3 Transportation

The excavated materials are required to be transported to the factory site or the point of treatment. With the wide variety of raw materials, cost of handling dictates the type of transportation that can best be utilised. If the lead is small, the material can be transported by mules. But if the quantity of materials to be transported is large, this method is not usually adopted. Motor truck or tractor trailer are being conveniently used. Rail transportation carries raw materials whereever track facilities are available. This method can handle a considerable quantity of material at a relatively low cost. These are the general methods of transportation. However, depending upon the terrain, other methods such as belt conveyor, aireal cable-way as at Surajpur (Haryana) etc. can be employed.

4.7.4 Proportioning Raw Materials

Proper combination of the ingredients of cement can be assured only by close control of the ratios of the various oxides involved. The following control ratios have been used quite successfully in raw mix proportioning and have served to assure control of the quality of Portland cement manufacture.

$$\text{Hydraulic Index } (SiO_2 + Al_2O_3) = 0.4 \text{ to } 0.6$$

$$\text{Silica modulus } \left(\frac{SiO_2}{Al_2O_3 + Fe_2O_3}\right) = \text{Not less than } 3$$

$$\text{Iron modulus } \left(\frac{Al_2O_3}{Fe_2O_3}\right) = \text{Not less than } 0.66$$

$$\text{Lea and Parker combination factor } \left(\frac{\text{CaO-free CaO}}{2.8SiO_2 + 1.18Al_2O_3 + 0.65Fe_2O_3}\right)$$

$$= \text{Not greater than 1.02 and not less than 0.66.}$$

4.7.5 Pulverization

Pulverization is the process of reducing cement raw materials from the maximum size made available by primary and secondary blasting in the quarry to the maximum size permissible in the feed to the clinkering kilns. This is usually accomplished in two stages ; namely

(*i*) Crushing, and

(*ii*) Grinding.

Crushing and grinding are also done in primary or preliminary and secondary or final stages or sometimes only in one stage.

The maximum size of the stone after the secondary blasting, controls to some extent, the type of primary crusher to be used. Important factors governing the choice of the primary crusher are crushability, moisture content and quality of the blasted rock, reduction ratio desired to meet the feed requirements of the secondary crusher, production rates needed, initial investment cost, operating and maintenance cost and installation space available. There are four types of crushers mostly used, namely :

1. Jaw crusher
2. Gyratory crusher
3. Hammer crusher
4. Roll crushers of single or double roll types.

Jaw crushers have low initial cost and consumes less power but handle less tonnage than other crushers having the same area of discharge opening. Gyratory crusher handles more tonnage but costs more, consumes more power and requires more headroom. Hammer crushers have a large reduction ratio than other types, thus making it possible to prepare material for grinding in one stage only. These crushers are suitable only for soft materials of limited maximum size. Single roll crushers need less headroom and power than jaw crushers of equal capacity but are not suitable for hard, abrasive materials.

All the four types of crushers, mentioned above, are adopted for use as secondary crushers usually with certain modifications to suit the reduction ratio needed.

Crushing is merely a preliminary stage, whereas grinding prepares the proper maximum particle size to assure thorough fusion and combination of the raw materials to form clinker. As such, grinding is a very important process in the manufacture of cement.

Primary grinders, usually called preliminators, are of two types :

1. Ball mills ; and
2. Ring roll mills.

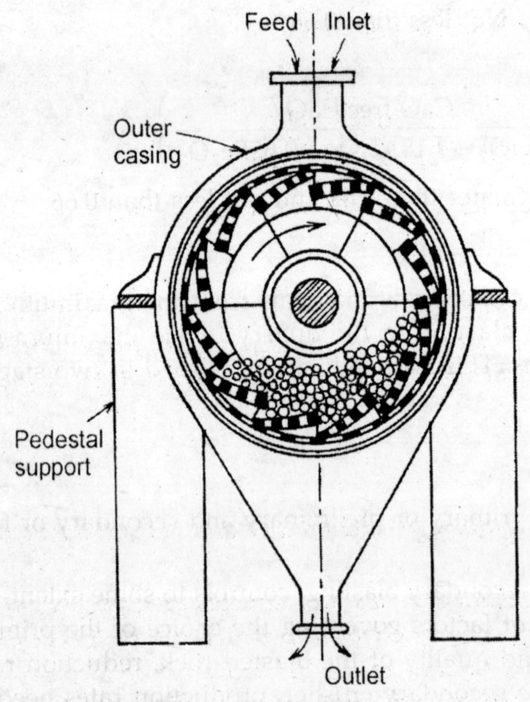

Fig. 4.2. Ball mill

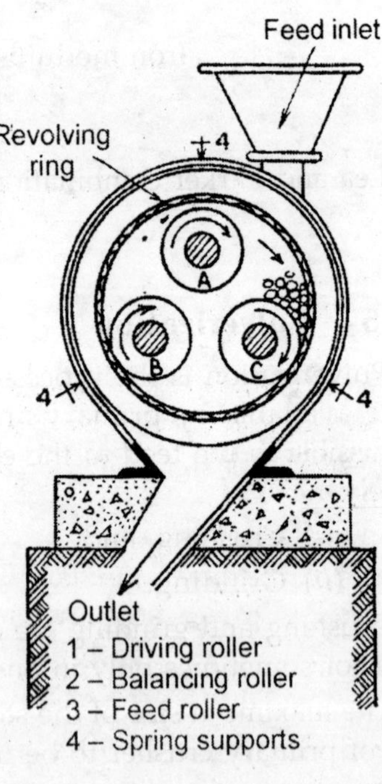

Fig. 4.3. Ring roll mill

Ball mill is a short cylinderical mil having length nearly equal to its diameter and loaded with 25 to 125 mm steel balls. Modern ball mills have large capacities and are capable of grinding 100 tonnes or more of crushed cement rock per hour to a fineness of approximately 45%, I.S. Sieve 75 micron. Ring roll mills consist of a ring shaped die against the inside surface of which, rolls ride under the pressure of centrifugal force. The rolls are suspended by means of hinged joints from arms mounted on a vertical operating shaft. These mills usually produce one-third than the modern large ball mills.

Both the ring roll type mill and the ball mill may be operated in close circuit with an external screen classifier, whereby the fines or undersize are sent for secondary grinding and the rejects or oversize are returned to the primary mill for further primary grinding. If sufficient fine particles are produced during secondary crushing ; they are first sieved and the oversize are sent to primary crushers.

For secondary grinding, tube mills are commonly employed. Tube mills are cylinders with steel balls from 12.5 to 30 mm diameter.

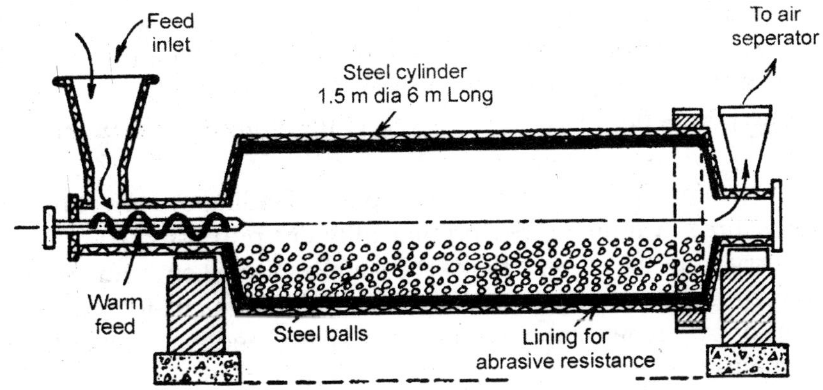

Fig. 4.4. Tube mill

They are longer than their diameter unlike those in ball mill and hence the name tube mill. Ordinary tube mill does not have division plate inside the mill to form compartments by means of which steel balls of different diameters can be kept separated. These do not give maximum grinding efficiency. As such compartmented tube mills are usually recommended.

Sometimes, inorder to avoid handling of raw materials between the primary and secondary stages of grinding, a combination mill is used. It has two compartments, the first compartment serving as a ball mill and having steel balls of 50 to 125 mm diameter and the second compartment as tube mill with balls of smaller diameter. The combination mill has the advantage of mechanical simplicity, reduced maintenance and lower power consumption.

The grinding mills rotate at a moderate speed of 15 to 20 revolution per minute. The raw materials alongwith the steel balls get lifted and fall in rapid succession, resulting in pownding or hammering action on the raw material by the balls. Due to the sliding of the balls, grinding effect is also produced. Thus by the combined effect of hammering and grinding, the material is ground to the required fineness.

4.7.6 Wet mixing

The method of grinding described above are confined to the manufacture of cement by dry process. Economical dry grinding requires drying of the raw materials before grinding. But clay or marl usually contains a high percentage of moisture and as such pre-drying is very expensive. In such cases, wet process of grinding is adopted. Water is added during grinding to form a mixture called slurry in a colloidal state.

The amount of water required is a function of the physio-chemical characteristics of the raw materials. Clay or argillaceous materials may require upto 65 percent water by weight of the slurry. Less quantity of water may create difficulties in the grinding mills such as (1) poor grinding efficiency or plugging of the screen plates (2) the various pumps and pipe line which convey the ground slurry from the mills to storage and blending tanks and finally to the kiln may be choked and (3) less efficiency in the storage and blending tanks by hampering the action of the mechanical and air agitators.

In wet process grinding, most of the grinding equipment discussed above can be used. Grinding efficiencies are higher in wet grinding. In this case, the separation of the coarse particles or oversizes is not done by screening but by entirely different means. Rake classifier or separators are used with primary ball mills, bowl classifiers or high speed centrifuges are also used. The slurry, thus produced, is conveyed to the storage tanks by means of pipes or pumps.

Slurry from the storage tanks is taken either to the blending or correcting tanks or silos. These are usually three in number, performing different purposes. The first silo is being filled with the slurry from the storage tank. The samples are taken from the second silo for analysis and correcting the same, if needed, by the adition of calculated quantity of slurry of either argillaceous or calcareous matter separately prepared and stored in independent tanks. The third silo contains the corrected slurry, to be drawn for feeding the kiln. If the correction of the slurry is to be done in two stages, an additional silo is required for rough adjustment of the slurry. The slurry is kept in the state of agitation by means of mechanical agitators and by compressed air blown up from the bottom. Silos have diameters ranging from 4.5 to 7.5 metres and height of 12 to 15 metres.

The water content from the slurry must be driven out to an appreciable value before being fed to the clinkering kilns. The more the water used, the more fuel will be consumed in evaporating water. For this purpose different types of dispersives and de-fluidizers have been used, but to little advantages. As such their use is limited.

The principal advantage in the wet process is more thorough mixing of the raw materials. This assures uniformity of kiln feed, better kiln operation and clinker quality. However the fuel consumption is more than the dry process and the length of the kiln is to be larger and therefore more expensive.

4.8 BURNING

There is no state in the manufacture of cement that equals in importance to the conversion of slurry into clinker, through the controlled combustion of fuel in the rotary kilns. The process is termed as *calcination*.

The principal objects of calcination of the slurry are :

(*i*) Evaporation of water.

(*ii*) Dissociation of carbonates of lime and magnesia.

(*iii*) Expulsion of alkalies.

(*iv*) Oxidation of ferrous to ferric oxides.

(*v*) Combination of lime and magnesia with silica, alumina, and ferric oxide to form the silicates, aluminates and ferrites which make up Portland cement.

Almost all the moisture is driven off at temperatures slightly exceeding 100°C. Dissociation of carbonates of lime take place at temperatures somewhat above 900°C and those of magnesium at temperatures between 800°C and 900°C. Silicates, alnminates and ferrites are formed at temperatures not less than 1100°C. It has been found that a temperature of 1550°C is necessary in order to ensure that all the lime has combined with the clayey constituents.

Fuel. Many factors affect the heat requirement, the important being :

1. Chemical and mineral composition of the kiln feed.
2. Fineness and particle size distribution of the feed.
3. Diameter, length, pitch and speed of rotation of the rotary kiln.
4. Efficiency of the combustion fuel.
5. Heat evolved from cooling clinker.
6. Radiation losses from the kiln.

The fuel used in the manufacture of cement is either pulverized coal or gas or oil. The coal is dried and pulverized, before use, in pulverizing mills. The air for drying is taken from the discharging end of the kiln. If needed, atmospheric air is also mixed to achieve the required temperature. Gas burning plants are the simplest to operate. This is directly blown into the kiln under pressure. Oil burning plants require boilers for pre-heating the oil to the required temperature and pumps to develop the proper pressure, usually 5 to 17.5 N/mm^2.

There are no standard burners for any of the fuels. Most of the cement factories develop their own burners.

4.8.1 Rotary Kiln

The clinkering process is carried out in rotary kilns. These are long and inclined steel cylinders. The length in the case of dry process varies from 30 to 100 metres. The wet process rotary kilns are longer, usually 50 to 150 m. The cylinders are made of 25 to 40 mm thick plates. The diameter ranges from 2 to 3.5 metres. The inclination of the kiln varies from 1 in 48 to 1 in 12; the feeding end being higher. The kiln is lined with refractory bricks. The speed of rotation of the kiln varies from 30 to 110 revolutions per hour. The lining with refractory bricks is either 100 or 200 mm. The 200 mm lining has longer life and is less troublesome. The steel cylinder is supported at intervals on rollers. One or more kilns may be arranged in parallel so as to operate independently.

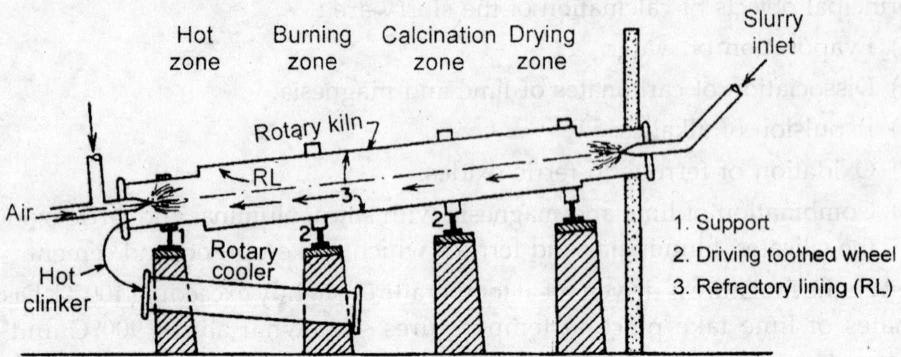

Fig. 4.5. Rotary kiln and cooler

Kiln room instrumentation has advanced to such an extent that the operator has to watch the instruments panel more frequently than the kiln itself. Typical installation would have the following instruments :

1. Recording speedometer or tachometer for kiln speed.
2. Recording optical pyrometer for lining temperature.
3. Oxygen and combustible recorders for showing efficiency of combustion.
4. Automatic draft controllers.
5. Temperature recorders for rear end temperatures.
6. Temperature recorders for preheated air from air quenching coolers.
7. Revolution counters.
8. Coal meters for pulverized coal.
9. Oil meters, indicating and recording.
10. Gas meters, indicating and recoding.

4.8.2 Stages in Burning

There are four distinct stages in the calcination of the raw materials for the manufacture of cement. The burning operation is largely endothermic. Heat is absorbed except in the cooling of clinker where exothermic reaction occurs and heat is evolved.

Drying zone. The raw materials either dry or the slurry is fed in the rotary kiln at the top. The fuel usually in the form of coal dust, is blown in under pressure from the lower end. A sufficient amount of air necessary for combustion is also blown in. In the drying zone, thick iron chains are provided. As the slurry gets sprayed against the surfaces of the hot chains, it dries and peels off in the form of flakes. The flakes fall on the bottom of the kiln. As the kiln, revolves at a slow speed, the material moves down slowly. The water gets evaporated in the drying zone.

Calcination zone. The flakes get calcined in this zone and carbon dioxide is driven off.

Burning zone. In this section, the argillaceous and the calcareous materials enter into chemical combination to form the silicates, aluminates and ferrites. The above compounds are believed to form in the burning zone of kiln at a temperature of about 1100°C at which point the melt or liquid begins to form. The reactions are completed at about 1200°C. At this point, about 20 to 30 percent of the mass is converted to a molten liquid.

Hot zone. The next step in the formation of compounds in the 'hot' zone of the kiln is the formation of di-calcium silicate witft continued heat and the presence of available lime, the dicalcium silicate takes an additional molecule of lime to form tricalcium silicate.

4.9 GRINDING OF CLINKER

Clinker is in the form of nodules of size 10 to 12.5 mm. This is required to be ground to produce cement. The effects of grinding are :

1. It produces a fine material in which all the ingredients are well distributed.
2. It has more surface area to come in contact and react with water. Therefore the setting property is accelerated. This enables, the development of high strength at early ages. Thus quick setting cement have to be ground to a greater fineness.
3. Fineness of grinding imparts workability and density to the mortars and concrete.
4. Expansions and contractions are reduced to a great extent.

There is a limit to the degree of fineness of grinding cement.. If this is exceeded, the fine particles will float in air during handling and cause an effect known as "blowing".

Factors of clinker grindability. Clinker grindability is influenced by the following factors :

1. Chemical combustion of the clinker.
2. Clinker burning.
3. Rate of cooling of clinker.
4. Moisture content of clinker.
5. Age and method of storing clinker.
6. Finish mill grinding temperatures.

Methods of grinding. As the size of the clinker is small, there is no need for crushing. The grinding operation consists of :

(*i*) Primary grinding ; and

(*ii*) Finish grinding.

The primary or preliminary grinding is accomplished in the ball mills or ring roll mills with certain modifications. The material is reduced so as to pass through I.S. Sieve No. 75 microns Factors to be kept in mind are physical and chemical properties of the cement, reduction in size and thorough distribution of the ingredients.

The finish grinding is done in tube mills and combination mills of various sizes. Proper compartmentation of tube mills and selection of suitable grinding media can help to a great extent to achieve desired results. The combination mill has the advantage of lower operating and maintenance costs but is less flexible in operation. It combines in one mill the preliminary and secondary grinding units.

4.10 RETARDER MIXING WITH CLINKER

If the material obtained after the grinding of clinker is mixed with water, it begins to set immediately and becomes hard. But from the practical point of view, some time is always required for mixing cement with water and aggregate, carrying the mix to the required location and laying the mix in position. To account for all this, it has been seen that about 15 to 30 minutes are required.. Thus, the cement should not start setting in this period and to achieve this, a compound called a retarder is added to the clinker during grinding. The function of the retarder is to increase initial setting time of cement or in other words to retard its setting. The retarder generally added is gypsum. About 3 to 5 percent by weight is mixed with the clinker before grinding. Alumina reacts with gypsum ((calcium sulphate) to form calcium sulpho-aluminate, which is insoluble and does not hydrate. Addition of excess quantity of gypsum is harmful as the proportion of SO_3 increases more than the permissible value.

PACKING CEMENT

Various types of containers have been used for packing cement. These include : Jute bags, cloth bags, multi-ply bags, wooden barrels, airtight steel drums.

In this country cement is commonly supplied in jute bags. Jute bags have made very satisfactory containers as their shape and size make them convenient to handle and they may be used repeatedly, probably making from three to five round trips from the factory to the cement user. However, the use of jute bags requires a considerable investment and expense for the cement manufacture. Large numbers are in stock and in transit. Returned bags must be sorted, cleaned; counted, and repaired before they are ready for the next trip and of course, there is a percentage of loss.

Tne factories will only accept bags which are fit for repacking and in order to preserve the bags as much as possible, the following instructions should be carefully carried out :

1. Do not cut the bags but open with a knife. Bags should be opened by pulling the loose end of the tying string.
2. Do not use iron hooks for moving bags around. They tear the bags and torn bags are useless for replacing.
3. When emptying cement from the bags, dust the bags thoroughly. Cement remaining in the corners and seams of the bags tends to turn it stiff with absorption of moisture from the air.
4. Do not allow the bags to come in contact with moisture. Wet bags deteriorate quickly. Stack bags neatly in a dry place. Do not use bags for curing. Old and unserviceable bags may be used for this purpose.
5. Do not use the bags for packing material other than cement such as coal, sand, stone, chips, grain, seeds, sugar, etc. These materials not only contaminate the cement, but often reacts chemically with it, altering its properties.
6. Each bag has a distinct brand mark and must go back only to that factory from which it came.

In recent years multiply paper bags have come to be largely used as cement containers in view of the rising cost and difficulty of supply of jute bags. Paper bags afford slightly more protection to the cement from the weather than jute or cloth bags.

Wooden barrels and steel drums, generally lined with paper, are mostly used for overseas shipments.

4.11 HOT CEMENT

When cement is ground, its temperature rises owing to the frictional heat generated in the ball mill, and the finer the cement the higher will be the temperature. Thus when taken to the silo, cement may have a temperature as high as 135°C. A large silo of hot cement has little chance to cool and therefore it is quite likely that when the cement is packed it will still have a fairly high temperature and reaches the consumers in a hot condition. Doubts have sometimes been expressed about the soundness and setting properties of such cement. It may be categorically stated that the fact of cement being hot is no indication of inferior quality. Tests indicate that hot cement has a somewhat shorter setting time than cool cement. Since this test is made on neat cement, the effect is probably very much less in mixed concrete. The crushing strength is not affected by the use of hot cement. The only effect of its use is to make the concrete a few degrees warmer than otherwise. In cold weather this is an advantage, but in hot weather it may contribute early stiffening of the concrete and later to shrinkage cracking on cooling. There should be no trouble, however, if care is taken to prevent rapid evaporation of water from the concrete.

4.12 BULK CEMENT

Recently, owing to the scarcity of jute bags, cement factories have instituted the practice of supplying cement in bulk, particularly to large consumers. Although this is a new development in India, bulk cement has been in supply in the United States since long where the existence of a ready mixed concrete industry has given it considerable impetus.

Bulk cement is loaded at the factories into ordinary or hopper-bottom railway wagons, barges, or specially fitted lorries. Tha cement is taken from the silos in the same way as the carrier from the convenient point on the conveying system. At the newer factories and in the United States, the cement is drawn directly from the silos through valves in the silo wall. Only about two thirds of the contents of a silo may be emptied by gravity in this way, but the procedure is rapid and satisfactory. The cement may or may not be weighed on the way to the wagon or lorry, but in either event, the weight of the cement may be obtained by weighing the carrier before and after loading.

The method of unloading bulk cement from railway wagon or other carrier varies from hand shovelling and packing into the consumer's own containers to the use of mechanically or pneumatically operated plant. The cement is shovelled by hand into the hopper and filled into a jute bag tied at the mouth of the hopper by operating a flap valve. Mechanical methods of unloading include the use of power scrapers, screw conveyors, bucket elevators, belt conveyors, etc., or a combination of these devices. With pneumatic system the cement may be pumped over considerable distance both horizontally and vertically. The cement is normally discharged into an elevated bin or the cement compartment of the hopper of a batching plant from which it is drawn into a scale hopper and weighed, then discharged into a truck or mixer.

Steel Containers. Special road-rail steel containers are also employed for conveying bulk cement. These may be mounted on flat railway wagons and can be transferred on to lorries. A hinged door at the top is clamped down after the cement has been poured in. At destination the container is lifted from the wagon or lorry by crane and the contents dropped in an elevated bin by opening two hinged doors which form its base.

STORAGE OF CEMENT

It is often necessary to store Portland cement, sometimes for a period of months. This is particularly true when transportation facilities must be used to their capacity and deliveries are uncertain.

Portland cement will retain its quality almost indefinitely if moisture is kept away from it. This has been shown by tests of cement kept for several years in tightly covered cans in which there was little air and no moisture could reach the cement. Similarly tests have been made of cement kept at cement factories in large air tight storage bins for several years. In neither case there was any deterioration. On the other hand cement exposed to air will absorb moisture slowly which will cause deterioration. An absorption of 1 to 2 percent of water has no appreciable effect, but further amounts of absorption retard the hardening of cement and reduce its strength. If the absorption exceeds 5 percent, the cement is, for all ordinary purposes, unsuitable. The more finely a cement is ground the more active it is and consequently the more rapidly it absorbs moisture from damp surroundings.

Every year and particularly during the monsoon, large quantities of cement are spoilt owing to neglect in keeping the cement absolutely dry. The methods of storage, therefore, merit careful consideration.

The best method of storage is in bulk and bins of loose cement 2 metres or more in depth can lie for longer than a year with no more damage than the formation of a crust about 50 cm thick which must be removed before the cement is taken for use. If prolonged storage of bagged cement is unavoidable it is better to empty the containers and stock the cement in as deep a heap as possible in a building of which the walls and floor are non-porous.

Bagged cement may also be kept safely for many months if stored in a waterproof shed with a dry floor well off the ground. The walls and roof are of waterproof construction and the few small windows are kept tightly shut. It is well above the ground. Drainage should be provided if necessary to prevent accumulation of water in the vicinity of the shed. The capacity of cement storage depends upon its floor area and the height to which bags are piled. When determining the size of shed required for storage of a certain number of bags, each bag may be counted as occupying 0.0506 cu m and not less than 0-293 sq m of floor space. A tonne of cement will, therefore, require about 10 cu m space. Usually piles about 14 bags high are the maximum that can be laid up economically. Since it is inadvisable to pile bags against the building walls, the allowance of 0.3 m all round should be made between exterior walls and the piles. Passages between piles should be provided for and be not less than 1 m wide.

Before a new warehouse is put into service, time should be allowed for the interior to dry thoroughly. Bagged cement can be piled directly on a dry floor. The bags should be placed closed together to reduce circulation of air. If the piles are to be more than seven or eight bags high, the bags should be placed in headers and stretchers, that is, alternately lengthwise and crosswise so as to tie the piles together and lessen the danger of piles overturning.

For long storage the tops and sides of the piles may be covered with *tarpauling*, waterproof paper or other covering as an extra precaution against moisture. Once the cement has been properly stored it should not be disturbed until it is to be used. There is no advantage in moving and restacking the bags to reduce warehouse pack as this practice only exposes fresh cement to the air.

When removing cement from storage, some bags should be removed three-tiers back rather all from one tier. If the piles are thus stepped back there is less chance of overturning. This is particularly true when the piles are high. It is also desirable to apply the "first in, first out" rule, that is remove the oldest cement first. Dead storage, that is an area where the cement remains in place for a long time while other consignments of cement come in and go out of the shed, should be avoided.

TYPES OF CEMENT

Cement can be divided into the following categories :

1. Portland Cement
   - (*a*) Ordinary Portland cement.
   - (*b*) Rapid hardening cement or high early strength cement.
   - (*c*) Low heat cement
   - (*d*) Sulphate resisting cement
   - (*e*) Portland blast furnace slag cement
   - (*f*) White Portland cement
   - (*g*) Coloured Portland cement
2. Natural Cement
3. High alumina cement
4. Super-sulphated cements
5. Special types of cements
   - (*a*) Masonry cement
   - (*b*) Expansive cement
   - (*c*) Oil-well cement
6. Portland cement with additives :
   - (*a*) Portland Pozzolana cement.
   - (*b*) Air entraining cement
   - (*c*) Hydrophobic cement.

A brief discussion follows :

4.13 PORTLAND CEMENT

The difference in properties of various kinds of Portland cement arises from the relative proportion in which the principal compounds are present in the cement and the fineness of the cement. The C_3S and C_2S constituents form 7 to 8 percent of all Portland cement and are most stable and contribute most of the eventual strength and resistance of the concrete to corrosive salts, alkalies and acids. The C_3S hydrates more rapidly than the C_2S and it therefore

contributes more to the early strength and heat generated and therefore the rise in temperature. The C_3S component is liable to decompose to hydroxides of calcium and aluminium on exposure to air and water and gets easily attacked by salts and alkalies. The C_4AF component is comparatively inactive and contributes little to the strength of heat of hydration of cement.

4.13.1 Ordinary Portland cement I.S. 269 : 1989 ; IS 8112 : 1989 ; and IS 12269 : 1987

This is used for normal work. All the discussion, given above relate to ordinary Portland cement. It is available in three grades i.e. 33 grade, 43 grade and 53 grade depending upon the crushing strength in MPa of 1:3 cement sand mortar cubes after 28 days curing.

Chemical properties laid down in IS 269 : 1989 are given in Table 4.3.

Table 4.3 Chemical Requirements for 33 Grade Ordinary Portland Cement

Sl. No. (1)	*Characteristics* (2)	*Requirement* (3)
(i)	Ratio of percentage of lime to percentage of silica, alumina and iron oxide, when calculated by the formula $\frac{CaO - 0.7\,SO_3}{2.8\,SiO_2 + 1.2\,Al_2\,O_3 + 0.65\,Fe_2\,O_3}$	Not greater than 1.02 and not less than 0.66
(ii)	Ratio of percentage of alumina to that of iron oxide, Min.	Not less than 0.66
(iii)	Insoluble residue, percent by mass, Max.	In case no flyash, silica fume, rice, husk ash and metakaoline is added-not more than 5.0
(iv)	Magnesia, percent by mass	Not more than 6.0 percent
(v)	Total sulphur content calculated as sulphuric anhydride (SO_3) percent by mass Max.	3.5
(vi)	Total loss on ignition	Not more than 5 percent
(vii)	Chloride content by mass Max.	0.1
(viii)	Alkali content	0.05 (for prestressed structures)

4.13.1.1. Physical Requirements

1. *Fineness:* Specific surface of cement should not be less than 225 m^2/kg.

2. *Soundness:* Unaerated expansion should not be more than 10 mm by 'Le Chatelier' method and 0.8 percent by autoclave test. If it fails, aerated sample shall not show more than 5 mm and 0.6 percent when tested by 'Le Chartelier' method and autoclave method respectively.

3. *Setting Time.* The setting time of the cements, when tested by the vicat apparatus shall conform to the following requirements:

(*a*) Initial setting time in minutes — Not less than 30; and

(*b*) Final setting time in minutes — Not more than 600.

4. *Compressive Strength.* The average of at least three mortar cubes (area of face 50 cm^2) composed of one part of cement, three parts of standard sand by mass and $\left(\frac{P}{4} + 3.0\right)$ percent (of combined mass of cement plus sand) water and prepared, stored and tested shall be as follows :

(*a*) 72 ± 1 hour; Min — not less than 16 MPa

(*b*) 168 ± 2 hours; Min — not less than 22 MPa, and

(*c*) 672 ± 4 hours; Min — not less than 33 MPa

Max. — 48 MPa

4.13.2 Rapid Hardening Portland Cement (IS 8041 : 1990)

It is similar to ordinary Portland cement but cement is ground finer with slight alteration in chemical composition. The extra rapid hardening type is obtained by grinding calcium chloride with rapid hardening Portland cements. The rapid hardening' property is due to the presence of more C_3S and less C_2S ordinarily as compared to ordinary Portland cements, and this gives a high early strength also. The calcium chloride added in the extra rapid hardening cements imparts the quick setting properties in addition to accelerating the process of hardening.

It should be so fine that the specific surface should not be less than 325 m^2/kg. The compressive strength of mortar cubes should not be less the 16 MPa after one day and 27 MPa after three days.

4.13.3 Low Heat Cements

Low heat cements are manufactured from raw materials with a low calcium content and a high silica content. In these cements the proportion of C_2S (Calcium silicate) is increased and the proportion of C_3S (Tricalcium silicate) is reduced as far as possible in addition to restricting the amount of C_3A (Tricalcium aluminate). Since a reduction in temperature retards chemical action of hardening the rate of evolution of heat will be less, thereby extending the evolution of heat over a longer period. Though due to this fact the rate of gain of strength will be slow, it is beneficial in resisting the formation of cracks in concrete.

4.13.4 Sulphate Resisting Cements

Sulphate resisting cements are similar to Portland cements but the amount of C_3A (Tricalcium aluminate) present is much less and silicates much more. The compound C_4AF (Tetra calcium alumino ferrite) is formed which is resistant to chemical agencies. These cements are useful in counteracting the effect of sulphates. As such it is useful in soils and water where sulphates exist. Although its ultimate strength is the same as that of ordinary Portland cement, its strength development proceeds at a slower rate. It generates less heat of hydration. Specific surface should not be less than 225 m^2/kg.

4.13.5 Portland Blast Furnace Slag Cements

Portland blast Furnace slag cements are manufactured by grinding together a Portland cement clinker and granulated blast furnace slag of high lime content obtained as a by product in the manufacture of pig iron, in such proportions that the content of granulated slag does not exceed 65 percent of the total weight. These cements which show a lower rate of hardening and a low heat of hydration can be used for concreting large masses in hot weather though they require more careful handling at extremes of temperature as they harden in cold weather rather slowly. These cements are more resistant to sea water, peaty or slightly acidic waters and sulphates. Their resistance to sulphates is generally attributed to the lesser content of free calcium hydroxide present when they set, as well as the less basic nature of the hydrated calcium silicates present. The specifications for strength requirements are the same as for 33 grade ordinary Poartland cements and these cements find extensive use in marine work. Indian Standards Specification No. 455: 1989 lays down the requirements for Portland blast furnace slag cement as follows :

(1)	**Chemical Requirement**	**Percent max.**
	Magnesium Oxide (MgO)	8.0
	Sulphur Trioxide (SO_3)	3.0
	Sulphide Sulphur (S)	1.5
	Loss on ignition	5.0
	Insoluble residue	4.0

(2) Fineness—Specific surface not less than 225 m^2/kg

(3) Total chloride content should not exceed 0.05 percent by mass for cement used in prestressed structures and long span reinforced concrete structures.

4.13.6 White Portland Cement

White Portland cements are manufactured from chalks and limestones containing low iron contents. The chemical requirements for white Portland cement, as stipulated in IS : 8042 : 1989 are given in Table 4.4.

Table 4.4 Chemical Requirements for White Portland Cement

Sl. No. (1)	*Characteristics* (2)	*Requirement* (3)
(*i*)	Ratio of percentages of lime to percentage of silica, alumina and iron oxide	Not greater than 1.02 and not less than 0.66
(*ii*)	Iron oxid, percent by mass	Not more than 1.0 percent
(*iii*)	Insoluble residue, percent by mass	Not more than 2.0 percent
(*iv*)	Magnesia, percent by mass	Not more than 5 percent
(*v*)	Total sulphur content calculated as sulphuric anhydride (SO_3), percent by mass	Not more than 3.5 percent

The compressive strength of mortar prepared with white cement should not be less than 90 percent of those specified for an ordinary Portland cement. The kilns are also fired by oil fuel instead of coal so that the contamination by coal wash is avoided.

4.13.7 Coloured Portland Cement

Coloured Portland cements consist of Portland cements with 5-10% or pigment, obtained by grinding the cement and pigment together. The more commonly used pigments are iron oxide (red yellow, brown and black), manganese dioxide (black and brown), chromium oxide (green), cobalt (blue), ultramarine (blue) and carbon pigments (black). The white and coloured cements may be used more for decorative purposes.

4.14 NATURAL CEMENTS

These are manufactured from naturally occurring cement rocks with composition similar to the artificial mix for the manufacture of Portland cements. Their properties closely resemble that of the natural cement rock from which they are manufactured.

4.15 HIGH ALUMINA CEMENTS

These cements are characterised by their dark colour, high early strength, high heat of hydration and resistance to chemical attack. They are manufactured by fusing or sintering a mixture of aluminous and calcareous substances and grinding to powder the resultant material. Indian standard specification No. 6452 : 1989 specifies that the total alumina content shall not be less than 32%, the specific surface should not be less than 225 m^2/kg and expansion not more than 5 mm. The rapid hardening properties of these cements are due to the presence of chiefly mono-calcium aluminate, in place of tricalcium silicate. Due to the formation of a ceramic bond in concrete on setting, these cements are useful as refractory materials also. Mixtures of Portland and high alumina cements are used when quick setting properties are required. As compared to special types quick setting cement these mixtures of Portland and high alumina cements are more useful as these mixtures can be stored for longer periods if they are kept dry. It is also seen that great care has to be exercised when these mixtures are used since their behaviour depends on various factors. The high alumina cement exhibits a very high resistance to sulphates. Usually it is seen that in Portland cements calcium sulpho-aluminate is formed by the interaction of aluminate compounds with sulphate solutions and sulphate resistance decreases with increasing alumina cement. But, in the case of high alumina cements, though the alumina content is very high, they show a higher resistance to sulphates. This has been explained due to the fact that expansion can occur only if a cement compound in the solid state reacts with a substance in solution. But if the compound itself goes into solution, it first reacts and then precipitates as solid, no expansion can occur. In the case of normal Portland cements, the free lime produced during setting renders the hydrated calcium aluminates insoluble. But in the case of high alumina cements this hydrated calcium aluminates pass into solution due to the absence of additional lime as very little free lime is liberated during their setting and hardening. The mono-calcium aluminate which is its principal constituent forms dicalcium aluminate and a small amount of mono-calcium silicate and free alumina in excess of that which can combine with any free lime during the hydration process.

The point to be noted is that while there is an excess of free lime over and above that which can combine with aluminates in Portland cements due to hydration of the tricalcium silicate, there is an excess of free alumina only in the case of high alumina cement. Hence this free alumina which is relatively inert imparts a high sulphate resistance, preventing the formation of calcium sulphate aluminate which produces expansion, the chief cause for destruction of concretes when used in conjunction with sulphates. There is no satisfactory and quick test for the resistance to chemical attack, but the resistance to sulphate attack can be found out by strength tests on concretes immersed for long periods in sulphate solutions. The quality of the cement also can be assessed from its chemical composition and from this an idea of its resistance to chemical attack can be estimated.

Following restrictions should be followed for its use in concrete:

1. It should not be used in locations where ambient temperature exceeds 18°C.
2. Accelerators like calcium chloride should not be used.
3. Steam curing or elevated temperature of curing should be avoided.
4. It should not be mixed with other types of cements.

4.16 SUPER-SULPHATED CEMENTS

These are made by inter grinding granulated blast furnace slag 80-85%, calcium sulphate 10-15% and about 5% of Portland cement. It has a low heat of hydration and resists chemical attack duet to the relatively low content of free lime released during hydration.

4.16.1 Application

Supersulphated cement has been successfully used in a variety of aggressive conditions, for emple, for marine works, mass concrete jobs to resist the attack by aggressive water, reinforced concrete pipes in ground water, concrete construction in sulphate bearing soils, and in chemical works under conditions involving exposure to high concentrations of sulphates or weak solutions of mineral acids. It has been used for the underside of bridges over railways and for concrete sewers carrrying industrial effluents. Its use under tropical conditions has also been recommended, provided the prevailing temperature is below 40°C. Although its use as a general purpose cement can be made with adequate precautions, it is not recommended for producing steam cured products. Production of this cement will also result in greater utilization of blastfurnace slag, an industrial by-product of steel in the country.

4.16.2 Chemical Requirements

Insoluble residue	4 percent, *Max*
Magnesium oxide	10 percent, *Max*
Sulphuric anhydride	6 percent, *Min*
Sulphide sulphur	1.5 percent, *Max*

4.16.3 Physical Requirements

(*i*) *Fineness.* It shall have a fineness (specific surface) of not less than 400 m^2/kg.

(*ii*) *Soundness.* Expansion not more than 5 mm by Le Chatelier method.

(*iii*) *Setting Time.* Initial setting time not less than 30 minutes, final setting time not more than 10 hours.

(*iv*) *Compressive Strength.*

(*a*) 72 h ± 1 hour not less than 15 MPa

(*b*) 168 h ± 2 hours not less than 22 MPa

(*c*) 672 h ± 4 hours not less than 30 MPa

4.17 SPECIAL TYPES OF CEMENTS

4.17.1 Masonry Cements

Masonry cements were manufactured to avoid the laborious process of mixing lime, cement and sand as and when required in order to prepare mortar for brickwork, as well as to avoid the presence of unslaked lime in the mixture of lime, sand and cement so prepared. Lime has been used in mortar since it makes the mortar easier to work.

Masonry cement can be used for all general purposes where mortars for masonry are required. Masonry cement is, however, not intended for use in structural concrete, for flooring and foundation work or for reinforced and prestressed concrete works.

Physical requirements: IS 3466 : 1988 has specified physical requirements as per Table 4.5.

Table 4.5 Physical Requirements for masonary cement

Sl. No.	*Characteristics*	*Requirements*
(*i*)	Fineness—Residue on 45-micron IS Sieve, Max percent (by wet sieving)	15 .
(*ii*)	Setting Time (by Vicat Apparatus)—	
	(*a*) Initial, Min	90 min
	(*b*) Final, Max	24 h
(*iii*)	Soundness:	
	(*a*) Le-Chatelier Max	10 mm
	(*b*) Autoclave expansion, Max	1 percent
(*iv*)	Compressive Strength—Average compressive strength of not less than 3 mortar cubes of 50 mm size, composed of 1 part masonry cement and 3 parts standard stand by volume, Min	
	7 days	2.5 MPa
	28 days	5 MPa
(*v*)	Air Content—Air content of mortar composed of 1 part masonry cement and 3 parts standard sand, by volume	6 percent
(*vi*)	Water Retention—Flow after suction of mortar composed of 1 part cement and 3 parts standard sand by volume, Min	60 percent of original flow

4.17.2 Expansive Cements

To counteract the effect of shrinkage of concrete during the hardening process, expansive cements were manufactured to be used with Portland cement. These expansive cements contain the expanding calcium sulpho-aluminate (3CaO $A1_2O_3$ 3Ca So_4). One of the composition for an expanding agent is a combination of kaolin (26-35%), slaked lime (43-55%) and Portland cement (about 31%). A stabiliser is also used, the commonly used one being the blast furnace slag which brings the expansion to the desired limit. The expansive action caused may be said to be made up of two parts namely the reaction of the calcium sulphate ($CaSO_4$) and C_3A (Tricalcium aluminate) of the Portland cement and the reaction of the $CaSO_4$ and the less basic aluminate in the Sulpho-aluminate clinker. The expansion continues only as long as the concrete is kept moist and careful control of curing is necessary. Its use in repair work enables the proper bonding of the new concrete with the existing concrete. This cement is also used in the method of prestressing. Its use with sea water is not recommended. The upper limit of expansion is 1 percent.

4.17.3 Oilwell Cements

Oilwell cements are used in petroleum industry for cementing the steel casing of oil wells. The type acquired varies according to the conditions but the setting time and rate of thickening must be retarded sufficiently. This is achieved by the adjustment of the composition of the cement and adding retarders to Portland cements. The proportion of Fe_2O_3 (iron oxide) is kept higher than that required to combine with all the $A1_2O_3$ (Alumina) to form $4CaOAl_2O_3Fe_2O_3$. Tetra calcium alumino ferrite). Hence the proportion of $3CaOAl_2O_3$ (Tricalcium aluminate) formed is small thereby increasing the setting time considerably.

4.18 CEMENTS WITH ADDITIVES

4.18.1 Pozzolonic Cements

Under the classification of cements with, additives, the most important one is the Portland pozzolonic cements prepared by grinding together Portland cement clinker and a pozzolona or by mixing hydrated lime and a pozzolona. A pozzolona is a siliceous material which itself possesses no cementitious properties but when in finely divided form it reacts in the presence of water with lime at normal temperature to form compounds having cementitious properties. The materials generally consist of calcined clays and shales, diatomaceous earth, opaline cherts, volcanic material such as tuffs, pumites and artificial materials such as blast furnace slag or fly ash. Pozzolonic activity is the property exhibited by the pozzolonas namely to combine with lime liberated during the hydration of cements to form cementitious compounds. There are various tests for determining this activity. The pozzolonas, are useful in counteracting the bad effects of free lime liberated during hydration of cement, which causes disruption of the concrete. The pozzolonas also interact with alkalies present in cement, thereby counteracting the effect of alkali aggregate reaction. Hence the pozzolonas are used as admixtures to cements wherever the aggregate used is known to be reactive. Other reasons for using pozzolonic cements are improved durability, combined with economy obtained in concrete constructed in marine, hydraulic and underground structures as also the reduction of heat evolved. It will be seen that all pozzolonic cements are rich in silica and alumina and low in alkalies. For the pozzolona to be effective, the silica it contains should be in the amorphous form and not crystalline. It should also be in the finely divided form for easy chemical interaction. This explains the difference in the behaviour of different types of pozzolonas.

Table 4.6 lists the chemical requirements of Portland pozzolona cement specified in IS 1489 : 1991 for calcined clay based. The setting time and compressive strength are the same as for ordinary Portland Cement. Specific Surface should not be less 300 m^2/kg.

Table 4.6: Chemical Requirements of Portland-Pozzolona Cement

Sl. No.	*Characteristics*	*Requirements*
(i)	Loss on ignition, percent by mass, Max	5.0
(ii)	Magnesia (MgO), percent by mass, Max	6.0
(iii)	Sulphuric anhydride (SO_3) percent by mass, Max	3.0
(iv)	Insoluble material, percent by mass, Max	$x + \frac{4.0(100 - x)}{100}$ where x is the declared percentage of pozzolona in the given Portland pozzolona cement

4.18.2 Air Entraining Cements

These are Portland or Portland blast furnace slag cements to which a small amount of air entraining agents such as natural resins, sodium salts of sulphonates, glues salts of fatty acids etc. are added during grinding. These materials reduce the surface tension of water and stabilise very small air bubbles in the mix in a manner analogous to the formation of a foam on soapy water. The entrainment of air increases frost resistance and to some extent imparts resistance to sulphate waters. It improves the workability and reduces the segregation and bleeding of the concrete mix. The amount of air entrained varies to some extent with the richness of the concrete mix, the type and grading of the concrete mix and aggregate and other factors. It is also seen that there is no advantage with respect to durability if the air content is increased to more than 6%, as it may cause a decrease in compressive strength than what is normally allowed. A number of indigenous and cheaper air entraining agents are being developed now-a-days.

4.18.3 Hydrophobic Cement

It is Portland cement with a special quality of not being affected by moisture in air or dampness. Hydro-phobic cement is prepared from Portland cement clinker in which water repellent chemicals are added during the final operation of grinding the clinker into a fine powder. The chemicals form a water repellent film around each particle of cement which prevents moisture from air or water to be readily absorbed to start the hydration process. However the film gets broken when the cement is being mixed in concete and more easily by the abrasion of the aggregate particles and normal hydration takes place.

This cement has been developed for the principal purpose of enabling normal Portland cement to withstand moisture attack during the long period of storage without deterioration or departure from the standard normally demanded by relevant specifications.

The advantages of the hydrophobic cement are:

1. It can be stored without being spoiled, even under very humid conditions.
2. It can be stored for a long time without deterioration and affecting its quality under unfavourable conditions.

As cement concrete made with hydrophobic cement entraps a small amount of air in the form of minute air bubbles, workability is increased which allows a better consolidation and therefore increases the density also. As the mix results in a better workability, reduction can be effected in the water content of the mix, resulting in higher strength.

The chemical requirements of hydrophobic cement are the same as for ordinary Portland cement. The specific surface should not be less then 350 m^2/kg. The compressive strength speficied in IS. 8043 ; 1991 are 15.69, 21.57 and 30.40 after 3, 7 and 28 days respectively.

TESTING AND SPECIFICATIONS OF CEMENT

In order to ensure that all cements manufactured and marketed are of standard quality, certain tests have been devised which may readily be applied in the laboratory. The object of these tests is to fix values of physical and chemical properties, strength characteristics, etc. so that inferior type may be detected. Properties of this type form the basis of I.S. specifications for cements.

Following are given the various tests performed on cements and their specifications.

4.19 CHEMICAL ANALYSIS

Ordinary and rapid-hardening cements should comply with the following chemical requirements :

(*a*) Ratio of percentage of lime to percentage of silica, alumina and ion oxide, when calculated by the formula.

$$\frac{CaO - 0.7\,SO_3 \text{ (Sulphuric Anhydride)}}{2.8\,SiO_2 + 1.2\,Al_2O_3 + 0.65\,Fe_2O_3}$$

should not be greater than 1.02 and not less than 0.66.

(*b*) Ratio of percentage of alumina to that of iron oxide should not be less than 0.66.

(*c*) Weight of insoluble residue should not be greater than 5%.

(*d*) Weight of magnesia should not be more than 6%.

(*e*) Total sulphur content, calculated as sulphuric anhydride (SO_3) should not be more than 2.5%, when tricalcium aluminate is 5% or less and 3% when it is greater than 5%.

(*f*) Total loss an ignition should not be more than 5%.

Low heat Portland cement should comply with the following requirements with regard to its chemical composition :

The percentage of lime after deduction of the amount necessary to combine with sulphuric anhydride present should not be more than 2.4 times the percentage of silica plus 1.2 times the percentage of alumina plus 0.65 times the percentage of iron oxide ; nor be less than 1.9 times the percentage of silica plus 1.2 times the percentage of alumina plus 0.65 times the percentage of iron oxide.

In all other respects, low heat Portland cement should comply with the requirements specified under *b* to *f* above.

4.20 FINENESS

The fineness of cement is tested in two ways :

(*i*) Sieve analysis and

(*ii*) Specific surface area test.

100 g of cement is continuously sifted for a period of 15 minutes on an I.S. test sieve No. 90 micron and the residue should not be more than values given in Table 4.7.

The sieve test is rather meaningless, for it does no more than measure the quantity of grit present in the cement and gives no adequate comparative measure of the real fineness. Special methods have therefore been devised which determine the particle size distribution or as a more convenient single figure, the specific surface that is the surface area per unit weight. This is expressed in square meteres per kilogram. I.S. specifications employ Blane method based on the measurement of the permeability to air of a bed of cement. The specifications are :

Table 4.7: Fineness requirements

Test	*Ordinary Portland*	*Rapid hardening*	*Low heat*
(*i*) After sieving the residue by weight on I.S. test sieve No. 90 micron, not to exceed.	1%	5%	—
(*ii*) Specific surface (m^2/kg) by air permeability method not less than	225	325	320

4.21 SETTING TIME

It is necessary that the setting time should be sufficient to allow the concrete after mixing to be deposited and worked into position. Two periods of set are distinguished, the initial set or interval between the mixing of water with cement known as gauging and partial loss of plasticity and the final set or time required for the gauged cement to acquire sufficient firmness to resist a certain definite pressure.

The Vicat apparatus, shown in Fig. 4.6 is used for determining the setting time. It is also used for determining the amount of water that is to be used for the strength, setting time, and soundness tests. For the purpose of arriving at the standard consistency of cement paste the plunger is fitted to the needle hole of the Vicat apparatus. Trial pastes of cement and water are mixed and placed in the mould. The plunger is then gently brought into contact with the surface of the paste and quickly released. The quantity of water required to produce a paste of standard consistency is that required for a paste which will permit the settlement of the plunger to a point 5.7 mm from the bottom of the mould.

The time of initial set is found by filling the mould with a paste formed by gauging with 0.85 of the water required to give paste of standard consistency. (In the case of high alumina cement, the cement is gauged with 22 percent by weight of water). The needle C is lowered to the surface of the cement paste in the mould and quickly released. This process is repeated until the needle ceases to pierce the test block beyond 5.0 ± 0.05 mm measured from the bottom of the mould. The period elapsing between the time when the water is added to the cement and the time at which the needle ceases to pierce the test block to a point 5.0 ± 0.05 mm measured from the bottom of the mould is the initial setting time.

To find the final setting time the needle *F* is used in place of needle C. It is 1 mm square but fitted at the end with an annular attachment hollowed out so as to leave a circular cutting edge 5 mm in diameter around the needle. The needle itself projects 0.5 mm below the cutting edge. The period elapsing between the time when water is added to the cement and the time at which the needle makes an impression on the surface of the test block while the attachment fails to do so will be the final setting time.

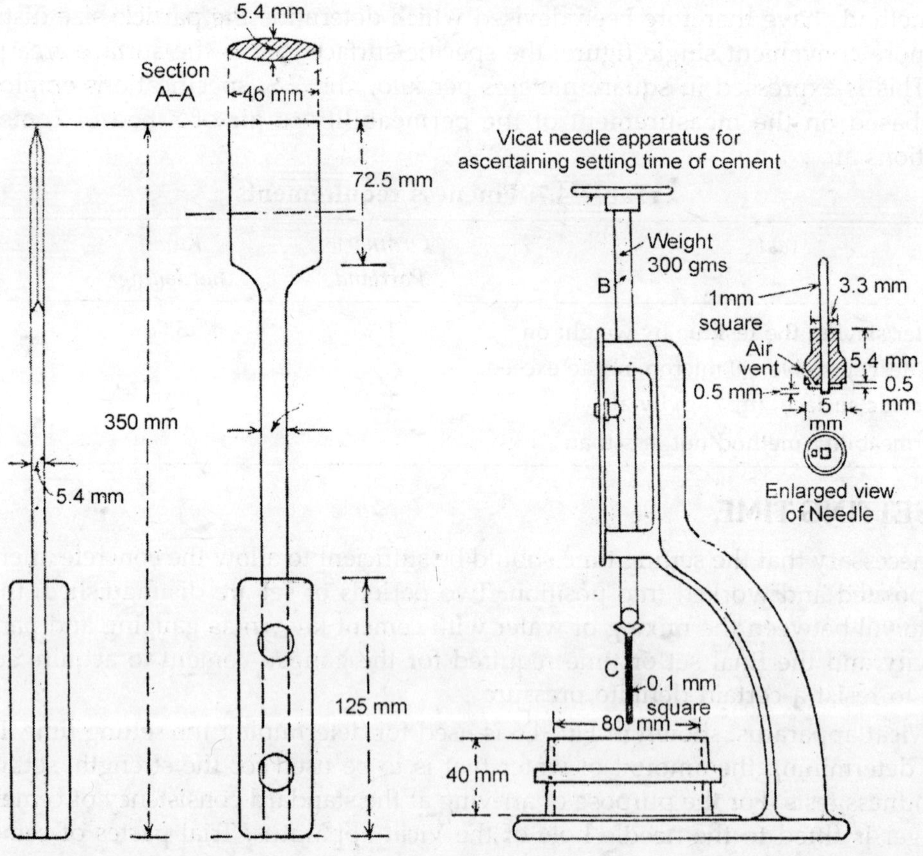

Fig. 40. Vicat apparatus

The initial setting time should not be less than 30 minutes and final setting time more than 10 hours for ordinary Portland, rapid hardenng, Portland pozzolona and low heat cements.

4.22 TENSILE STRENGTH

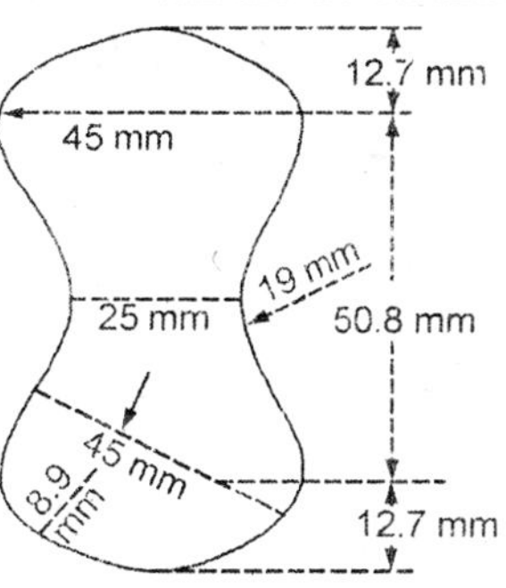

Fig. 4.7. Briquette for tensile strength

This test is provided for ordinary and rapid-hardening Portland cements. The tests are made on briquettes shaped like a figure eight and with a cross section across the middle of 25 mm. A mixture of one part of cement to three parts of standard sand by weight is used, the quantity of water derived by $\frac{P}{4} + 3.0$, where P is the percentage of water required to produce a paste of standard consistency which is determined by the use of the Vicat apparatus. The briquettes are kept at a temperature of 14 to 20°C in an atmosphere of at least 90 percent relative humidity for 24 hours after gauging when they are removed from the moulds and immediately submerged in clean, fresh water maintained at a temperature between 14 to 20°C and left there until taken out just prior to breaking. The requirements for ultimate tensile strength of the briquettes are given in Table 4.7.

Table 4.7: Indian Standard Specification for the Tensile Strength of Cement

	Ordinary Portland $N/mm^2\left[\frac{(kg/cm^2)}{10}\right]$	*Rapid hardening* $N/mm^2\left[\frac{(kg/cm^2)}{10}\right]$
(*a*) 1 day (24 hour), not less than	—	2
(*b*) 3 days, not less than	2	2
(*c*) 7 days, not less than	2.5	

4.23 COMPRESSIVE STRENGTH

This test is provided for ordinary, rapid hardening and low heat cements. It is carried out on mortar cubes having area of face 5000 mm^2 and prepared from a mixture by weight using the following quantities of materials :

Cement	185 g.
Sand	555 g.
Water	75 g.

The conditions for carrying out the test are similar to those for the tensile strength test except that the mortar in the cube is compacted on a standard vibration machine. Because of this, the compressive strength test is not affected by personal error as much as the tensile strength test. The crushing strengths specified are summarized in Table 4.8.

Table 4.8: Indian Standard Specifications for Compressive Strength of Cement

	Ordinary Portland *N/mm² (MPa)*	*High Alumina* *N/mm² (MPa)*	*Super Sulphated* *N/mm² (MPa)*
(*a*) 1 day, not less than	—	30	—
(*b*) 3 days, not less than	16	35	15
(c) 7 days, not less than	22	—	22
(*d*) 28 days, not less than	33	—	30

4.24 SOUNDNESS TEST

If during the manufacture of cement, lime is present in excess, it is liable to remain uncombined and be overburnt in the kilns. The overburnt particles of lime slake very slowly and expand. This expansion may take place after the setting action is completed and cause disruption of the mass. The soundness test is designed to accelerate the slaking of such particles, if they exist, by the application of heat, thereby discovering in a very short time defect which might not be discovered until sometime after the cement had been used.

QUESTIONS

1. List five major ingredients of cement, alongwith their range of percentage composition. Briefly state the functions performed by these ingredients.
2. What role is played by the following, when present in cement:
   (*a*) Calcium sulphate
   (*b*) Soda and potash
   (*c*) Sulphuric anhydride
   Give the important constituents of Portland cement clinker.
3. State the stages of Portland cement manufacture. What considerations govern the selection of site for raw materials?
4. List the stages for the preparation of raw materials for cement manufacture. Briefly discuss these stages.
5. What is meant by the term 'pulverisation'? How is it carried out?
6. What is calcination? State its principal objectives. What factors affect the heat requirement for burning of slurry?
7. Draw a neat sketch of a rotary kiln for clinkering process. What instruments are installed in a kilm room to control the process of clinkering? Give the stages in the calcination process.
8. What is the effects of grinding clinker? On what factors is clinker grindability influenced? Briefly explain the methods of grinding.
9. Why is gyprum added to clinker before ginding? Write shorts notes on:
   (*a*) Packing of cement; and
   (*b*) Storage of cement

10. What are the various types of cements? How is Portland cement further categorised? What grades of ordinary Portland cement are manufactured? Discuss their salient features.
11. List atleast four types of Portland cements and describe them.
12. Write short notes on:
    (*a*) High alumina cement (*b*) Super sulphated cement
    (*c*) Special types of cements
13. Give the utility and characteristics of the following Portland cement with additives:
    (*a*) Portland flyash cement (*b*) Portland pozzolona cement
    (*c*) Air-entraining cement (*d*) Hydropholic cement
14. What tests are performed on cement? Give their specifications as per Indian standard codes.
15. Describe the following tests on cement
    (*a*) Chemical analysis (*b*) Fineness test
    (*c*) Setting time test
16. How will you determine the compressive strength and transverse strength of cement?
17. Describe 'Le Chatelier' apparatus to test the soundness of cement. Give the specifications for the soundness of various types of cement.

OBJECTIVE TYPE QUESTIONS

1. The main ingredient in cement is:
   (*a*) Lime (*b*) Silica
   (*c*) Alumina (*d*) Iron oxide
2. The percentage of lime in cement should be:
   (*a*) 40 to 50% (*b*) 50 to 60%
   (*c*) 60 to 70% (*d*) 70 to 80%
3. The percentage of iron oxide in cement should not exceed:
   (*a*) 2% (*b*) 4%
   (*c*) 6% (*d*) 8%
4. The alkali of Portland cement commonly referred is:
   (*a*) Calcium oxide (*b*) Sodium oxide
   (*c*) Iron oxide (*d*) Magnesium oxide
5. Which one of the following is present upto the extent of one-fourth of cement:
   (*a*) Magnesia (*b*) Alumina
   (*c*) Iron oxide (*d*) Silica
6. Which of the following has the maximum percentage composition in cement:
   (*a*) Tetra-calcium alumino-ferrite (*b*) Tri-calcium silicate
   (*c*) Tri-calcium aluminate (*d*) Di-calcium silicate

7. The clinkering temperature, at which all the lime combines with clayey constituents is:
(*a*) 1350° C (*b*) 1450° C
(*c*) 1550° C (*d*) 1650° C
8. Silicates, aluminates and ferrites in cement are formed at a temperature not less than
(*a*) 1100° C (*b*) 1000° C
(*c*) 900° C (*d*) 800° C
9. Dissociation of carbonates of lime take place at a temperature above
(*a*) 600° C (*b*) 700° C
(*c*) 800° C (*d*) 900° C
10. Gypesı'm is added to clinker before grinding in amounts varying from:
(*a*) 1 to 3% (*b*) 3 to 5%
(*c*) 5 to 7% (*d*) 7 to 9%
11. Gypsum is added to:
(*a*) Retrard setting time (*b*) Accelerate setting time
(*c*) Reduce burning temperature (*d*) Reduce grinding time
12. Dry grinding can be done in:
(*a*) Ball mills (*b*) Ring roll mills
(*c*) Tube mills (*d*) All these mills
13. Strength is imparted to cement by:
(*a*) Lime (*b*) Alumina
(*c*) Silica (*d*) Magnesia
14. The clinkaring temperature is lawered by the presence of ______ in excess quantity
(*a*) Silica (*b*) Alumina
(*c*) Lime (*d*) Iron oxide
15. The initial setting time of Portland cement should not be less than:
(*a*) 20 minutes (*b*) 25 minutes
(*c*) 30 minutes (*d*) 35 minutes
16. The final setting time of Portland cement should not be more than
(*a*) 8 hrs (*b*) 10 hrs
(*c*) 12 hrs (*d*) 16 hrs
17. The silica modulus $\left(\frac{SiO_2}{Al_2O_3 + Fe_2O_3}\right)$ shoule not be less than
(*a*) 1 (*b*) 2
(*c*) 3 (*d*) 4
18. The iron modulus $\left(\frac{Al_2O_3}{Fe_2O_3}\right)$ should not be less than
(*a*) 0.55 (*b*) 0.66
(*c*) 0.77 (*d*) 0.88

19. Clinker is in the form of nodules of size
(*a*) 10 to 12.5 mm (*b*) 12.5 to 15.0 mm
(*c*) 15.0 to 17.5 mm (*d*) 17.5 to 20.0 mm

20. The compressive strength of white Portland cement should not be less than ________ per cent of 33 grade Portland cement.
(*a*) 95 (*b*) 90
(*c*) 85 (*d*) 80

21. For white Portland cement, the weight of insoluble residue should not be more than:
(*a*) 0.5% (*b*) 1.0%
(*c*) 1.5% (*d*) 2.0%

22. The cementing property in cement is imported by
(*a*) Silica (*b*) Iron oxide
(*c*) Lime (*d*) Magnesia

23. The percentage of flyash in Portland Pozzolona cement should be between:
(*a*) 5 to 15% (*b*) 10 to 25%
(*c*) 15 to 35% (*d*) 20 to 45%

24. The initial and fiinal setting times of flyash based cement and ordinary Portland cement are:
(*a*) Less (*b*) Same
(*c*) More (*d*) None of these

25. Granulated slag percentage in Portland blast furnace slag cement should not exceed:
(*a*) 65% (*b*) 60%
(*c*) 55% (*d*) 50%

26. In high Alumina cement the specific surface in m^2/kg should not be less than:
(*a*) 225 (*b*) 250
(*c*) 275 (*d*) 300

27. The setting times for Masonry cements are specified as
(*a*) 30 min/10 hrs (*b*) 45 min/15 hrs
(*c*) 60 min/20 hrs (*d*) 90 min/24 hrs

28. The compressive strength of masonry cement at 28 days should not be less than
(*a*) 2.5 MPa (*b*) 5 MPa
(*c*) 7.5 MPa (*d*) 10 MPa

29. Rapid hardening in cement is due to:
(*a*) more C_2S (*b*) more C_3A
(*c*) more C_3S (*d*) more C_4AF

30. Low heat cement has:
(*a*) Low calcium content (*b*) High silica content
(*c*) Restriction of the amount of C_3A (*d*) All the above

31. The soundness of cement is tested by:
   (*a*) Le-Chatelier apparatus (*b*) Vicat apparatus
   (*c*) Specific surface area test (*d*) Compressive strength test
32. For Portland cements, the expansion should not exceed ________ mm after 7 days aeration
   (*a*) 3 mm (*b*) 5 mm
   (*c*) 7 mm (*d*) 10 mm
33. The compressive strength of ordinary Portland cement after 7 days should not be less than
   (*a*) 16 MPa (*b*) 22 MPa
   (*c*) 33 MPa (*d*) 40 MPa
34. The compressive strength of high alumina cement after 24 hours curing should not be less than
   (*a*) 30 MPa (*b*) 35 MPa
   (*c*) 40 MPa (*d*) 45 MPa
35. The final setting time is tested by:
   (*a*) Le-chatelier method (*b*) Vicat apparatus
   (*c*) Air-permeability method (*d*) None of these

Answers

1. (*a*)	**2.** (*c*)	**3.** (*c*)	**4.** (*b*)	**5.** (*d*)	**6.** (*b*)
7. (*c*)	**8.** (*a*)	**9.** (*d*)	**10.** (*b*)	**11.** (*a*)	**12.** (*d*)
13. (*c*)	**14.** (*b*)	**15.** (*c*)	**16.** (*b*)	**17.** (*c*)	**18.** (*b*)
19. (*a*)	**20.** (*b*)	**21.** (*d*)	**22.** (*c*)	**23.** (*c*)	**24.** (*b*)
25. (*a*)	**26.** (*a*)	**27.** (*d*)	**28.** (*b*)	**29.** (*c*)	**30.** (*d*)
31. (*a*)	**32.** (*b*)	**33.** (*b*)	**34.** (*a*)	**35.** (*b*)	

FURTHER REFERENCES

Bureau of Indian Standards,
I.S. Code Nos. 269, 8041, 455, 6909, 8042, 6452, 3466, 8043, 1489.

AGGREGATES [FINE AND COARSE]

5.1 CATEGORIES OF AGGREGATES

Aggregate is a general term applied to those materials which are inert *i.e.* chemically inactive. Broadly speaking agregates are further sub-divided into two categories, namely:

(*i*) Fine aggregates such as sand, surkhi, cinder etc.

(*ii*) Coarse aggregates such as gravel, shingle, crushed stone aggregates etc.

The most important functions of fine aggregates are to assist in producing workability and uniformity in the mixture. Fine aggregate also helps the cement paste to hold the coarse aggregate particles in suspension, thus promoting plasticity in the mixture and preventing the possibility of segregation of the paste and coarse aggregates.

Coarse aggregates are used mainly for providing bulk to the concrete. They are frequently used in two or more sizes to increase the density of the mix.

Characteristics of Aggregates

Aggregates in general should be hard, strong, dense, durable, clean and free from adherent coating. Flaky and elongated aggregates should be avoided, as for as possible. Below are given the important properties of aggregates:

5.2 SOUNDNESS

An aggregate should be physically sound *i.e.* it must be sufficiently strong and capable of resisting weathering action without disruption or decomposition. Mineral and rock particles which are soft and porous cleave easily or swell when wet, do not stand up to exposure and their use reduces strength or leads to premature deterioration by promoting weak bond between aggregates and cement or by inducing cracking and spalling. Shales, friable sandstones, some micaceous rocks, clay rocks and some very coarsely crystalline rocks are examples of physically unsound aggregate materials which may be inherently weak or may deteriorate through saturation, alternate wetting and drying, freezing, temperature changes, or by the disruptive forces developed as a result of growth of crystals in the cleavage planes or pores.

5.3 STRENGTH AND RESISTANCE TO ABRASION

An aggregate should and usually does have sufficient strength to develop the full strength of the cementing matrix. When wear resistance is important, for example in such structures as heavy-duty factory floors, roads and aerodrome runways, the aggregate should be hard and tough. Quartz, quartzite and many dense volcanic and siliceous rocks are well adapted for making wear-resistant surfaces.

5.4 VOLUME CHANGE

Volume change in aggregate resulting from wetting and drying is a common source of damage to mortar or concrete. Shales, clays and some rock nodules are examples of materials which expand by absorption of water and shrink on drying. Internal stresses in the concrete may also develop when the change in volume of aggregate due to temperature variations is substantially different from that of the cement paste. Instances of cracking and spalling are generally attributed to this cause.

5.5 PARTICLE SHAPE AND SURFACE TEXTURE

The external characteristics of any mixture of mineral aggregate include a wide variety of physical shape, colour and surface condition. The physical shape of an aggregate has an important bearing on the workability while the surface texture has a marked effect on the strength of bond with the cementing matrix.

Table 5.1: Particle Shape of Aggregates

Classification	*Description*	*Example*
Rounded	Fully waterworn or completely shaped by attrition.	River or sea-shore gravels; desert, sea-shore and wind blown sands.
Irregular	Naturally irregular or partly shaped by attrition and having rounded edges.	Pit sands and gravels, dug flints, cuboid rock.
Angular	Possessing well-defined edges formed at the inter-section of roughly planner faces.	Crushed rocks of all types such as talus.
Flaky	Material, usually angular of which thickness is small relative to the width and/or length.	Laminated rocks.

The particle shape of an aggregate may be classified as rounded, irregular, angular or flaky. Table 5.1 gives a simple definition and examples of each of these groups. Experience indicates that angular aggregates do not produce as smooth and workable a mixture as materials having more rounded and smoother particles grading, other conditions being similar.

An excess of thin, flat, elongated, flaky or splintery particles in concrete aggregates, is objectionable as it materially decreases the workability for a given water-cement ratio and necessitates more highly sanded mixes and consequent use of more cement and water. Specifications, therefore, generally limit the quantity of such particles.

The surface texture of aggregates may have a marked effect on the strength of bond between aggregates and cement paste. This is chiefly a matter of mechanical anchorage, sometimes due to roughness and often resulting from adhesion caused by suction of the paste into the surface pores of the aggregates. Some types of seemingly smooth aggregates may, therefore bond more strongly than ones with a rougher surface texture. The surface texture can be glassy, smooth, granular, crytalline, pitted or honeycombed and porous.

5.6 SPECIFIC GRAVITY

The specific gravity of aggregates is of direct importance only in those cases where design or structural considerations requires that the mortar or concrete have minimum or maximum weight. When lightness is required, natural or artificial aggregates of low unit weight are normally used.

Speicfic gravity is a useful quick indicator of suitability of an aggregate if used with caution and checked by other methods of examination and testing. Low specific gravity frequently implies porous, weak and absorptive material but numerous exceptions to this criterion prevent its application without confirmation by other means.

5.7 HARMFUL IMPURITIES

Aggregates should contain no harmful material in such quantity as to affect adversely the strength or durability of the mortar or concrete or in the case of reinforced concrete to attack the steel bars. The common harmful impurities present are dirt, silt, clay, mica, coal, chemical salts, iron pyrite and organic matters. The effects of these impurities are to:

(*i*) make the concrete unsound *i.e.* expand

(*ii*) decrease strength and durability

(*iii*) cause unsighty appearance

(*iv*) complicate mixing operation

(*v*) break down of aggregates by weathering agencies

(*vi*) prevent the development of maximum bond between aggregates and cementing material

(*vii*) react chemically with the cementing material.

5.8 ALKALI AGGREGATE REACTION

In general aggregates are inert substances but some of them react chemically with the alkalies released by the setting of cement in mortar or concrete. This results in excessive expansion, cracking and deterioration of the mortar concrete. On account of this, serious failures have been reported.

Opal (Amorphous, hydrous silica) is the most conspicuous aggregate reacting in this manner. Other such aggregates are schist, olivine and felspar rocks and kaolitic materials. It has been observed from tests that alkali-aggregate reaction is reduced in intensity or sometimes altogether eliminated by the limitation of alkalies (Na_2O and K_3O) in cement to 0.5 to 0.6 percent.

FINE AGGREGATES

Aggregates finer than 4.75 mm, *i.e.* passing I.S. Sieve No. 4.75 mm and coarser than 0.150 mm *i.e.* retained on I.S. Sieve No. 150 micron are classified as fine aggregates. These are chemically inactive materials and are added in mortar or concrete to serve different purposes.

Examples of fine aggregates are:

(*i*) natural sand
(*ii*) artificial or crushed stone sand
(*iii*) surkhi
(*iv*) cinder or coal ash.

A brief description of these follows:

5.9 NATURAL SAND

Sand comprises of grains of silica (SiO_2), formed by the decomposition of sandstone under the action of weathering agencies like water, wind, frost, etc. Sand, in the natural state, is available from three sources:

(1) *Pit sand.* Sand from the pits consists of sharp, rough and angular silica grains. It contains impurities like clay, hornblende, $CaCO_3$, felspar etc. which easily decompose by the atmospheric action and acids in the rain water. It is considered necessary to clean and wash pit sand, before putting it to use. Clean pit sand forms an excellent inert material for mortar.

(2) *River-sand* is available in the banks or beds of rivers. Due to the action of the moving water, the sand grains are fine and rounded. It is free from salt and clay impurities and thus should be preferred to pit sand. It has light colour. It lacks in adhesion on account of round and smooth surfaces. Owing to its fineness, it is mostly used in plastering. It finds use also in making mortar.

(3) *Sea sand* is also rounded due to wave action. It has fine and light brown coloured grains. It contains impurities like alkaline salts and organic matter. These salts absorb moisture from the air and are sources of dampness and efflorescence. Moreover the setting is also retarded. As a rule, sea sand should not be used for building purposes, especially as a mortar. However, it is being put to use in the south for outside plaster work, because of its thin thickness and exposure to the sun.

5.9.1 Qualities of Good Sand

The essential qualities of good sand are:

(*i*) It should be clean and sharp.
(*ii*) It should be well graded so that the voids are minimum. This will give a dense mortar or concrete mixture.
(*iii*) It should consist of hard and durable silica grains.
(*iv*) When rubbed between thumb and the fingers, it should not stain. It should be gritty to the touch.
(*v*) It should be free from flaky and flat particles such as those of mica.
(*vi*) It should be free from salts, which cause dampness and efflorescence.
(*vii*) It should be free from organic matter, which get decomposed and are thus a source of weakness. They also do not allow proper adherence between sand and the cementing material.

(*viii*) Neither too fine nor too coarse sand should be used. If sand is too fine, it will require much greater amount of cement or lime for coating its surfaces and also cracking may occur. On the other hand coarse sand will give a rough finish to the plastered surface; but it has greater strength. Thus for masonry work, coarse sand should be preferred and for plaster work, somewhat fine sand would give a better finish.

5.9.2 Bulking of Sand

The moisture content has an important effect on the volume of sand. It is a recognised fact that when dry sand is moistened, it increases in volume known as bulking of sand. This is due to the fact that sand particles are surrounded by a film of water, which tend to hold the surrounding particles apart, thereby resulting in an increase in volume.

There is a limit to the bulking of sand; which has been estimated to be 30% at about 10% moisture content. Further increase in water content, decreases the volume. When sand is fully saturated, it occupies the same volume as that of dry sand; because the particles come to their original position. Thus there is no bulking for a fully saturated sand.

As delivered and used on a job, sand invariably contains moisture. The volume of a given weight of sand is increased by moisture for out of proportion to the quantity of moisture. Thus the effect on the volume of sand must be taken into account when batching has tc be done by volume, to avoid inaccurate proportioning in mortars and concrete.

The amount of bulking varies with the nature of sand. It is less for coarse sand and more with fine sand.

5.9.3 Fineness Modulus of Sand

The fineness modulus of any particular variety of sand is an index of its coarseness or fineness. It is computed by adding the cumulative precentages by weight of materials retained on I.S. Sieves No 4.75 mm, 2.36 mm, 1.18 mm, 600 microns, 300 microns and 150 microns and dividing the sum by 100. It is worked out in Table 5.2.

Table 5.2: Determination of fineness modulus of sand

I.S. Sieve	*Percentage Retained*	
	Natural sand	*Crushed Stone sand*
4.75 mm	3	5
2.36 mm	12	25
1.18 mm	35	40
600 micron	60	65
300 micron	80	80
150 micron	90	95
Fineness modulus	$\frac{280}{100} = 2.8$	$\frac{310}{100} = 3.1$

Sand may be classified as fine, medium and coarse in accordance with its fineness modulus as follows:

Sand	Fineness Modulus
Fine	2.20—2.60
Medium	2.60—2.90
Coarse	2.90—3.20

It may be seen from above that fineness modulus is an index number which is roughly proportional to the average size of particles in sand, i.e., the coarser the particles, the higher the fineness modulus. But it does not give any idea of the gradation i.e. it cannot distinguish between a single sized sand and a graded sand. However it is useful in connection with normally graded aggregate. It is used for specification and record purposes and for controlling grading and uniformity.

5.10 ARTIFICIAL OR CRUSHED STONE SAND

At places where river sand, free from deleterious materials, is not available at reasonable cost, artificial sand is manufactured by crushing sandstone rock.

Artificial sand is produced in a rod mill either by dry or wet process. The mill is charged with coarse aggregates and high carbon steel rods. The coarse aggregates are crushed due to tumbling action in the mill, when it revolves.

In the wet process, water is also added in the mill. The liquid from the rod mill is dicharged into a sump, from where it is pumped into a sand separator. Due to the formation of a vortex in the separator, fine sand passing I.S. Sieve No. 150 micron and clay are washed away with water into the overflow and sand is allowed to drop down.

5.11 SURKHI

Surkhi is the name given to powdered burnt brick or clay and used with mortars as a pozzolona or as fine aggregate. Most authorities specity that surkhi should be prepared from slightly underburnt clay.

A practical test to see whether the clay is not under baked is to verify whether it has lost the power of plasticity when mixed with mortar. Recent researches have indicated that different clays have their own definite temperatures of calcination at which they produce most satisfactory surkhi. These temperatures are different from the temperatures at which they form the best bricks. For big works, therefore, it will be advantageous to burn clay to predetermined temperatures for production of surkhi rather than powder the burnt brick for this purpose as is usually the current practice. It is manufactured by grinding burnt bricks or bats, no underburnt bricks or bats should be used.

Surkhi should be clean and free from admixture of any clay, dust and foreign materials. The materials shall be ground to a fineness varying from I.S. Sieve No. 3.16 mm to 1.58 mm. In case where surkhi is used in conjunction with coarse sand in mortars to obtain pozzolonic condition the finer grinding should be perferred. In cases where surkhi is used partly to contribute to pozzolonic action, the coarser grinding may be preferred. Powdering surkhi to desired fineness presents enough dry condition for sieving. While installation of driers will be justified in works of big magnitude, the wet grinding of mortars helps considerably in reducing the commercial surkhi to a standard of fineness which is efficient for all practical purposes.

5.12 CINDER OR FLY-ASH

Furnace ashes or clinker can be used as aggregate in lime mortar. The properties of such ashes and clinker vary greatly and little control exists over their quality. These aggregates sometimes contain material that can cause trouble such as unburnt fuel or too high proportions of compounds containing sulphur. These defects are likely to accentuate when the material is used, in conjunction with cement.

The cinder used should be clean and ground in mill and screened through an I.S. Sieve No. 3.16 mm. The stipulation regarding fineness given in the case of surkhi apply to cinder as well. Cinder can be used as a fine aggregate in mortars and is also pozzolonic like surkhi. If unburnt carbon and soluble sulphates are present to an excessive degree in cinder, the mortars should not be used in very wet situations. As a practical guide, the quantity of unburnt carbon may be limited to 20% and quantity of the sulphates limited to the percentages specified for surkhi.

COARSE AGGREGATES

Coarse aggregate means an aggreeate most of which is retained on I.S. Sieve No. 4.75 mm and does not contain fine aggregates more than a prescribed limit. The maximum size of the aggregate depends upon the type of concrete, in which the aggregate is being used. For ordinary concrete work, a maximum size of 75 mm is usually recommeded.

Coarse aggregates may be any of the following varieties:

(1) Natural picked gravel or shingle stone.

(2) Crushed gravel or stone.

(3) Artificial aggregate such as broken-brick or air cooled blast furnace slag.

(4) Special type aggregates, *e.g.*, light-weight aggregates.

5.13 NATURAL GRAVEL OR SHINGLE-STONE AGGREGATES

These aggregates are available in the river and stream beds and are carried by the flowing water from high reaches of the river and deposited in plains. These aggregates are usually round, strong, dense and durable. These aggregates should be free from adherent coatings of dirt, clay, fine silt, etc.

They are available in varying sizes and should be properly graded for a particular use.

5.14 CRUSHED GRAVEL OR STONE AGGREGATE

If natural coarse aggregates do not conform to the grading specifications for a job, stones may have to be crushed to obtain coarse aggregates. This is particularly necessary on large construction projects such as dams, multi-storeyed buildings, etc. For this purpose a crushing and screening plant has to be installed; which comprise of the following units:

(*i*) Feeder.

(*ii*) Crushers; primary, secondary and tertiary.

(*iii*) Screening unit.

(*iv*) Conveyors for transporting the aggregates.

The various types of crushers used are:

Primary—Jaw, gyratory, hammer mill.

Secondary—Cone, hammer mill.

Tertiary—Roll, rod mill, ball mill.

5.15 ARTIFICIAL AGGREGATE

There are two types of artificial aggregates used in ordinary construction, namely broken brick and air cooled blast furnace slag.

(*a*) *Broken Brick:* This aggregate is suitable for plain but not for reinforced concrete work. Its porosity makes it unsuitable for waterproof construction. It should be thoroughly wetted before use. Broken brick concrete is lighter than broken stone concrete. Owing to its poor resistance to wear, broken brick should not be used for wearing surfaces. Overburnt brick aggregate, called *jhama* is some-times used in reinforced concrete work when the working stresses are low and the structural dimensions are generous.

(*b*) *Air-cooled Blast Furnace slag:* This is a by-product in the reduction of iron ore in the blast furnace, and is cooled slowly in air.

It is different from blast furnace slags treated in the molten state to produce light weight products such as expanded slag. Air cooled blast furnace slag is chiefly used as coarse aggregate, in combination with well-graded sand. Though this slag is extensively used in Europe and America, its use is not common in India mainly because of the fear of corrosion of the reinforcement. However, carefully selected slag produces a strong and durable concrete which has a high fire resistance. When the concrete is dense and impermeable corrosion does not occur.

5.16 LIGHT-WEIGHT AGGREGATE

These are usually by-products of industries or aggregates processed from natural sources or natural materials. They have a high percentage of voids and are therefore light. As concrete made with light weight aggregates does not possess the strength and soundness of ordinary concrete, its use in limited to parts protected from the weather which do not require great strength. The workability of concrete with light weight aggregates is greatly increased with air entraining agents and this has made the use of many light weight aggregates practicable.

Some properties of light-weight aggregates are given in Table 5.3

Table 5.3: Properties of Light-weight Aggregates

Aggregate	*Unit weight kg/cu m*	*Bulk specific gravity*	*Water absorption*	
			By weight g/kg	*By volume kg/cu m*
Vermiculite	160	1.35	128	300
Diatomite	500	1.44	75	385
Perlite	144	0.86	153	225
Slag	480- 960	1.13-2.26	5-23	48-160
Ash	670-930	1.74-2.10	8-15	24-96

Pumice	700-790	1.46-1.66	29-43	240-270
Shale	960-1220	1.74-2.90	7	80
Slate	625-1270	1.29-2.20	8-15	90-110
Clay	850-1060	1.65-1.98	16-21	176

STORAGE AND HANDLING OF AGGREGATES

Aggregates have to be stored and handled carefully so that they remain free of contamination with dirt, the aggregation is prevented and "their moisture content remains unaltered.

Before the aggregates are received at the site, the space for their storage must be selected carefully keeping in view the position of the mixer and convenience in handling. A clean, hard and dry patch of ground should be selected for storage. If such ground is not available a platform of planks or corrugated iron sheets, or a brick or lean concrete floor should be made. For large works, timber or steel bins may be used for storage. Fine aggregates as well as the different sizes of coarse aggregates must be stored separately.

Prevention of segregation or the separation of the various sized particles during storage and handling is of utmost importance. To-avoid rolling down of coarser particles, successive consignments should not be dropped on the same place each time to form a pyramid. Aggregates should be placed in layers, each layer being not thicker than that of lorry load. Aggregates should not be dropped from a height.

To avoid excessive variation in the moisture content, the piles of aggregates must be kept large in area and low in height (1 to 1.5 m). They must be allowed to stand for at least 24 hours before use.

The correct and incorrect methods of handling, storage and batching aggregates are shown in Fig. 5.1.

Correct

Methods which place material in the pile in individual units not larger than a truck load and which do not permit the aggregate to run down the slopes at the edge of the pile.

Incorrect

Methods which permit the aggregate to roll down the slopes as it is added to the pile.

Stockpiling of Screened Aggregate

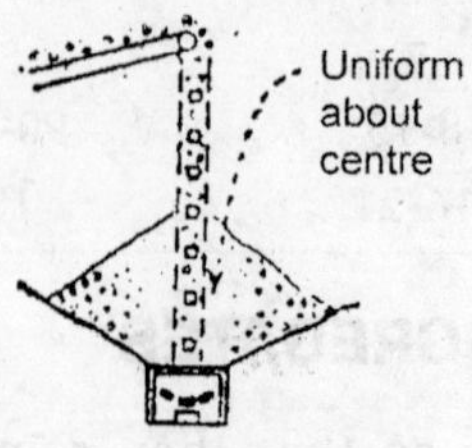

Correct

Chimney surrounding material falling from end of conveyor telt to prevent wind from separating fine and coarse materials. Openings provided as required to discharge materials at various elevations on the pile.

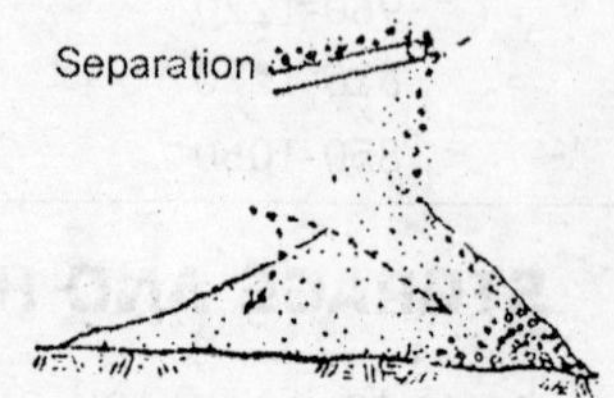

Incorrect

Free fall of material from high end of stacker permitting wind to separate fine from coarse material.

Unfinished or Fine Aggregate Storage (Dry Materials)

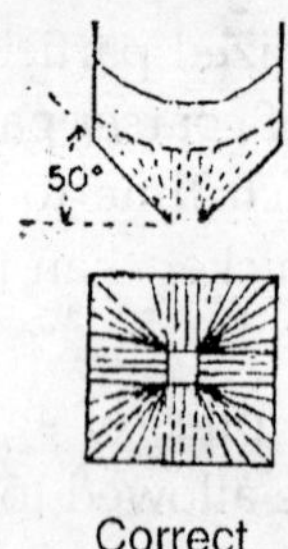

Correct

Full bottom sloping 50° from horizontal in all directions to outlet with corners of bin properly rounded.

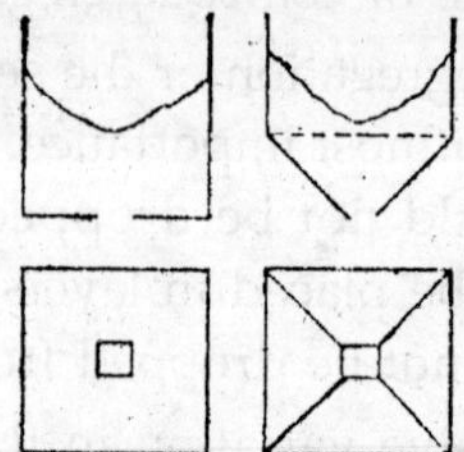

Incorrect

Flat bottom bins or those with any arrangement of slopes having corners or areas such that all material in bins will not flow readily through outlet without shovelling.

Filling of Aggregate Bins

Correct

Material drops vertically into bin directly over the discharge opening permitting discharge of more generally uniform material.

Incorrect

Chuting material into bin on an angle. Material falling other than directly over opening not always uniform as discharged.

Filling of Aggregate Bins

Fig. 5.21. Correct and incorrect methods of handling, storing and batching of aggregates

TESTING OF AGGREGATES

Methods of testing aggregates have been standardized in Indian Standard specifications. The following tests are usually carried out:

(1) Sieve analysis

(2) Determination of clay, silt, and fine dust

(3) Determination of organic impurities

Control tests

(*a*) Aggregate crushing value

(*b*) Bulk density or unit weight

(*c*) Determination of voids

(*d*) Determination of suitability of aggregates by reference to the strength of test pieces of concrete made with them

(*e*) Additional tests such as the abrasion and impact tests though not normally required, may be called for in special cases.

5.17 SAMPLING

In all tests, it is important that the samples of aggregates be representative of the materials as they will be when batched and mixed. A reasonable amount of local variation must be expected. For example a single handful of sand in one sample does not in itself justify rejection of an entire wagon load of aggregate.

Samples may be obtained from stockpiles and bins. The greatest accuracy in sampling is, however, obtained during loading or unloading. The main sample is made up of several small samples taken at different places in a stockpile or bin or at intervals during loading or unloading.

When taking samples from the surface of aggregates heaped in a stock and bin, the coarse material which tends to roll to the edges and the fine material which tends to remain near the centre must be avoided. At each selected point of sampling the dry surface material should be scrapped off for a depth of several mm, the material under-neath being taken, at the same time taking care to avoid segregation.

Whenever possible, samples of sand should be taken damp, as dry sand segregates badly and the very fine material may be blown away or left at the bottom of the container. Sand may be sampled by means of a sampling tube driven into the pile at several different points. In taking samples to determine moisture content, it should be remembered that the bottom of a pile or bin of wet aggregate which has stood for a few hours will be better than the top.

5.18 SIEVE ANALYSIS

Sieve analysis of aggregates is a means of controlling gradation and of checking the grading requirements of the specifications and they provide information which may aid in determining the proper proportion of the fine and coarse aggregates to be used in a mix.

A sample of sand for sieve analysis, after being dried, should weigh at least 500 g. To avoid segregation, the sample should preferably be reduced to the desired size before being dried. The sample may be heated gently but preferably air dried, as excessive heating may cause breakage of some of the particles. A sample of coarse aggregate for sieve analysis should weigh from 2 kg for a maximum size of 10 mm to 25 kg for a maximum size of 60 mm. It is generally not necessary to dry the coarse aggregate for sieve analysis.

If sieving is carried out by hand each sieve is shaken separately over a clean tray for a period of not less than two minutes. The shaking should be done by a combined lateral and vertical motion of the sieve, accompanied by a jarring action so as to keep the sample moving continuously over the surface of the sieve. Material less than 20 mm should not be helped through the sieve by hand. In any laboratory where a considerable amount of sieving is to be done, it is advisable both in order to save time and to ensure greater accuracy, to use a sieving machine which can deal with a set of several sieves at a time.

5.19 SILT CONTENT

The extremely fine mineral material (clay, silt, dust or loam) occuring in most aggregates requires relatively large amounts of mixing water and tends to work to the surface of the concrete and cause cracking due to shrinkage upon drying. It also tends to interfere with the bond between the aggregate particles and the cement water paste. Indian Standard limits the amount of such material to 5 percent by mass for natural sand or crushed graval sand or crushed stone sand and 1 percent for coarse aggregate.

Three methods are available for the determination of the silt content. Two of which are laboratory methods and one a field method.

Decantation Method. This method consists of washing the silt and clay through I.S. Sieve No. 75 micron, and finding the difference in weight between the original sample and that retained on the sieve. The difference is assumed as clay and silt. An objection to this method is that any coarse meterial lost during the test counts as silt and clay.

Sedimentation Method: The sample is sieved through a 2.36 mm I.S. sieve, the material passing being set aside. The residue is treated with sodium oxalate, the washed liquid poured in a 2.36 mm sieve, allowed to rest, and the clear liquid decanted. The total material passing the 2.36 mm sieve is subjected to the sedimentation test. If, however, the dry material passing the No. 2.36 mm sieve exceeds 50 gm in weight, two sedimentation tests are necessary one of 30 gm of this dry material and the other on the wash liquid.

Field settling test. This method of test, is intended as a guide to the quantity of silt, etc., contained in an aggregate. A sample of the aggregate is placed without drying in a 200 ml measuring cylinder. The size of the sample should be such that if fills the cylinder upto the 100 ml mark. Clean water is then added up to the 150 ml mark, the mixture shaken vigorously and the contents allowed to settle for three hours. The height of the silt visible as a layer above the sand is measured and expressed as a percentage of the height of the sand.

While large amounts of clay and silt are objectionable, a small quantity of clean, fine sand particles is desirable for workability. Specification permit upto 10 percent of fine aggregate passing the I.S. Sieve No. 300 micron.

5.20 ORGANIC IMPURITIES

A fraction of even one percent of certain kind of organic material in sand will delay or prevent the hardening of mortar or concrete and will reduce the strength greatly. Organic matter in sand usually occurs in the form of decayed vegetable matter; it will be seldom found in washed sand.

The calorimetric test for the presence and approximate amount of organic material is carried out. Its principal value is to furnish a warning that further tests of the sand are necessary before it is finally approved for use.

The test sample should preferably be slightly damp. An excess of surface moisture weakens the testing solution, but if the aggregate is dried, some of the organic material may be lost in handling or may be burnt. The sample is immersed for 24 hours in a 3 percent solution of sodium hydroxide in a clear glass bottle. An approximate idea of the final result can be gained by inspection. The colour of the liquid is noted and compared with a standard colour solution prepared according to Indian standard or with the standard colour chart. If the solution is clear or light yellow, the sand is satisfactory and free from organic matter. If the solution is in the range from light yellow to brown, its quality is doubtful and is may be necessary to make strength test of mortar containing the sand, before it is finally approved for use.

Some sands contain a small quantity of coal or lignite which gives the liquid a very dark colour. The quantity may not be sufficient to reduce the strength of mortar or concrete appreciably and the sand may otherwise be acceptable. Here again, mortar strength tests of the sand in question should be made to determine the exact amount of the impurities present.

5.21 CONTROL TESTS

The control tests for moisture absorption and specific gravity are quite common and details need not be given here.

5.21.1 Aggregate Crushing Value

A quantity of aggregate graded between given sieves is placed in a cylinder, a piston is inserted and a compression load varying between 10 and 160 tonnes depending on the size of the aggregate is applied. The aggregate is then resieved on the appropriate sieve and the aggregate crushing value is calculated from the proportion of material crushed sufficiency to pass the sieve. The test is intended io provide a measure of the hardness of an aggregate. The curshing value should not exceed 30 percent in the case of aggregate to be used for wearing surfaces such as heavy-duty floors, roads, etc. and 45 percent in all cases.

5.21.2 Bulk Density or Unit Weight

The bulk density of a material is the weight of the material held by a container of unit volume when filled and compacted under defined conditions. It is normally expressed in kg per cubic metre or N/m^3.

The weight is obtained by filling loosely a suitable container of known volume with the dry aggregate. The method of filling the box can make an appreciable difference to the result obtained in this test.

The weight per cubic metre is used in translating the proportions determined by weight from the sieve analysis, into proportions by volume as used in many cases, on the site. It is sometimes necessary to convert proportions by weight into the equivalent proportions by volume of aggregate or vice versa. If the aggregate contains an appreciable amount of material below a No. 4.75 mm sieve, it is then necessary not only to find the weight per cubic metre dry, but also to make tests for bulking. The weight per cubic metre at any moisture content can be calculated from the dry density and the percentage bulking.

5.21.3 Voids

A method sometimes used in proportioning concrete is based on the void content of the aggregates. Although this method is not reliable, a measurement of voids may be useful in certain cases. This is suitable for fine and coarse aggregate measured separately. It cannot be used for all-in one material owing to the marked segregation occuring when all-in aggregate is inundated, or for aggregates of high porosity. As with measurements of the weight per cubic metre, the void content, will vary according to the method of filling the gauge box.

5.21.4 Strength of Concrete Test Pieces

If further confirmation of the satisfactory nature of the aggregates as tested by the above methods is required, recourse may be had to compression tests of concrete made with the materials in question. Details are given in chapter 7.

5.21.5 Abrasion and Impact Tests

These tests are carried out on the coarse aggregates.

Abrasion test is done in the Los Angeles abrasion testing machine. A 5 kg sample of aggregate along with 12 steel balls of about 50 mm diameter are rotated in a 700 mm diameter steel cylinder filled with an internal shaft. It is rotated for about 500 revolution at the rate of 30 to 33 r.p.m. The result is expressed as a percentage by weight of the fines passing a particular sieve.

The test is also carried out in the abrasion testing machine shown in Fig. 1.7 About 100 gm of 12.5 mm chipping are held in a shallow tray and pressed against the rotating steel disc by a weight of 2 kg for 500 revolutions at 28 r.p.m. Standard sand being fed continuously on to the disc. The result is expressed as a percentage loss in weight of the chipping by abrasion.

Impact test: About 350 gm of 12.5 to 10 mm aggregate is subjected to 15 blows from a height of 380 mm in the impact testing machine. The result is expressed as the percentage fines passing No. 60 micron I.S. Sieve.

QUESTIONS

1. Define aggragate. What are the different categories of aggregates and what functions are performed by them in a mortar or concrete mix.
2. What important characteristics should an aggregate possess. Discuss briefly.
3. Describe the following types of aggregates and give examples of each one of them.

   (*a*) Rounded (*b*) Irregular
   (*c*) Angular (*d*) Flaky.

4. List the common harmful impurities present in aggregates and list the effects of these impurities. What is alkali-aggregate reaction and its ill effects in mortar or concrete mix? How can this be prevented?
5. Classify fine aggregates on the basis of its size. Give examples. State the sources from which natural sand is available.
6. Give the qualities of good natural sand. What do you understand by bulking of sand?
7. What does fineness modulus of sand indicate? How can you determine it? Give the limits of fineness modulus for fine, medium and coarse sands.
8. Write short notes on:
   (*a*) Crushed stone sand (*b*) Surkhi
   (*c*) Cinder or fly-ash
9. Define coarse aggregate. What are the different varieties of coarse aggregates. Give details briefly.
10. Write short notes on:
    (*a*) Light weight aggregates and their properties.
    (*b*) Storage and handling of aggregates.
11. List the various tests which are performed on aggregates. Describe two important tests.
12. How will you collect a sample of aggregates for performing tests. Give details of sieve analysis.
13. Give details of tests on aggregates to determine inorganic and organic impurities.
14. How will you perform the following tests on aggregates:
    (*a*) Specific gravity and absorption of aggregates
    (*b*) Crushing strength test
    (*c*) Unit weight test (*d*) Void test
15. What tests do you perform to determine the suitability of aggregates by making test specimens of mortar and concrete.

OBJECTIVE TYPE QUESTIONS

1. Which of the following is not a fine aggregate
   (*a*) Sand (*b*) Shingle
   (*c*) Surkhi (*d*) Cinder
2. Coarse aggregates increases :
   (*a*) Density of the mix (*b*) Workability and uniformity
   (*c*) Plasticity (*d*) prevention of segregation
3. Aggregates should not be :
   (*a*) Dense (*b*) hard
   (*c*) elongeted (*d*) free from adherent coating
4. Crushed rock aggregates are:
   (*a*) rounded (*b*) Irregular
   (*c*) flaky (*d*) Angular
5. Pit sands are :
   (*a*) flaky (*b*) angular
   (*c*) irregular (*d*) rounded

6. Fine aggregates should pass .... I.S. Sieve
   (*a*) 600 micron (*b*) 1.18 mm
   (*c*) 2.36 mm (*d*) 4.75 mm
7. Fine aggregates should be coarser than
   (*a*) 75 micron (*b*) 150 micron
   (*c*) 300 micron (*d*) 600 micron
8. The fineness modulus of fine sand varies from
   (*a*) 2.20 to 2.60 (*b*) 2.60 to 2.90
   (*c*) 2.90 to 3.20 (*d*) 3.20 to 3.50
9. The fineness modulus of coarse sand should lie between :
   (*a*) 2.60 to 2.90 (*b*) 2.90 to 3.20
   (*c*) 3.20 to 3.50 (*d*) 3.50 to 3.80
10. The fineness modulus of sand is checked with the help of .... numbers of I.S. Sieves.
   (*a*) 4 (*b*) 5
   (*c*) 6 (*d*) 7
11. Surkhi is made from :
   (*a*) Crushed sand stone (*b*) Sea sand
   (*c*) Cinder (*d*) burnt clay
12. Coarse aggregates are those which are retained on I.S. Sieve :
   (*a*) 2.36 mm (*b*) 4.75 mm
   (*c*) 10 mm (*d*) 12.5 mm
13. The grading requirements are checked by :
   (*a*) sieve analysis (*b*) specific gravity
   (*c*) unit weight (*d*) bulk density
14. The silt content in the aggregates is determined by :
   (*a*) Decentation method (*b*) Sedimentation method
   (*c*) Field settling method (*d*) All of these methods
15. The bulking of sand is ... for coarse sand than fine sand
   (*a*) less (*b*) more
   (*c*) same (*d*) None of these

Answers

1. (*b*)	**2.** (*a*)	**3.** (*c*)	**4.** (*d*)	**5.** (*b*)	**6.** (*d*)
7. (*b*)	**8.** (*a*)	**9.** (*b*)	**10.** (*c*)	**11.** (*d*)	**12.** (*b*)
13. (*a*)	**14.** (*d*)	**15.** (*a*)			

FURTHER REFERENCES

Bureau of Indian Standards, New Delhi
I.S. Code Nos. 383

MORTARS

Mortar may be defined as an intimate mixture of cementing material, *e.g.*, cement, lime etc. fine aggregates or inert materials like sand, surkhi cinder etc. and water.

Uses. Mortar is used to bind brickwork or stone masonry, blocks or other units in building construction. Besides being binding materials, mortars serve to different degrees, as impermeable materials offering protection against moisture penetration by filling the joints. It is also used in the form of plaster to cover the walls. It acts as a matrix in concrete. The usefulness of mortars is judged from the properties that they are required to possess as a finished end product.

DESIRABLE PROPERTIES OF MORTARS

Mortars should possess the following properties:

6.1 WORKABILITY

Mortars should be easy to work so as to reduce labour in their application and to ensure proper filling of the joints by spreading. The three main factors that affect the workability of mortars are bulk density, water retentivity and good trowelling property. Of these, the bulk density is a factor of minor importance and is usually ignored.

Water retentivity is the ability of mortars to retain water against suction and evaporation. In general, water retentivity is a measure of flow and workability of mortars. As water is the sole lubricating material in the mortar, more the retentivity the more workable will be the mortar under practical conditions. If the mortar loses water to the brick too rapidly, it will stiffen quickly and intimate contact with bricks is vitiated and the adherence to bricks will be uneven. In such cases initial bond is poor but this is further reduced by volume changes in hardened mortars through successive wetting and drying.

In laboratory, water retentivity is measured by determining the flow of mortar before and after removing some of the water by suction.

Mortars of low retentivity when used on non-absorptive base will undergo segregation and form puddles. After sometime volumetric changes occur in the hardened mortars and the brick or masonry units are found separated from the mortars forming cakes.

It has been observed that whatever may be the conditions of the bricks (poor, middle or high rate of water absorption) lack of water retentivity of mortars results in poor bonding properties.

Workability in connection with building mortars exists in a mix in which the viscosity of the matrix is such as to allow distortion without segregation. A lean cement mortar with coarse sand which is usually unworkable cannot be made workable in the sense, by adding more water. Addition of more water may increase its flow, but does not increase its workability because the process of adding water leads to segregation. Some authorities therefore, specify that for use in masonry the mortars should have a water retention of not less than 70%.

The property of water retentivity and plasticity imparts good trowelling properties. Mortars of good trowelling properties spread and travel like butter. Non-plastic mortars do not fill the interstices of the rough faces and the header ends of the bricks. They tend to bridge across these shallow indentations, thereby reducing the extent of bond and further provide a path for moisture penetration.

Good working qualities in a mortar may be obtained by the admixture of fat lime in a finely divided state. The higher the proportion of the lime used, the better are working qualities. In this respect even a rich cement mortar of 1 : 3 mix is inferior to lime mortar.

6.2 RATE OF STIFFENING

The stiffening of mortars, which can be controlled to some extent by the initial wetness of bricks immediately upon being placed, is due to the loss of water through suction or evaporation. Development of early strength of mortars is due to continuation of the stiffening process. Mortars made of fat lime have no set and hardening is due to slow carbonation. Such mortars harden slowly on any impervious building unit. With more porous units this difficulty is less. Lime mortars and cement lime mortars with high lime content are particularly suitable to porous units when they set dry. These mortars are sufficiently water retentive to make intimate contact with the units.

The process of hardening is a chemical change which produces loss of plastic properties and acquiring of elastic properties. The rate of early hardening should be rapid enough to allow a normal rate of brick laying and should be such that the plastic properties are retained till the initial shrinkage of mortars is over. In such cases, there is less danger of cracks or opening of joints. Mortars, specially lime mortars continue to harden for year long after they have ceased to be plastic. Addition of cement to lime mortars increases their hardening rate and this hardening is less dependent on weather.

6.3 VOLUME CHANGES

Mortars undergo three types of distinct changes ; namely

(*i*) Compacting on a porous base,

(*ii*) Shrinkage during the early hardening, and

(*iii*) Volume changes subsequent to hardening.

The first two occur once during the life history of mortars, the third occurs over and over again and is a cyclic process. The mortars expand on becoming damp or wet and shrink when they dry out. This cyclic system and alternate expansion and contraction continues over a period of years. The first type of volume change cannot take place unless there is contact with the porous base. The other two types do not need contact of mortars with a solid object.

Compacting on a porous base is essentially unidirectional and occurs in the direction of application of pressure. This compacting on a porous base decreases with increase in water retentivity of the mortar and conversely, in a porous base such compaction occurs usually before any appreciable hardening or cementing action of the mortar. It is very rapid at the first instant of contact and takes place within a very few minutes. It is attended by a rapid loss of workability due to stiffening of the mortar. This stiffening is distinct from the hardening through cementing action. The process renders it impossible to make good contact between the mortar bed and the next course of bricks or to slush the mortar in the joints. It also prevents the mortar in the horizontal bed from slumping into and filling the vertical joints left open by poor workmanship. The unidirectional compacting tends to pull the mortar away in the cross joints from the bricks, the mortar being too stiff to flow under the pressure of bricks laid above it to cause any renewal of contact. This action is prevented by high water retaining capacity of mortars as obtained in lime cement mortars.

Shrinkage during early hardening is through cementing action and occurs only once in the life history of the mortar. A mortar which hardens slowly is more or less plastic when most of this type of shrinkage is completed. Any possible damaging effect of this type of shrinkage is dependent as much and probably more on the rate of hardening as on the magnitude of the shrinkage. It has been shown that this type of volume change does not damage either the mortar or the bond between bricks regardless of the magnitude of the shrinkage of mortar during early hardening.

Volume change subsequent to hardening is characterised by expansion on wetting and shrinkage on drying. It is cyclic and frequently rapid. It takes place in rigid mortar that has an elastic limit. This type of change is characteristic of dense mortars of outstanding hydraulic properties and is lowest in lime mortars. In cases of joints of masonry where initial bond is poor, these volume changes tend to destroy the bond totally. Where the bond is good cracks occur perpendicular to the plane of contact of brick and mortar. In an external wall these cracks heal up slowly when mortar becomes wet but they open up again while drying. It has been observed that the early expansion in cement mortar is slow while that in lime or high lime-cement mortars is rapid. In the latter case the early expansion on wetting results in sealing of cracks caused by previous drying thus reducing water penetration.

It is generally observed that high volume changes are characteristic of mortars with low water retentivity and this undesirable combination results in poor initial bond and subsequent destruction of the bond between the masonry and the mortar.

6.4 EXTENSIBILITY

Extensibility is defined as the maximum linear deformation (stretching) that a hardened mortar may undergo before it fails in tension. It may be expressed in mm per mm, cm per metres, etc. This property becomes significant when it is considered in its relation to the magnitude of volume changes subsequent to hardening.

6.5 RESISTANCE TO MOISTURE PENETRATION

Two major causes of water leakage in masonry are lack of intimate contact between bricks and mortars and lack of permanency of adhesion of such mortars to bricks. The former is dependent on high water retentivity of the mortar and the latter on volume changes subsequent to hardening. These aspects of mortars have already been discussed in previous paragraphs. The term adhesion or bonding is applied to the ability of the mortar to take hold on a solid surface. The conditions for bonding can be altered by the wetness of the unit. In so far as resistance to rain penetration is concerned it is the extension of the bond rather than the strength of the bond that is the criterion. Good bonding capacity is dependent on the general adaptability of the mortar, that is, it enables the mortar to adhere uniformly and completely at all points where the unit and the mortar meet under widely different conditions. The lack of sufficient contact between mortar and brick at the internal plane results in small capillary cracks which induces capillary suction sufficien for rain water to penetrate directly into the wall through the entire length of the bricks in contact with mortar. A good and complete bond prevents this capillary transmission.

The above arguments lead to the inference that a high water retaining, low-volume change mortar is to be preferred to reduce moisture penetration in walls. Investigations carried out reveal that moisture penetration in masonry of certain buildings was caused largely by separation cracks between bricks and mortar. Practically all structure that did not leak had high lime content mortar. The adverse effects of the rain penetration on strength and druability are dealt with in the following paragraphs.

6.6 COMPRESSIVE STRENGTH

It is usual to specify strength of mortar in terms of compressive strength or tensile strength. These are not always the critical factors in the design of masonry walls for strength. A strong wall is one having adhesion of mortar to units throughout. The weakest link in the masonry is at the plane of contact of bricks and mortar when the latter has poor bonding power. The strength also depends on the degree of saturation of the wall and its water tightness which in turn is obtained with mortar of good adhesion and bonding power.

The compressive strength of mortars is usually determined in the laboratory on 50 mm cubes. In actual practice the thickness of mortar is 10 mm. The strength in compression of the mortar in the adjoint is thus several times greater than its laboratory value. Even a mortar of 7 N/mm^2 can sustain in compression the weight of several storeys of wall superimposed on it in masonry construction.

6.7 TENSILE STRENGTH

Over a period of years Portland cement mortars increase slightly in strength. An increase of 40% over 28 days strength has been observed. The corresponding increase in the case of lime mortars (Laboratory specimens) has been as much as five times while stored in air. Strength of mortars depends on conditions of curing. In practice the masonry should be kept moist for at least 10 days after erection to avoid rapid drying out of mortars with consequent reduction in strength.

6.8 BONDING EFFICIENCY

The bonding efficiency of a mortar may be defined as the ratio of compressive strength to tensile strength. Certain investigations found the bonding efficiency of Portland cement mortar as varying from 11 to 20%. The average value for cement lime mortar 1 : 2 : 9 is 26%. The following figures are the bonding efficiencies for some mortars tested in India.

Table 6.1: Bonding efficiencies of Mortars

Mortar		Bonding Efficiency percentage
Lime : Sand (Coarse)	1 : 2	25
Lime : Surkhi	1 : 2	33
—do—	1 : 3	48
Lime : Cinder	1 : 2	24
—do—	1 : 3	25
Lime : Surkhi : Sand	1 : 1 : 1	33
Lime : Cinder : Sand	1 : 1 : 1	14
Cement : Sand (Fine)	1 : 3	12
—do—	1 : 7	18
Cement : Sand (Coarse)	1 : 3	12
—do—	1 : 7	12

6.9 DURABILITY AND EFFLORESCENCE

All building materials contain at least some trace of water soluble substances. With the entrance of moisture into the masonry, these salts get dissolved and on evaporation they get deposited on the surface of the masonry. This leads to unsightly efflorescence. This condition accompanies excessive saturation and the cycle of drying and wetting of the wall.

Under persistently damp conditions, soluble sulphates from the bricks come in contact with mortar and a chemical reaction between the sulphate and the cement or hydraulic lime takes place. This is accompanied by marked increase in volume which may cause the mortar to crumble and brickwork to expand. Conditions favourable to this reaction are found in very wet circumstances. Dense strong mortars are attacked slowly by sulphates but they are also not immune.

For most normal brickwork in superstructure, no special consideration need be given to the question of sulphates. But there are special cases where high resistance to sulphates attack may be essential. This may occur when the sub soil water is high and contains considerable soluble sulphates. In such cases a mortar containing special cements like high alumina cement etc. instead of Portland cement is suitable. Lime should be used with high alumina content. But to reduce their rate of hardening, ground chalk, limestone dust or finely ground surkhi may be used as additives.

TYPES OF MORTARS

In practice the above requirements cannot be fully attained in one single mortar composition, therefore, the selection of mortar should be done to suit a particular type of work. The various types of mortars are:

6.10 CEMENT MORTAR

Cement mortars consist of cement and sand. These harden rapidly, attain high strength at early age, possess high shrinkage on setting and a high degree of rigidity. Sometimes lime is added to increase workability, reduce shrinkage and the rate of stiffening of cement mortar.

6.11 LIME MORTARS

Lime mortars contain lime and sand. Usually 2 to 3 volumes of sand are added to one volume of dry hydrated lime. High calcium lime is used in the form of putty which imparts better workability. Lime mortars have high water retentivity, plasticity and capacity for autogenous healing. Since these harden by drying and by carbonation brought about by the absorption of carbon dioxide from atmosphere, the development of strength is rather slow. Hydraulic or semi-hydraulic limes have the property of setting under water and are used when a situation demands an early gain in strength and higher ultimate strength.

6.12 LIME-POZZOLONA MORTARS

In these mortars lime and pozzolona are the main constituents. Lime may be either non-hydraulic or semi-hydraulic. Surkhi and cinder are the two pozzolonas that are commonly used. Sometimes part of mortars possess hydraulic properties and attain strength higher than that of lime sand mortars.

6.13 COMPOSITE MORTARS

Composite or gauged mortars contain both cement and lime in varying proportion. Lime should be either non-hydraulic or semi-hydraulic. These are known to possess good properties of both cement and lime mortars. They have good strength, workability, water retention and bonding properties. Use of these mortars results in a more watertight masonry as shrinkage cracking is considerably reduced.

6.14 AERATED CEMENT MORTARS

These contain a small amount of entrained air by incorporating a small quantity of air entraining agent in a mix. The workability of a lean cement-sand mortar is thus improved and bleeding and segregation are reduced.

6.15 MUD MORTAR

For cheap construction, mud mortar obtained by mixing clay with sand is used. This type of mortar is liable to shrinkage cracks and has low strength. It is mostly used in temporary and village house construction.

The clay is broken up into fine powder and then mixed with clean water and allowed to mature for at least two days during which time it shall be worked up at intervals with man's feet and spade. Depending upon the composition of the clay, adequate quantities of sand and chopped straw may be added to obtain the necessary consistency and freedom from shrinkage. The mud mortar when ready should be of such consistency that it will readily slide of the face of a trowel but must not be so wet that the mortar falls apart into large drops. Sometimes, bitumen and kerosene are added to make mortar more resistant to rain attack.

MATERIALS FOR MORTARS

Mortars used in building works usually have one or a combination of the following ingredients: cement, lime, mud, sand, surkhi, cinder or ash or special admixtures like air entraining agents and water. Desirable properties of these materials are discussed briefly in the subsequent paragraphs.

Clay

Stiff clay should be used for preparation of mud mortar. The clay must be free from grass roots, stone and other foreign matter.

Sand

Sand serves the following functions in mortars:

1. It prevents shrinkage and cracking of the mortar in setting.
2. It sub-divides the rnortar into thin films to provide air passage for carbon dioxide from the atmosphere and thus helps in setting of fat lime mortars.
3. It provides the requisite surface area for the film of binding material to adhere and spread.
4. It does not improve the tensile strength of mortar but increases its crushing strength.
5. It reduces the amount of cementing material required for a given volume of mortar and being cheaper than the cementing material, reduces the cost of the mortar.

Sand used for mortar work shall be inert, hard, insoluble and free from deleterious materials like acid, alkali, and animal or vegetable organic matter. Coarse sand is preferable to fine sand in high strength mortars even though such mortars are likely to become harsh with lean mixes. Fine sand makes the mortar more workable but causes greater shrinkage. The fineness modulus should be in the range of 1.15 to 2.95. A sand with fineness modulus of 2.0 to 3.0 is considered suitable for lime sand mortars made from fat lime while sand of fineness modulus between 1.5 to 2.5 would serve the purpose in the hydraulic limes, though coarser and well graded sand would be preferable.

Neither the grading nor the size and shape of sand is of special importance for normal work. But a well graded sand from fine to coarse yields the best results and greater strength is obtained with a sharp grain than with a rounded one.

It is not always possible to obtain a well graded sand at site and the best sand that is available has to be used where graded sand is not available.

Grading

(*a*) The particle size grading of sand for use in mortars should be within the limits as specified below by 1S.2116 : 1980.

IS Sieve Designation	*Percentage Passing by Mass*
4.75 mm	100
2.36 mm	90 to 100
1.18 mm	70 to 100
600 micron	40 to 100
300 micron	5 to 70
150 micron	0 to 15

(*b*) Various sizes of particles of which sand is composed should be uniformly distributed throughout the mass.

Cement

Cement should conform to the standard specification of Portland cement. Cement used should be fresh and must not contain lumps and show signs of hydration or setting. The effect of storage of cement for long periods in ordinary godowns is to reduce its strength considerably.

Lime

Lime may be fat, semi-hydraulic or hydraulic and should conform to I.S. 712—1984. Non-hydraulic and semi-hydraulic limes should be converted to putty and the actual weight of the hydrated lime which the putty contains may be determined by the following formula :

$$W = \frac{G}{(G-1)(W_p - 1)}$$

Where, W = weight of dry hydrated lime in gms per cm^3 of putty.

G = Speicific gravity of hydrated lime (usually 2.25 for fat lime and 2.75 for semi-hydraulic lime)

W_p = Weight of putty in gms per cm^3.

Surkhi

It is important that surkhi used in mortar should not contain too much sulphates.

The acid soluble sulphate content should not be more than 0.5 percent except in the case of internal walls where it may be as with soluble salts within these limits. In such cases and where the quantities of acid soluble sulphates are not known or cannot be determined, surkhi should be used with caution and must be used only in a situation where great dampness does not prevail.

Water

Water fit for drinking purposes is normally suitable. It should be clean and free from deleterious matter.

METHODS OF MIXING MORTARS

6.16 MIXING OF CEMENT MORTARS

Dry cement and aggregates should be first thoroughly mixed in the required proportions. Water should then be added and mixing continued until the mortar has the right consistency. Any mortar not used within two hours of mixing should be thrown away. The partly set or dried cement mortar should never be tempered for use. Cement and sand used for mixing should conform to the specifications stated earlier. Good potable water should be used. Only sufficient water should be added to enable the mixture to work up to the required consistency. The work in which the mortar is used should be kept moist at least for seven days.

The porportioning of the materials is done by:

(*i*) weight, and

(*ii*) volume of packed cement and loose sand.

The first method is considered most accurate and should be adopted wherever possible. The second method is convenient and is normally used in the field.

Cement should be measured by volume, three bags being roughly taken as 1.00 cubic metre. The quantity of sand specified is for dry sand. Thus it is necessary to find out the dry volume of any sample of wet sand. For this purpose, an ordinary graduated measuring jar may be used. The sample of sand may be filled up to any desired mark, say 100 c.c. Water shall then be filled into the jars so that the sand is completely immersed. The volume of sand is reduced due to saturation. Let the new volume be 90 c.c. The bulking of the sand to be allowed, therefore, is 10% of the wet sand *i.e.* to obtain one measure of dry sand, 10/9 measure of the wet sand should be used.

6.17 MIXING OF LIME MORTAR

The lime mortar should consist of slaked lime and sand each complying with its respective standard specifications, mixed in the proportions specified. The lime used in this country mostly is quick lime. The slaked lime is sieved through an I.S. Sieve No. 300 micron for use in ordinary mortar works. The slaked lime should be run into putty before being used on work. The putty is prepared only from non-hydraulic and semi-hydraulic quick lime. The lime should be run into a putty as soon as possible after delivery.

The process of running quicklime directly into putty has been described in Chapter 4.

Where hydrated or slaked lime (Class *B* or C) is received it should be mixed thoroughly with water by adding it to water in suitable container. It should be stirred to a thick consistency. This should be left undisturbed for not less than 36 hours. Extra water which will be separated on top may then be stirred to a thick consistency. This water which will separate on top may then be siphoned. The putty formed should be protected from drying out.

Where hydraulic quicklime (Class A) is used it should be slaked or hydrated and the screened slaked lime should be mixed with the fine aggregate. In proportioning lime mortar, the weight of hydrated lime in putty is determined by the formula given in article.

Alternately the volume of putty may for the sake of convenience, be taken as equal to the volume of dry slaked lime ; thus in a mortar of 1 : 3 proportions, one volume of lime putty should be used to three volumes of dry sand.

All mortars should be used as soon as possible after grinding. As a rule it should be used on the day on which it is made and in no case should it be used after 72 hours. If hydraulic lime (Class A) is used the ground mortar should not be kept unused for more than 4 hours after grinding. The ground mortar in all cases should be kept damp and should never be allowed to go dry. This is ensured by covering the ground mortar with wet bags or by any other suitable means.

A mortar and a plaster mix are one and the same. For undercoat there are no special differences. For finishing coat of plaster, certain precautions must be taken. The sand must be fine, all finer than I.S. Sieve No. 400 micron and surkhi finer than I.S. Sieve No. 200 micron.

Strength and quality of the mortar or plaster depend, inter alia, on the method adopted for mixing. There are essentially two methods of mixing lime mortars, *viz.*

(*a*) Manual mixing on a platform or masonry tank, and

(*b*) Grinding in a mortar mill.

Manual Mixing. Measured quantities of slaked lime in the powdered form (or putty) and aggregate are placed on a water tight masonry platform or in a small tank and the ingredients are first mixed by turning over 2 or 3 times with a spade. Water is added and mixing is further continued till a mortar of uniform consistency is obtained. For large works generally two tanks are provided for the purpose of alternate use so that while mortar already prepared in one tank is being used, mortar in the other tank could be mixed.

Mortar Mill Mixing. The bullock driven mill shown in Fig. 6.1 popularly termed the *chakki* in Northern India is too well known to require any description. In this mill slaked lime which is generally in the form of putty and aggregate is added in the requisite proportions along with water and grinding is done till a mortar of uniform consistency is obtained. As grinding proceeds, the mixture is continuously racked essentially at angles and sides so as to ensure uniformity of mix. Grinding is done generally for 80 to 180 revolutions of the stone roller depending upon the nature of lime and aggregate used and the quality of mortar desired. Time required for mixing of mortar by this method varies from 1 to 3 hours depending upon the number of revolutions specified and the size of the mill, etc.

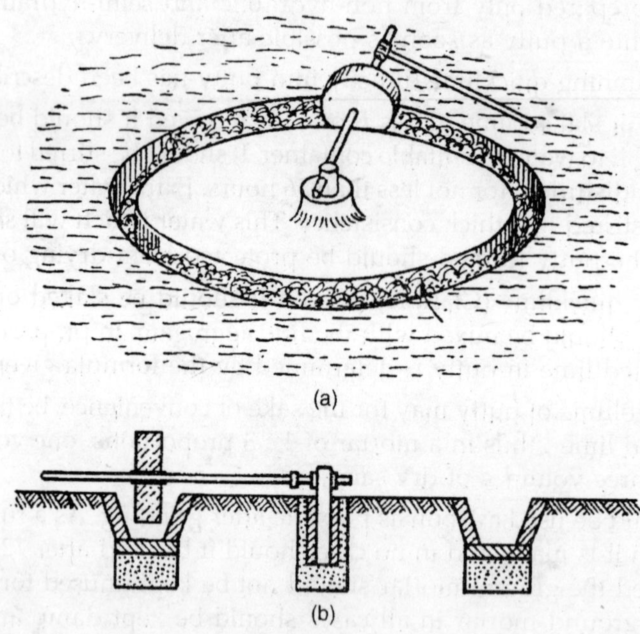

Fig. 6.1. Mortar mill

If a mechanically powered mixer of roller and pan type shown in Fig. 6.2 is employed the result obtained is equally satisfactory and the time taken is much less. In fact, in case of construction of buildings enmass, powered mortar mixers are preferable because of less space requirements and greater speed achieved.

Lime mortar made by mixing without grinding does not attain maximum workability and strength. Grinding or good mixing for preparation of lime mortars is of very great importance and is essential for all high class work where strength and quality are required. Grinding brings lime hydrate and particles of aggregate into intimate contact with each other, increases workability, decreasing the water required and thus the water lime ratio, thereby contributing to additional strength.

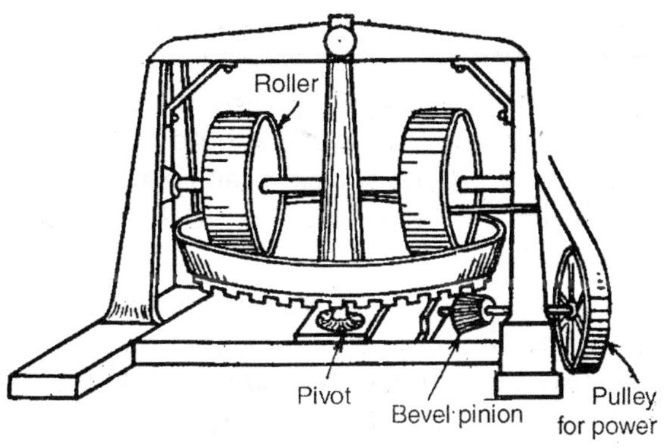

Fig. 6.2. Pan mill

Keeping Mortar after Mixing. Lime mortar should be used as soon as possible after mixing and mortars made from fat lime and semi-hydrated lime should be used on the same day; mortars made from eminently hydraulic lime should be used within 4 hours. Mortars prepared from dry slaked fat lime, however, improve in workability if kept for a few hours before use. The ground mortar in all cases should be kept damp and should never be allowed to go dry. This is ensured by covering the ground mortar with wet bags or by any other suitable means.

6.18 MIXING OF LIME POZZOLONA MORTARS

Surkhi Mortar. The same procedure as for lime mortar should be followed for surkhi mortar except that surkhi in the specified proportions is to be added in place of sand. It should be noted that lime mortar made with surkhi is hydraulic and must be used within 24 hours of being ground.

Lime-Cinder, Lime-Surkhi-Sand and Lime-Sand-Cinder Mortars. The same specifications as for lime mortar and surkhi mortar shall apply to these mortars as well.

6.19 MIXING OF COMPOSITE MORTARS

Cement Lime Mortar. Cement lime mortar is prepared by adding the desired quantity of cement to lime mortars. When, lime mortar made from fat lime is required to be improved with regard to its initial setting, cement is added in small batches so that any mortar in which cement has been added is utilised in work within two hours of addition of cement. Thus to prepare a 1 : 2 : 9 mix of composite mortar, lime mortar using lime and sand in the proportion of one volume of lime to 4½ volumes of sand is prepared. Immediately before use one volume of cement should be added to 9 volumes of this mortar.

When hydrated lime is used in mortars, it is desirable to soak the lime overnight for 24 hours before using it in the mortar. This procedure is found to increase the plasticity and workability of the mortar.

Though best results are obtained by grinding lime mortars in mortar mills, this procedure is often cumbersome and to some extent less important in the case of cement-lime mortars. In the absence of mechanical mixers intimate mixing of cement to the wet lime mortar may also present difficulties. In such cases, any one of the following methods may be used:

(*a*) Cement slaked lime and sand are first mixed dry very thoroughly on a platform or in a masonry tank in the required proportions. Water is then added and further mixing done till mortar of uniform consistency is obtained.

(*b*) Cement and sand should be mixed dry on a platform or in a tank in the requisite proportions. Lime should then be added to it in the form of putty and dissolved till uniform consistency is obtained. This method is better when the proportions of lime in the mortar is not much.

(*c*) Lime and sand are first mixed separately for the entire day's requirements into a stiff paste and kept in stock. Cement should then be mixed with lime mortar in small batches in accordance with the requirements. When lime mortar has been prepared by grinding in a mortar mill, this is the only method which can be adopted for making of cement lime mortar. As in the case of cement mortars all masonry using lime or composite mortars shall be kept wet for a period of at least seven days to prevent the mortars drying out and losing strength in its early hardening days.

6.20 MIXING OF AERATED MORTAR

Aerated mortar is mixed mechanically. The predetermined amount of air-entraining agent is added to the mixing water in the mixer. Cement is next added and the mixer started tc obtain a slurry. Finally sand is introduced. Sufficient quantity of water should be added to obtain desired workability. The mixer should have a revolving paddle which cuts into the mortar and entrains air.

6.21 PRACTICAL HINTS FOR PREPARING GOOD MORTAR

(*a*) The quantity of water added in preparing mortar should be just sufficient to give good workability. The use of too much water should be avoided as it reduces strength, increases shrinkage and impairs the bond between the brick and the mortar.

(*b*) Partially set or dried cement mortar should never be tempered for reuse.

(*c*) Lime should never be added to high alumina cement mortar.

(*d*) Cement should not be used with hydraulic lime mortars.

(*e*) Lime should be completely slaked.

(*f*) Magnesium limes are not suitable for use in mixes containing cement since they contain overburnt magnesia, some of which may not hydrate when lime is slaked. It slakes slowly and expands on hydration which may cause defects or failure in the building.

(*g*) Fresh and clean water should be used.

ADDITION OF POZZOLONIC AND AIR-ENTRAINING MATERIALS TO MORTARS

Researches have shown that the addition of very finely ground pozzolona like stone dust or fly-ash reduces the capillarity of the mortar specially in the case of lean cement mortars. It also reduces the early strength of the mortar to some extent even though the strength improves considerably at later stages. In some of the tests carried out it was found that a replacement of 20 percent of cement by surkhi in 1 : 6 cement mortar with graded sand resulted in a reduction of 30 percent of strength and 15 percent of permeability at the age of 8 days. Where high early strength is not desired this reduction in strength is not harmful. The replacement with surkhi or other pozzolona is, therefore, advantageous to reduce permeability.

It has been stated earlier that cement mortars of even the densest variety are permeable and porous to various degrees. The permeability is reduced and the water penetration resistance increased by the addition of integral water proofer. Several proprietary brands are available in the market. These have, however, limited application in building mortar, but are used in damp proofing of basements and floors in water retaining structures, etc.

Recently workability aids have been placed on the market by various manufactures. Addition of these to cement mortars produces the same beneficial effects as obtained by the addition of lime. This reduces shrinkage cracking, increases bond and impermeability and does not adversely affect the tensile strength very much. These admixtures are air entraining agents. They entrain small stable air bubbles in the mortars. The presence of these air bubbles imparts cohesiveness and plasticity to the mix. By addition of these aids, the mortar becomes fatty and spreads and trowels like butter. In places where coarse sand is available and good lime is not available, the addition of these agents make the mortar workable and enables the mortar to work into the joints without segregation. These materials should be used to the manufacturers specifications.

RECOMMENDED MIXES

The selected mortar should impart to masonry adequate strength and resistance to rain penetration. The strength of masonry increases upto a certain limit with the increase in the strength of mortar, beyond which there is practically no gain in strength. Also the strength of masonry is not proportional to the strength of the mortar and a rich mix does not necessarily increase the strength of the masonry. For example the strength of medium strength brick masonry built in 1 : 6 cement sand mortar is about 90 percent of that in 1:2:4 cement-lime sand mortar even though the strength of the cement and mortar is only about 60 percent of that of cement lime sand mortar.

In masonry, rain water generally penetrates through the mortar joints. For a masonry to be water tight, extent of bond (area of contact between mortar and the building unit) is more important than the bond strength. The former depends primarily on the water retaining capacity of a mortar and suction capacity of the building unit. The bricks in India have generally a high suction capacity and tend to draw water from the mortar before it hardens. Mortars with high water retention capacity are therefore preferable. Composite mortars are superior to cement mortars in this respect and increase water tightness of the masonry.

Strong mortar should not be used in masonry as it will cause cracking in the unit due to initial drying. Most of the shrinkage cracking of walls occurs during the first drying of the building. Compared to other mortars, lime-pozzolona mortars possess a higher pre-hardening and if drying is quick they may crack. However, once they have hardened their resistance to cracking wilL be greater than that of cement mortar because of their lower modulus of elasticity and reduced volume changes.

Aerated mortars may be used when either the quality of available lime is poor or transportation of lime increases the cost.

6.22 PROPORTIONS OF MORTARS FOR VARIOUS PURPOSES

Recommended mortar mixes for the various parts of the building are given below:

(*a*) *Masonry in Foundation and Plinth*

(*i*) Cement mortar 1 : 6

(*ii*) Cement lime mortar 1 : 1 : 8, 1 : 1 : 7 or 1 : 1 : 6.

(*iii*) Where good quality lime and surkhi is available and sub soil water is not within 2.5 metres of foundation level, following lime mortar may be used for loadings upto 44 metric tonnes/sq. m :

1 lime : 2 sand

1 lime : 2 surkhi,

or 1 lime : 1 surkhi : 1 sand.

(*iv*) Where sub-soil water is within 2.5 metres below the foundation level and the soil contains high percentage of soluble sulphates, 1 : 3 cement mortar should be used.

(*b*) *Masonry in Super-structure*

(*i*) Cement mortar 1 : 6 or cement lime mortar 1 : 1 : 8 for single and two storeyed constructions.

(*ii*) Cement lime mortar 1 : 1 : 7 for three storeyed constructions.

Cement lime mortar 1 : 1 : 6 for 4 storeyed constructions.

(*iv*) Cement lime mortar 1 : 2 : 9 (1 cement : 2 lime : 9 sand) for loading upto 80 metric tonnes/sq m.

(*v*) For cavity and half brick walls and for loading upto 110 metric tonnes/sq m use cement mortar 1 : 3 or cement lime mortar 1 : 1 : 6 and

(*vi*) For arch work-cement mortar 1 : 3

(*c*) *Internal Plaster*

(*i*) Cement plaster 1 : 6 (1 cement, 6 sand)

(*ii*) Cement lime plaster 1 : 1 : 7 (1 cement, 1 lime, 7 sand)

(*iii*) Cement lime plaster 1 : 2 : 9 (1 cement : 2 lime : 9 sand)

(*iv*) Lime plaster 1 : 2 (1 lime : 2 sand)

(*v*) Lime surkhi plaster 1 : 2 (1 lime : 2 surkhi).

(*d*) *External Plaster*

(*i*) Cement plaster 1 : 6 (1 cement, 6 sand)

(*ii*) Cement lime plaster 1 : 1 : 7 (1 cement, 1 lime, 7 sand)

(*iii*) Cement lime plaster 1 : 2 : 9 (1 cement : 2 lime : 9 sand)

QUESTIONS

1. Define mortar and state its uses. List the properties which should be possessed by mortar.
2. What factors affect the workability of mortar? Discuss in detail.
3. What volume changes take place in mortar? Briefly explain these.
4. Write short notes on the following in connection with mortars :
   (*a*) Rate of hardening or stiffening
   (*b*) Resistance to moisture penetration
   (*c*) Durability and efflorescence
5. Define bonding efficiency. On what factors does it depend? Give the limits for common types of mortars.
6. List the various types of mortars and state their suitability.
7. What are the various materials used in the preparation of mortars and state the functions performed by each one of them.
8. How is lime mortar mixed? Give berief details of the methods used mixing.
9. State practical hints for preparing good mortar. How is mixing of composite mortar done?
10. In what way the addition of pozzolonas and air-entraining agents help to modify the properties of mortars.
11. Which type of mortar will you recommend for use in masonry work? Give justification.
12. What proportions of mortor are normally recommended for the following purposes:
    (*a*) Masonry in foundation and plinth
    (*b*) Masonry in super-structure
    (*c*) Internal plaster; and
    (*d*) External plaster.

OBJECTIVE TYPE QUESTIONS

1. Which of the following does not form a part of mortar?
   (*a*) Cement (*b*) fine aggregate
   (*c*) coarse aggregates (*d*) water
2. Surkhi acts as a .... in mortar
   (*a*) fine aggregate (*b*) coarse aggregate
   (*c*) cementing material (*d*) None of these
3. The most important function of mortar in masonry work is to
   (*a*) increase compressive strength
   (*b*) increase lateral strength
   (*c*) bind masonry
   (*d*) increase protection against moisture penetration

4. Workability of mortar depends upon :
(*a*) bulk density (*b*) water retentivity
(*c*) Trowelling property (*d*) All of these

5. The process of stiffening or hardening of mortar results in a
(*a*) physical change (*b*) chemical change
(*c*) gain of plastic properties (*d*) loss of elastic properties

6. A strong wall is one which has more :
(*a*) compressive strength (*b*) tensile strength
(*c*) adhesion of mortar to units (*d*) water tightness

7. Which of the following types of mortars has the maximum bonding efficiency :
(*a*) Cement : Sand (coarse) 1 : 3
(*b*) Cement : Sand (fine) 1 : 3
(*c*) Lime : Sand (coarse) 1 : 2
(*d*) Lime : Surkhi 1 : 3

8. Which of the following types of mortar has the least bonding efficiency
(*a*) Lime : Cinder : Sand 1 : 1 :1 (*b*) Lime : Surkhi : Sand 1 : 1 : 1
(*c*) Lime : Surkhi 1 : 2 (*d*) Lime : Cinder 1 : 2

9. Lime is added to cement mortar to
(*a*) decrease workability (*b*) Reduce shinkage
(*c*) Increase rate of stiffening (*d*) attain high strength at early age

10. Cement mortar possesses properties of :
(*a*) slow hardening (*b*) low shrinkage
(*c*) low degree of rigidity (*d*) high strength at early age

11. Lime martars have :
(*a*) high water retentivity (*b*) high plasticity
(*c*) better workability (*d*) All of these

12. The compressive strength of lime surkhi mortar increases much higher than lime sand mortar due to :
(*a*) non-hydraulic properties (*b*) Semi-hydraulic properties
(*c*) hydraulic properties (*d*) None of these

13. Aerated cement mortars have :
(*a*) Improved workability (*b*) Low bleeding
(*c*) Low segregation (*d*) All the above

14. The specific gravity for fat lime is normally taken as :
(*a*) 2.25 (*b*) 2.35
(*c*) 2.45 (*d*) 2.55

15. For preparing good mortar, lime should never be added to :
(*a*) Ordinary Portland cement mortar (*b*) high alumina cement mortar
(*c*) Low heat cement mortar (*d*) Portland blast furnace slag cement mortar

16. For good results, cement should not be used with :

(*a*) Fat lime mortars (*b*) Semi-hydraulic lime mortars

(*c*) hydraulic lime mortars (*d*) All these

17. The cement mortar for foundation and plinth are of the proportion :

(*a*) 1 : 4 (*b*) 1 : 5

(*c*) 1 : 6 (*d*) 1 : 8

18. When sub-soil water is within 2.5 m below the foundation level, and soil has high percentage of soluble sulphates, the cement mortar should have the proportions

(*a*) 1 : 2 (*b*) 1 : 3

(*c*) 1 : 4 (*d*) 1 : 6

19. The cement lime mortar for 3 storeyed construction has the following proportions

(*a*) 1 : 1 : 5 (*b*) 1 : 1 : 6

(*c*) 1 : 1 : 7 (*d*) 1 : 1 : 8

20. For internal plaster, lime surkhi plaster recommended is :

(*a*) 1 : 2 (*b*) 1 : 3

(*c*) 1 : 4 (*d*) 1 : 5

Answers

1. (*c*)	**2.** (*a*)	**3.** (*c*)	**4.** (*d*)	**5.** (*b*)	**6.** (*c*)
7. (*d*)	**8.** (*a*)	**9.** (*b*)	**10.** (*d*)	**11.** (*d*)	**12.** (*c*)
13. (*d*)	**14.** (*a*)	**15.** (*b*)	**16.** (*c*)	**17.** (*c*)	**18.** (*b*)
19. (*c*)	**20.** (*a*)				

FURTHER REFERENCES

Bureau of Indian Standards, New Delhi.
IS Code Nos. 712, 2016, 2116.

CONCRETE

7.1 DEFINITION

Concrete is an intimate mixture of the following materials:

(*i*) Cementing materials like lime, cement, etc.

(*ii*) aggregates, fine and coarse

(*iii*) water

(*iv*) any approved admixture, if necessary, for giving a particular property

(*v*) steel reinforcement to make it strong in tension, (for reinforced and prestressed concrete only).

ADVANTAGES OF CONCRETE CONSTRUCTION

Concrete, as a material of construction, has taken a prominent place in recent years. There have been great advances in its design and manufacture. It is considered today as the most important material in the field of construction. Wherever there is construction work in progress, whether in the building of small houses, big offices or large industrial buildings, concrete will be found in evidence in some form or the other. Concrete is also found in use where there are large projects being carried out such as construction of dams, power houses, spillways for reservoirs and even for the lining of canals. In fact, concrete has become synonymous with civilization and progress in the fast moving world of today. The obvious advantages of concrete are:

1. Concrete is very strong. It is considered as an artificial stone and is used as a substitute for stone.
2. It is very durable and is not affected, to a great extent, by the weathering agencies.
3. It is plastic, when freshly prepared and can be cast to any desired shape.
4. The quickness and ease of construction in the case of concrete work is a great advantage.
5. It can be given various finishes, which is not possible with any other material.
6. It is free from defects and flaws, usually present in other materials such as stone, wood, etc.
7. It can be manufactured to the required strength economically.
8. Cost of maintenance of concrete construction is less.

9. It can be used for all types of works, massive work where great weight is required, in small sections where strength is required, for architectural purposes, for water proofing and for a number of special uses where no other material would be suitable.

CONCRETE MATERIALS

7.2 CEMENTING MATERIALS

The cementing materials such as cement and lime have been discussed in detail in earlier chapters.

The function of a cementing material in concrete is two folds, firstly it binds the fine and coarse aggregates together and secondly it fills up the voids in between the fine and coarse aggregates.

7.3 AGGREGATES

Both fine and coarse aggregates are added in concrete. The most important characteristics of aggregates are the quality and gradation. The essential qualities have been discussed in chapter 5.

The grading of fine aggregates is expressed in terms of percentage passing or retained on I.S. test sieves No 10 mm, 4.75 mm, 2.36 mm, 1.18 mm, 600 micron, 300 mictron and 150 microns. The grading of fine aggregates has a very marked effect on the workability, uniformity and finishing qualities of concrete. Experience has shown that either very fine or very coarse sand, or sand having a large deficiency or excess of any fraction is generally undesirable and a material having a smooth grading curve will produce more satisfactory concrete. The allowable grading limits for sand depend to some extent on the shape and surface characteristics of the particles. A sand composed of smooth, rounded particles may give satisfactory results with coarser grading than would be permitted for sand made up of sharp angular particles with rough surfaces. It is not hard to visualize the interlocking position taken by angular particles in close contact and the contrast between such particles and smooth rounded particles with respect to freedom of movement in the fresh concrete. It is also evident that roughness of the surfaces of the grains increases internal friction. The grading limits for fine aggregatest are given in Table 7.1 for different zones in the country.

Table 7.1: Grading of Fine Aggregates in zones I to IV

IS Sieve Designation	*Percentage Passing for*			
	Grading Zone I	*Grading Zone II*	*Grading Zone III*	*Grading Zone IV*
10 mm	100	100	100	100
4.75 mm	90–100	90–100	90–100	95–100
2.36 mm	60–95	75–100	85–100	95–100
1.18 mm	30–70	55–90	75–100	90–100
600 micron	15–34	35–59	60–79	80–100
300 micron	5–20	8–30	12–40	15–50
150 micron	0–10	0–10	0–10	0–15

The grading of coarse aggregate may be varied through wider limits than that of sand without appreciable effect on the workability of concrete. As a rule, it is less difficult to provide satisfactory grading of coarse aggregate than of sand. It is practicable to make size separations of the coarser material by screening at comparatively small expense and to recombine the separated sizes in the required proportions. In the screening of aggregates however, each sized product usually contains undersize and oversize material. Oversize is due to screen wear or the use of screens whose effective openings are somewhat larger than specified. The amount of undersize is increased and of oversize decreased by breakage and attrition during handling operations. As over and undersize material in coarse aggregate increases the size range of the fraction and thereby tends towards increased segregation, its presence defeats to that extent the purpose of size separation. The Indian standard specification limits the amount of oversize and undersize material in 40 mm, 20 mm and 10 mm sizes.

The gradings of coarse aggregates for single sized of a graded coarse aggragates is given in Table 7.2. The grading for coarse aggregates for use in mass concrete is set in Table 7.3.

Table 7.2: Grading of coarse aggregate for Single sized Aggregates & Graded Aggregates.

IS Sieve Designation	*Percentage Passing for Single-sized Aggregate of Nominal Size*						*Percentage Passing for Graded Aggregate of Nominal Size*			
	63mm	40mm	20mm	16mm	12.5mm	10mm	40mm	20mm	16mm	12.5mm
80 mm	100	–	–	–	–	–	100	–	–	–
63 mm	85 to 100	100	–	–	–	–	–	–	–	–
40 mm	0 to 30	85 to 100	100	–	–	–	95 to 100	100	–	–
20 mm	0 to 5	0 to 20	85 to 100	100	–	–	30 to 70	100	100	100
16 mm	–	–	–	85 to 100	100	–	–	–	90 to 100	–
12.5 mm	–	–	–	–	85 to 100	100	–	–	–	90 to 100
10 mm	0 to 5	0 to 5	0 to 20	0 to 30	0 to 45	85 to 100	10 to 35	25 to 55	30 to 70	40 to 85
4.75 mm	–	0 to 5	0 to 5	0 to 5	0 to 10	0 to 20	0 to 5	0 to 10	0 to 10	0 to 10
2.36 mm	–	0 to 5	–	–	–	0 to 5	–	–	–	–

Table 7.3: Coarse Aggregates for Mass Concrete

Class and Size	*IS Sieve Designation*	*Percentage Passing*
Very large, 160-80 mm	160 mm	90–100
	80 mm	0–10
Large, 80-40 mm	80 mm	90–100
	40 mm	0–10
Medium, 40–20 mm	40 mm	90–100
	20 mm	0–10
Small, 20-4.75 mm	20 mm	90–10
	4.75 mm	0–10
	2.36 mm	0–2

Where all-in aggregates are used, the grading limits as set forth in the Indian standard specification are given in Table 7.4.

Table 7.4: Grading of All-in Aggregates:

IS Sieve Designation	*Percentage Passing for All-in Aggregates of Nominal Size*	
	40 mm	*20 mm*
80 mm	100	—
40 mm	95–100	10
20 mm	45–75	95–100
4.75 mm	25–45	30–50
600 micron	8–30	10–35
150 micron	0–6	0–6

7.4 WATER

Functions of water in concrete. Water has two functions in a concrete mix, firstly, it reacts chemically with the cement to form a cement paste in which the inert aggregates are held in suspension until the cement paste has hardened and secondly it serves as a vehicle or lubricant in the mixture of fine and coarse aggregates and cement and this facilitates the making and placing of concrete.

Specification for water. According to I.S. : 456—2000, (Reaffirmed 2005) the water to be used in concrete work should be clean and free from injurious amounts of deleterious materials.

Quantity of water for making Concrete. The quantity required is as follows:

(*a*) *for mixing concrete.* The water requirement for an ordinary concrete mix is generally 5% by weight of aggregate plus 30% by weight of cement, necessary variation being made for the required consistency of the concrete and the surface water present in the aggregates. For vibrated concrete, the quantity can be reduced by 20%. The exact quantity of waer required can be determined as described in the part dealing with the design of concrete mix.

(*b*) *For other purposes.* For washing aggregates, curing, etc., about 120 litres per cu m of concrete work may be assumed.

7.5 ADMIXTURES

Admixtures are substances other than cement, aggregate and water which are added to concrete in order to improve its qualitites either during its placing or in its subsequent performance. An admixture may be added to the mixing water, to the batch before or during mixing, or if its presence in concrete significantly afects the properties of the concrete, it may be grounded with the clinker during the manufacture of cement.

Purposes of Admixtures. The imprtant modification obtained in the properties of concrete by the use of admixtures are the following:

(*a*) Improvement in workability.

(*b*) Reduction in bleeding.

(*c*) Retardation or reduction of heat evolution.

(*d*) Retardation of initial stiffening or increased setting time.

(*e*) In grout mixtures, improvement of penetration and pumpability and the reduction of segregation.

(f) Acceleration of the rate of development of strength at early ages.

(*g*) Increase in durability and in resistance to deterioration under special conditions of exposure.

(*h*) Decrease in capillary flow of water.

(*i*) Decrease in permeability to liquids.

(*j*) Increase in the bond between reinforcement and concrete.

(*k*) Prevention of shrinkage or creation of slight expansion in concrete for certain special uses.

Classification of admixtures. Admixtures can be broadly divided into the following categories according to the kinds of materials constituting them or to the characteristic effects produced by them:

(*a*) Workability agents.

(*b*) P ozzolonic materials.

(*c*) Retarders.

(*d*) Accelerators.

(*e*) Water repelling agents.

(*f*) Air-entraining agents.

(*g*) Gas forming agents.

(*h*) Natural cementing materials.

(*i*) Miscellaneous.

Commercial admixtures may contain materials belonging to two or more of the above groups. For example, a workability agent having a retarding action may be combined with an accelerator to compensate for the retarding action. Certain admixtures may also have properties of more than one group.

The various classes of admixtures, their composition, the effects they produce on concrete, and the quantities in which they are to be used are summarised in Table 7.5.

7.6 REINFORCEMENT

The tensile strength of plain concrete being negligible, reinforcement of concrete is required wherever concrete is subjected to tensile stresses. Reinforcement also serves to limit the size of shrinkage cracks occuring during the setting of the concrete and also of cracks caused by changes due to temperature variations.

An ideal reinforcement should be able to develop a good bond with concrete so that the stresses can be easily transferred from one material to the other. It should have sufficiently high tensile strength, a high modulus of elasticity which is essential for an economical use of the concrete, and a temperature coefficient of expansion and contraction the same as that of concrete so that thermal stresses are not developed. Steel which fulfills almost all the above requirements is therefore in common use as a reinforcing material.

Types of Steel Reinforcement. The commonly used types of steel reinforcement are:

(*a*) Mild steel and medium tensile steel bars and hard-drawn steel wires.

(*b*) Structural steel sections.

(*c*) Hot rolled mild steel and medium tensile steel deformed bars.

(*d*) Cold twisted steel bars.

(*e*) Hard-drawn steel wire fabric.

Table 7.5: Admixtures

Classification of admixtures	*Purposes for which required*	*Composition of admixture*	*Quantity to be used*
1. Workability agents.	1. Increase in workability. 2. Increase in strength.	1. Finely divided minerals— (*a*) bentonite clay ; diatomaceous earth (*b*) Fly ash; finely divided silica, clay, fine sand ; hydrated lime ; talc ; pulverised stone. 2. Water-reducing agents, sulphonated organic compounds; carbohydrate salts. 3. Aerating agents.	3 to 5 percent by weight of cement. Upto 20 percent by weight of cement, to be determined by test 2 to 4 percent
2. Pozzolonic materials.	1. Reduction in heat of hydration. 2. Resistance to sea water and sulphate bearing soils. 3. Slight imrovement of workability. 4. Reduction of bleeding and segregation.	1. Natural volcanic ash, fly ash. 2. Artificial-heat treated diatomaceous earth, heat-treated raw shales or clay. 3. Surkhi ground to the same fineness as cement.	10 to 30 percent by weight of cement.
3. Retarders.	1. Retard the setting times of cement. 2. Remedy premature set of certain cements.	Carbohydrate derivatives; calcium lignosulphonates.	Fractions less than 1 percent by weight of cement.
4. Accelerators.	1. Increased rate of early strength development. 2. Early removal of forms. 3. Reducing curing poriod.	1. Calcium chloride. 2. Tri-ethanolamine soluble carbonates and fluosilicates. 3. Alumious cements.	2 to 3 percent by weight of cement. to be determined by test.

	4. Emergency repairs. 5. Concreting in cold weather.		
5. Water repelling agents.	1. Decreased capillary rise of moisture. 2. Increased impermeability. 3. Reduction of efflorescence.	1. Stearates of calcium ammonium, aluminium and sodium butyl sterate. 2. Soaps, oils, waxes.	1 to 0.2 percent by weight of cement.
6. Air-entraining agents.	1. Protection against frost attack. 2. Improved workability. 3. Reduced bleeding.	Natural wood resins and their soaps, fats and oils, their fatty acids and soaps; alkali salts of sulphonated or sulphated organic compounds.	0.005 to 0.05 percent by weight of cement.
7. Gas forming agents.	1. Reduced bleeding. 2. Improved bond. 3. Improve grount penetration and permeability.	Unpolished aluminium powder.	0.005 to 0.02 percent by weight of cement.
8. Natural cementing materials.	1. Increased workability 2. Reduced bleeding, segregation and heat of hydration.	Natural cements, hydraulic limes; water-quenched blast furnace slag, mixtures of blast furnace slag and lime.	Substituted for 10 to 25 percent by weight of cement.
9. Miscellaneous.	1. Increased resistance to abrasions. 2. Colour effects.	Admixtures other than those classified above.	

TYPES OF CONCRETE

The various types of concrete are:

(1) Mud concrete.
(2) Lime concrete, lime-surkhi concrete.
(3) Cement concrete—plain and reinforced.
(4) Plum concrete.
(5) Light weight concrete.
(6) Air-entrained concrete.
(7) Pre-stressed concrete.
(8) Pre-cast concrete.
(9) No fines concrete.
(10) Vacuum concrete.
(11) Waterproof concrete.

7.7 MUD CONCRETE

In the case of mud concrete, the coarse aggregate is 40 mm brick ballast. 40 mm gauge brick ballast is one which will completely pass through a ring of 40 mm internal diameter and not more than 20 percent should pass through 50 mm ring. It should also be ensured that not more than 15 percent of material passes 20 × 20 mm square mesh screen.

The fine aggregate and the binding material is soil having clay and sand content as required for brick earth, crushed into fine powder and freed from grass roots, kankar and other such matter. No soil shall be used which contains efflorescent salts nor shall soil be taken from a locality where there are white ants.

Mixing is done on a platform. 40 cubic metre of mortar should be used per 100 cubic metre of ballast. Necessary quantity of water is then added with a hose and the whole mass is turned over and over until thoroughly mixed.

Mud concrete is laid in layers and consolidated. Each succeeding layer is placed before the previous one dries out. No water should be sprinkled before laying the succeeding layer nor any curing is necessary.

7.8 LIME CONCRETE

Materials used. The various materials for making lime concrete are:

Lime. Fat lime is seldom used for mass concrete. It may be used on small jobs *e.g.* filling spandrels of arches etc. The reason is that it does not set in the interior places *i.e.* where it doss not come in contact with air. If fat lime has to be used, it must be converted first into a very fine powder by slaking and should be passed through a fine mesh sieve. Then it is mixed with the aggregate and kept wet until it is converted into a hydrate.

In all cases, hydraulic lime is mostly employed because it has the property to set under water. If necessary, it may be ground or slaked to convert it into powder. It should be passed through a Sieve No. 150 micron to remove the unslaked particles. Lime taken from a kiln is not mixed with aggregates as there is a danger of unslaked lime being present in the concrete. It is mixed with aggregates only after slaking. Any one of class A or class B lime may used.

Fine Aggregate. The fine aggregate used is either sand or surkhi. Sand should pass through I.S. Sieve No. 300 micron and not more than 10 percent should pass through I.S. Sieve No. 150 micron. Surkhi should pass through I.S. Sieve No. 300 micron.

Coarse Aggregate. The coarse aggregate consist of either brick ballast or broken stone or shingle. The nominal size of the coarse aggregate may be 40, 20 and 16 mm. Normally 40 mm size aggregate is used for mass concrete. If the aggregate is brick ballast, it is soaked by heavily sprinkling with water for 3 hours before the layer of surkhi and lime is added.

Proportion. Different proportions are used for different works.

Ordinary lime concrete used in foundations and as a base for pavings of various types, consists of one part of lime mortar and 3 parts of coarse aggregate. In metric units, 40 cubic meters of mortar is used per 100 cubic metres of coarse aggregate. The proportions of lime to sand or lime to surkhi is either 13 : 26 or 16 : 24.

Fine lime concrete, used for roof terracing, is prepared by mixing 47 cubic metres of lime mortar with 100 cubic metre of 20 mm size coarse aggregate. The mix of the lime mortar is 1 part of lime and 3 parts of surkhi or sand.

Mixing lime Concrete

(*a*) *Dry method.* Mixing is done on a clean water tight masonry platform of sufficient size to provide ample mixing area. Slaked lime and fine aggregate are spread on the platform in layers not exceeding 50 mm. The whole is turned over until thoroughly mixed. A mortar mill may be used for dry mixing on important works.

The coarse aggregate is washed and stacked to a depth of 150 mm on a separate platform and thoroughly soaked by heavily sprinkling water. Dry mortar, from the other platform, is taken and required quantity spread on the coarse aggregate. The whole is turned over backward and forward with a shovel until thoroughly mixed. Then water is added and the mixture again turned over and over again with the shovel, till the ingredients are thoroughly mixed.

(*b*) *Wet method.* In this case, the mortar is prepared by adding water. Mixing is done either on a platform or in a mortar mill. Lime and aggregate are mixed in the required proportions and water added in calculated quantity and the whole thoroughly mixed. The mortar so formed, must be used within 2 days of its mixing. If shell lime is used, the mortar should be used within 24 hours. The coarse aggregates are placed to an even thickness, usually on the platform covering not more than half its length. Lime mortar in the specified proportion, is then evenly spread over the coarse aggregate. The whole is thoroughly incorporated, using just sufficient water applied with a sprinkler to enable the mortar to adhere to each piece of the aggregate by turning it over backwards and forwards several times (not less than 4 times) and gradually transferring it to the vacant half of the platform until all particles of aggregate are covered with mortar. No more concrete is mixed than can be laid in position the same day of mixing.

Laying and Consolidation of Lime Concrete

(*a*) *Ordinary lime concrete* is laid (not thrown from a height) in layers not exceeding 15 cm in thickness. It is thorougly consolidated with 5.5 kg rammers. Square rammers are used for consolidating edges. Consolidation is complete when a skin of pure mortar covers the surface and completely hides the aggregate and a stick dropped endways from a height rebounds with a ringing sound. The mixing, placing and ramming should go on continuously when once started. No concrete should be placed later than two hours before work is stopped for the day.

The next day, the layer laid on the previous day should be swept or washed clean before the next is laid. Where joints in layers are unavoidable the end of each layer should be sloped at an angle of 30°. When vertical joints occur in an upper and lower layer, they must be atleast 600 mm apart horizontally.

(*b*) *Fine lime concrete* is laid to the required levels and throughly consolidated by beating with flat wooden piece called *"thapi" and* continually tested during the process with straight edge. The beating is carried out with two rows of labourers placed across the entire width of roof, sitting as close as they can, slowly traversing its length, beating all the time with wooden *thapies* and continued for not less than 3 days until the mortar is almost set and the *"thapies"* rebound-while beating is being done. The surface is continually sprinkled with a mixture consisting of 3.2 kg of *gur* and 1.8 kg *bael fruit* to 115 litres of water. As soon as the beating has been completed, clean water should be sprinkled on the surface to soften it and the mortar that comes to the surface is well rubbed and finished smooth with a steel trowel.

Curing of lime concrete. In the case of ordinary lime concrete, the surface is kept wet for a period of not less than 10 days. No brick work or masonry should be laid on the concrete for at least 7 days after laying.

Fine lime concrete is covered with 50 mm of fine sand which should be kept thoroughly saturated with water for a fortnight. Alternatively, the surface may be divided into squares by earthen bunds which should be kept flooded with water.

7.9 PLAIN AND REINFORCED CEMENT CONCRETE

Plain cement concrete. It is prepared by mixing cement, fine and coarse aggregates and water in definite proportions.

Cement concrete vs Lime concrete. Cement concrete requires much greater care in its preparation than lime concrete for the following reasons:

1. Cement concrete sets quickly, within half an hour of its mixing. This is also affected by atmospheric moisture. Addition of a little extra quantity of water alters the properties altogheter, specially the strength is much reduced. This is not the case with lime concrete which sets very slowly. Slight variations in the time of preparing the mortar and mixing of concrete and the quantity of water do not affect the strength of lime concrete to a great extent.
2. Cement is costlier than lime. As such it must be used with care.

Reinforced cement concrete. As plain cement concrete is very strong in compression and very weak in tension, it has to be reinforced in structures subjected to tensile forces. The reinforcing material is. usually steel, the variety and shape of which vary according to different uses of concrete. Thus compression is taken by concrete and tension by steel. Such a concrete is called "Reinforced cement concrete".

Proportioning of concrete. The properties and individual requirements of the different constituents of concrete have been considered earlier. It will now be appropriate to discuss their proper proportioning in order to ensure that the concrete produced meets the needs of the job in hand. The aim should be a judicious choice of the constituents and their correct proportioning so that all the reasonable requirements of the structure for serviceability, safety and appearance are satisfied to the desired degree.

In order to understand the principles governing the proportioning of concrete, it is necessary to consider the relative importance of the factors influencing the quality of concrete. Concrete has to be satisfactory in two stages *i.e.,*

(*i*) When being placed in position and

(*ii*) When it ultimately becomes hard.

The proportioning is controlled by both these stages.

In general cement, water, fine aggregate and coarse aggregate should be so proportioned that the resulting concrete has the following properties:

(*a*) *Strength.* When hardened the concrete should have sufficient strength to bear the imposed stresses safely with the required factor of safety.

(*b*) *Durability.* The concrete must be durable with proper regard to the weathering conditions to which the structure will be subjected.

(*c*) *Workability.* While fresh, the concrete must be workable enough for economical and uniform placing, but not excessively fluid.

(*d*) *Economy.* The cost of preparation of the concrete should be minimum possible. Particularly use of cement should be minimum with due regard to the first three properties.

When specifying concrete mixes, fixed proportions of cement, fine aggregate and coarse aggregate which have been found by experience to be suitable for the particular types of structures, are usually specified. This arbitrary method of preportioning has many drawbacks.

On the other hand, the specifications may stipulate that the concrete must develop a certain strength and resistance to external agencies and may not indicate the proportions. In this case the proportions have to be arrived at by a consideration of the properties desired of the concrete. This is a more rational method.

7.10 PLUM CONCRETE

Plum concrete consists of a specified concrete mix containing large embedded stones. It is used in massive piers, gravity abutments, heavy footings, retaining walls, etc. The proportions of plums (stones) should not, unless otherwise stated, exceed 20 percent of the total volume of the plum concrete. The plums are usually not less than 200 mm in diameter. Thsy are either of good quality stone or broken pieces of cement concrete dismantled from some existing work. These should be free from earth or clay or disintegrated matter and must not have sharp corners or soft materials embedded in them.

During concreting, first 600 to 900 mm of concrete of the nominal mix is laid at the bottom. While the top layer of this concrete is still wet, the plums are laid so that they are slightly embedded in the wet concrete. Normally these plums will sink in sufficiently under their own weight. Complete submergence should be avoided and subsequent layer is at least twice that of the largest plum. The plums should be placed so that the clear distance between any two is not less than the greatest width or thickness of either of the plums, the clear distance between any plum and the face of the work should not be less than the width of the plum. Plums should be washed and well wetted before placing. The surface of the plums concrete is generally left uneven and not smoothed up.

7.11 LIGHT WEIGHT CONCRETE

This type of concrete is used where lightness of a structure is desired. It incorporates certain lightweight aggregates in place of the usual aggregates. Another type of lightweight concrete is obtained by adding gas-forming or foaming agents. The introduction of such agents into concrete during mixing produces a cellular structure on hardening and reduces the weight of the concrete. The light weight aggregates used may be foamed blast furnace slag, bloated clay aggregate, cintered fly ash aggregate and cinder aggregate. The grading requirements for light weight combined aggregates for concrete are listed in Table 7.6.

Table 7.6: Grading Requirements for Lightweight Combined Aggregates for Concrete Masonry Units

Sl.No.	*Size Designations*	*Percentage (By mass) Passing IS Sieves*						
		20 mm	*12.5 mm*	*10 mm*	*4.75 mm*	*2.36 mm*	*1.18 mm*	*300 micron*
(1)	(2)	(3)	(4)	(5)	(6)	(7)	(8)	(9)
(*i*)	Fine aggregate (4.75 to 0 mm)	—	—	100	85–100	—	40–80	10–35
(*ii*)	Coarse aggregate (12.5 to 4.75 mm)	100	90–100	40–80	0–20	0–10	—	—
	(10 to 2.36 mm)	—	100	80–100	5–40	0–20	—	—
(*iii*)	Combined fine and coarse aggregate (10 mm to 0)	—	100	90–100	65–90	65–65	—	10–25

7.12 AIR ENTRAINED CONCRETE

The purposeful entrainment of air in concrete is one of the standing developments in concrete technology in recent years. This air-entrainment is accomplished either by the use of air-entraining Portlant cement or by adding directly to the concreter, at the time of mixing, some air-entraining agents. Air-entrained concrete has higher resistance to frost action, sulphate attack, the internal stresses caused by the heat of hydration, and the disruptive forces caused by frequent wetting and drying and heating and cooling. It has greater workability, plasticity and cohesiveness. It increases water-tightness, reduces bleeding and segregation and permits a reduction in the water-cement and fine aggregate content of the concrete.

7.13 PRE-STRESSED CONCRETE

Prestressed concrete is an improvement over reinforced concrete. Prestressing is the technique of introducing forces in a structural member so as to produce stresses, opposite to that which will occur during the service. Generally this force is applied by means of high tensile steel wires or bars, which are stretched and anchored by suitable means.

Prestresses concrete has many advantages over reinforced concrete, important of which are:

1. In the design of reinforced concrete members, concrete in the tension zone is considered inactive. In prestresses concrete, the whole section can be arranged to be active in compression.
2. Tensile cracks can be completely eliminated in prestressed concrete so that the members can be made more durable than reinforced concrete members.
3. The effect of dead load and shear forces can be reduced by proper prestressing. Thus the sections of the members are thin, as compared to those of R.C.C. members.
4. There is saving in the materials used.

7.13.1 Materials used

The principal materials used in prestressed concrete work are high strength concrete. Mild steel cannot be used because of losses due to shrinkage and other causes. Hence high tensile steel capable of sustaining a working stress much higher than the above losses is needed for all prestressed concrete work. Use of high strength concrete (33 to 53 N/mm^2) reduces the required area of the member, increases the modulus of elasticity of the material and allows earlier prestressing of the member.

7.13.2. Method of prestressing

There are basically two systems of prestressing namely pre-tensioning and post-tensioning. The prefixes 'pre' and 'post' denote the time at which tensioning of wires is carried out with regard to concreting the member. Thus pre-tensioning means initial stressing of the wires between external abutments and placing the concrete around wires ; the wires being released after the concrete has set and attained the specific strength. These wires are bonded to concrete and the prestress is transferred to concrete without use of external anchorage devices.

Post tensioning means stressing and anchoring the wires that pass through perforated holes in the member. The wires are tensioned by jacking against the members and the holes are generally grouted after final stress operation.

7.14 PRE-CAST CONCRETE

Pre-cast concrete, as the name implies, is manufactured before hand in factories or at the site. This is a concrete which is not directly placed in the formwork like the green concrete or in-situ concrete.

The obvious advantages are:

(*i*) Greater control can be exercised in the factory where facilities are readily available.

(*ii*) The work has not to be stopped in unsuitable weather.

(*iii*) Cost is less as large scale production is done.

(*iv*) Formwork and shuttering required for precast work is very less.

The disadvantages are:

1. Additional costs are involved in transporting the precast units from the factory to the site of work.
2. During handling and transporting, there is risk of damage and breakage of units.
3. More space is required for storage and curing of units.
4. Additional staff has to be employed for supervision, etc.

Uses of Pre-cast Concrete. Precast concrete members can be used for building construction to facilitate construction and economise in cost. Walls of hollow block masonry are advantageous as they are easy to construct, are cheaper and have great thermal insulation effect. They considerably save mortar as compared to brickwork as the joints are lesser. Internal plastering is reduced as a good surface can be obtained with one coat of material only instead of two.

These blocks are built with concrete, forms of which can be reused a number of times. Special types of forms are adopted keeping in view the types of hollows, etc. which are to be left in the blocks.

Precast concrete door and window frames can also be built. Steel bars of about 5 mm to 10 mm in diameter run through them. Suitable hard wood blocks are used for fixing hinges, etc. to them. The concrete is vibrated on a table vibrator as it is poured into the moulds, thereby forming a durable mix.

Precast concrete can be used for roofing elements also. Precast reinforced concrete battens can be used for flat roofs instead of wooden sections. They are also cast in special types of forms.

7.15 NO-FINES CONCRETE

In this type of concrete, fine aggregates are not added. The concrete consists of only cement and coarse aggregates. The nominal size of the coarse aggregate used is 20 mm. It should be ensured that not more than 5 percent pass through 10 mm sieve. The proportion generally used is 1 : 8, with a water-cement ratio of 0.4. The aggregates are made wet before use and excess of water is avoided. The advantages are:

(*i*) It has low thermal conductivity than the plain concrete.

(*ii*) It weighs less than plain concrete.

(*iii*) It is cheaper in cost than the plain concrete.

(*iv*) It does not segregate. As such formwork need not be closely jointed.

(*v*) It has little capillary action.

It is used for walls of residential buildings, particularly 2 to 3 storeys high. No-fines concrete is laid in a manner similar to the plain concrete except that greater care is taken in compaction.

7.16 VACUUM CONCRETE

All the water added to concrete is not used in the hydration of cement, but extra water has to be added for its workability, *i.e.* for laying and compaction. Thus if after placing the concrete in position, water in addition to that required for hydration is removed the water-cement ratio will be reduced and the strength of concrete correspondingly increased. To remove this water, the freshly laid concrete is subjected to partial vacuum. This is done by placing a suction mat, *i.e.* filter cloth with an impermeable and airtight covering of rubber, wood or metal at the back. The suction mat is connected to vacuum pump. Water is pumped out of concrete and partial vacuum created. The strength is increased considerably especially if the concrete is vibrated. Vacuum concrete is dense and less liable to shrinkage on drying. This concrete is used with advantage for laying concrete pavements, floor and roof slabs, deck floor of bridges, piles, etc.

7.17 WATER PROOF CONCRETE

Normal concrete is a porous material and as such it is not water tight. To make the concrete water proof, certain water repellent cements or water proofing compounds are added. These are proprietary materials. Greater attention should be paid to the concrete construction operation in this case. The water proofing agents act either as pore fillers or water repellent agents or wetting agents to help compaction by increasing workability or as catalysts to assist hydration of cement.

CONCRETE CONSTRUCTION PRACTICE

For obtaining good concrete great care has to be taken at every stage of the concrete construction procedure. The formwork, which holds the concrete till it sets has first to be erected with great care. The reinforcement has to be bent and placed as desired by the designer. The concrete, carefully proportioned has to be properly mixed, transported and placed in the forms. This has to be followed by careful compaction, finishing and curing of the concrete. Precautions have also to be taken to ensure that proper construction and expansion joints are provided in the concrete structures so that unsightly cracks are avoided. Certain important aspects of concrete construction have been dealt with below:

7.18 FORMWORK

Requirements of Formwork. Green concrete, being a plastic material, requires some kind of formwork to mould it to the required shape and also to hold it till it sets. As the cost of formwork may be as high as 25% or even more of the cost of the concrete structure, it is essential that formwork is designed properly so that it is economical and at the same time efficient.

The essential requirements of formwork are:

(*i*) It should be strong enough to take the dead and live loads imposed upon it during construction.

(*ii*) It should be sufficiently rigid so that any bulging, twisting or sagging due to the loads is minimized.

(*iii*) It should be as tight as possible so that loss of mortar and consequent honeycombing are avoided.

(*iv*) It should be well aligned. All its lines should be true and all its surfaces plane. This ensures a pleasing appearance of the structure without any extra cost of finishing.

(*v*) It should be easily removable without any damage so that its repeated use is ensured.

(*vi*) It should be designed to be as economical as possible, consistent with its other requirements.

Material for Formwork. Formwork is usually constructed of timber. However, plywood and steel forms are also in use. Soft woods are usually chosen as they are cheaper, easier to work, lighter, and do not warp so much when wet. Wooden forms are bound to absorb moisture when used, and therefore, kiln dried wood is not to be preferred as it is likely to swell and get loose shape. The selection of the kind of wood for formwork depends on the local market, but the commonest woods used are Chir, Deodar, Sal, Shisham and Kail, etc.

Where appearance of finished concrete is of little or no importance or is purposely kept rough, clean sawn timber may be used, but in view of the greater ease of fitting, the use of planned timber will often prove to be economical in both time and labour. Sometimes, when the concrete surfaces are required to be very smooth, the formwork is lined with waterproof building paper, grainless wood, plywood sheets, metal or similar material.

Stripping of formwork

All formwork should be removed without such shock or vibration as would damage the concrete. Before the bottom forms and posts are removed the concrete surface should be exposed where necessary in order to ascertain that the concrete has sufficiently hardened. Proper precautions should be taken to allow for the decrease in the rate of hardening that occurs with all cements in cold weather.

In no circumstance should forms be struck until the concrete reaches a strength of at least twice the stress to which the concrete may be subjected at the time of striking. Where possible the formwork should be left longer, as it would assist the curing. In normal circumstances, that is at temperatures above 20°C, forms may be struck as indicated in Table 7.7. For rapid hardening cement, 3/7 of the periods indicated in table will be sufficient in all cases, except for vertical sides of slabs, beams and columns which should be retained for at least 24 hours.

Table. 7.7: Stripping Time for Formwork

Part of Formwork	*Stripping time*
Vertical sides of slabs, beams and columns.	48 hours
Bottoms of slabs upto 4.5 m span	7 days
Bottoms of slabs above 4.5 m span, bottoms of beams upto 6m span, and arch rib bottoms upto 6m span	14 days
Bottoms of beams over 6 m span and arch rib bottoms above 6 m span.	21 days

7.18.1 Special types of Formwork

In addition to the usual timber formwork, steel forms, moving-forms and climbing forms are also used in concrete construction.

(*a*) *Steel Forms.* Steel forms are manufactured as proprietary articles by several firms. The forms usually comprise standard flat or curved panel units made of steel sheets reinforced with angle sections and are clipped together by keys, wedges or other similar devices. Horizontal and vertical stiffening is provided by bearers of angle sections which are clipped to make up panels. Flat panels are made in a variety of sizes but a common size is 60 mm with a range of narrower sections for use as closures.

(*b*) *Moving Forms.* Where long lengths of concrete works of constant cross section have to be constructed, moving forms may be used with advantage. In such cases a complete section of formwork is made up and carried on some suitable formwork of travelling carriage running on railway lines or other suitable tracks.

(*c*) *Climbing Forms.* Climbing forms are used for tall structures. These consist essentially of a narrow band of formwork encircling the structure which can be raised as the work proceeds. They may be made up at site or specially manufactured. In most types the height of concrete lift is restricted between 0.25 to 1 m but certain types allow for continuous pouring in order to produce joint free structure.

7.19 REINFORCEMENT

7.19.1 Concrete cover over Reinforcement

Reinforcement should have adequate concrete cover to prevent its deterioration. The thickness of such cover, exclusive of plaster or other decorative finish should be as follows:

(*a*) At each end of a reinforcing bar, not less than 25 mm, nor less than twice the diameter of such rod or bar.

(*b*) For a longitudinal reinforcing bar in a column, not less than 40 mm, nor less than the diameter of such rod or bar. In the case of columns of minimum dimensions of 200 mm or under, whose bars do not exceed 12 mm diameter, 25 mm cover may be provided.

(*c*) For a longitudinal bar in a beam, not less than 25 mm, nor less than the diameter of such rod or bar.

(*d*) For tensile, compressive, shear or other reinforcement in a slab, not less than 13 mm, nor less than the diameter of such reinforcement.

(*e*) For any other reinforcement, not less than 13 mm, nor less than the diameter of such reinforcement.

For all external work, for work against earth faces, and also for internal work where there exist particularly corrosive conditions, the cover of the concrete should be increased by 13 mm beyond the figures given above, depending upon the degree of corrosiveness.

7.19.2 Splicing and Welding of Reinforcement

As a rule a joint in a reinforcing bar should be avoided. If this is not possible, the lapping of the bars should be done at least to a distance equal to 40 times the diameter of the bar. Laps in reinforcement bars should always be staggered. In place of lapping welding of bars may be done.

7.19.3 Bar Bending Schedules

To enable reinforcement to be cut and bend accurately as well as economically bar bending schedules must be prepared and supplied by the designer. The schedule normally contains six columns as follows:

(*a*) Location of bar,

(*b*) Number of bars,

(*c*) Diameter of bars,

(*d*) Length of bars,

(*e*) Dimensioned sketch of bars, and

(*f*) Remarks.

7.20 BATCHING OF CONCRETE

After fixing the proportions of the different ingredients of concrete for a particular work the materials involved, namely cement, aggregates and water will have to be measured correctly in batches for mixing. This process is called batching. The batching may be done either by volume or by weight. When concrete is batched by volume there is always the danger of variation between one batch and another unless extra care is taken to ensure that the quantities are correct. Especially bulking of fine aggregate will have to be taken into account in volume batching as otherwise more cement will be consumed, and the mix will become under sanded and harsh.

Batching of concrete by weight is more dependable for the above reasons and should preferably be adopted for controlled concrete as well as in all important works.

For large works and for a good quality of concrete, it is essential to have weighing machines with measuring devices for weighing the concreting ingredients correctly. Many types of weigh batching plants are available. These are swinging hopper types, horizontal hopper types and simple weigh batchers. They can be completely mobile or partially mobile and the hopper may be elevated or may have bottom discharging arrangement. The capacity of the plants should be sufficient to give about 7 cu m of concrete per hour. For bigger works, fixed batching plants are used. They can be fed with bulk cement instead of cement available in sacks. They are economical only in large scale construction as they need lot of time for erection.

7.21 MIXING OF CONCRETE

7.21.1 Hand Mixing

The simplest method used for concrete mixing is by manual means. It is done only where the quantity of concrete is insufficient to warrant the installation of machinery, the working area is less or when the distance from the mixing position to the placing area is not excessive, etc. Whenever small quantity of concrete is to be mixed, good concrete can be obtained by hand mixing. This is not possible if concrete has to be mixed in greater quantities by hand mixing. The usual procedure for mixing concrete manually is to carry out this operation on a level, clean and less absorbent platform. The top of this platform must completely be covered with sheet metal or similar covering which will not flake under the action of shovels. The aggregates and cement are stored adjacent to this platform in a proper manner. Coarse aggregate is laid evenly in a layer over the platform and the fine aggregate is spread over it. The cement layer is then spread over uniformly at the top. In mixing the materials care should be taken that each shovel contains some quantity of each of these materials. A suitable method of mixing is that the whole mass is mixed thoroughly 2 to 3 times while in dry state and again 2 to 3 times when water is added.

7.21.2 Machine Mixing

Mechanical mixing is adopted these days to a great extent. The following types of mixers are usually adopted:

1. *Non-tilting types.* These are made in sizes from about 0.15 to 5 cubic metres. In most cases the drum rotates in one direction and the concrete is discharged by a hopper. The direction of rotation of the drum may be reversed to discharge the concrete.
2. *Tilting Drum Mixers.* These are generally made only in smaller sizes but a few larger machines upto 3 cu m of capacity have been made which are capable of producing about 60 cu m of concrete per hour. (Fig.7.1)
3. *Pan Mixers.* These mixers are efficient but have small capacity.
4. *Continuous Mixers.* These mixers measure the ingredients, by volume and once they have been set, their measuring is done more or less automatically. The power consumption of these mixers is small and they are capable of producing 10 to 20 cu m of concrete per hour.
5. *Truck Mixers.* These are used when the mixed concrete is to be carried to a long distance. They are used in conjunction with a central batching plant. Water is carried in a separate tank so that it can be mixed in the truck when it is near the deposition area.

The ingredients in dry state are filled into hoppers or skips attached to the mixer. Coarse aggregates are placed first at the bottom followed by fine aggregates and cement. These materials get admitted into the drums in the reverse order. No material should be put in the skip unless they have been mixed and used immediately. The materials are loaded into the skip after being correctly measured.

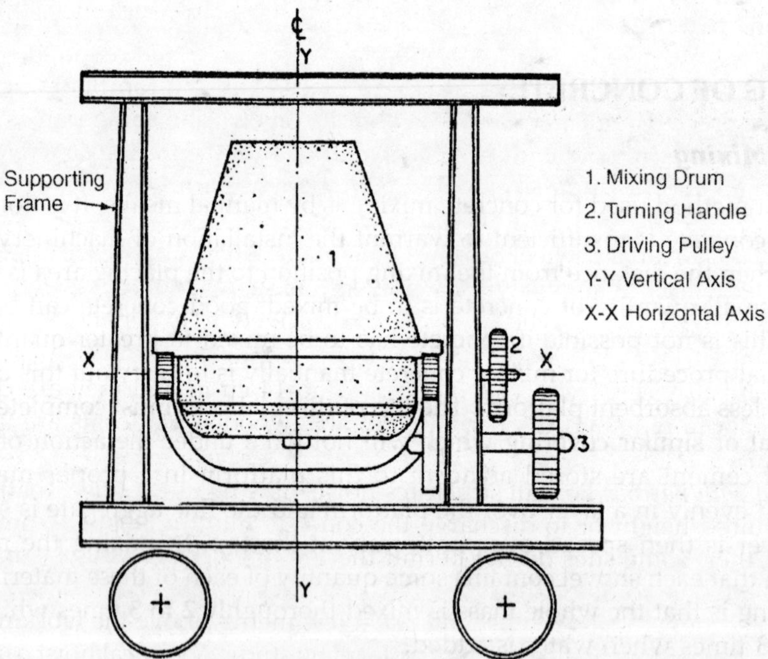

Fig. 7.1. Tilting drum mixer

After all the materials have been put in the drum mixing should continue until the mixture is uniform. The strength of concrete is affected by the duration of the mixing process. About one minute is necessary to give good strength and mixing upto 1½ minutes also improves the quality of concrete. There is no advantage in mixing beyond two minutes. The time between charging of the materials into the drums and discharging of the concrete is not less than 1½ minutes. Some machines have a device for discharging the mixed materials automatically after they have been mixed for a sufficient time. Sometimes it is specified that the materials should be mixed in a dry state before adding water. This may not be practicable .with large batching and mixing plants. When the mixing plant is to be closed for more than one hour it should be cleaned.

7.22 TRANSPORTING CONCRETE

After mixing the concrete, it has to be delivered to the final position which it has to take in a building unit. The need of transporting concrete at the earliest possible time is essential because concrete should be placed and compacted before the initial set of cement takes place. While transporting, it should be ensured that .the containers are water-tight and that the loss of water and cement is avoided to a great extent. Segregation of the particles must be avoided at all costs as this would result in bad concrete. If it is unavoidable during transit, concrete should be remixed before being placed.

Good concrete may get spoiled by mishandling during transportation from the mixer to the point of incorporation. This is more true for areas where greater lead of concrete is essential.

The various methods adopted for transporting concrete are:

1. *Use of Iron pans which are carried by Men.* These pans have small capacity and are usually of such a size that one man can lift and walk easily over the required distance. For raising concrete over some height so as to place it on the top floors of a building, ladders are erected and men stand at various levels. The same pan is transferred from one person to the other quickly. This method of transporting concrete is adopted only for small and medium size buildings in India as it works out to be the cheapest method.
2. *Wheel Barrows.* Ordinary wheel barrows hold about 0.05 cu m of concrete. The wheels may be fitted with pneumatic tyres to prevent the segregation of concrete on account of the series of jerks which the wheel barrow is likely to experience while travelling over uneven surface.
3. *Concrete Carts.* These vary in capacity from 7 to 14 cu m and they are designed for easy pushing, tipping and discharging. Suitable wooden platforms are laid for their easy travel.
4. *Hoists used with Barrow.* For tall structures hoists may be used either to lift the barrow to the required height or to discharge the concrete into the hopper from which it is carried further. Some sites do not permit the use of external hoist, *e.g.* in the case of steel buildings. In confined areas the hoist must be erected within the building and built up in sections. The top most will rise a few metres above the working level. The guide rails are held up against each floor level by studs. A vertical hoist can frequently be arranged in multi-storeyed buildings within the space allowed for stairs or lift wells. Otherwise temporary openings have to be left in the floor panels to allow for the hoists. Cranes may be used with hoist mainly to lift concrete from the site of the mixer to the building where the concrete is to be placed.

5. *Lorries.* Lorries are used whenever the distance is great. Such a case occurs when concrete is to be deposited at a place where it cannot be mixed due to non-availability of space. Central mixing plant placed at some distance from the site of work is erected. Special lorries suitably designed are used for transporting concrete. The concrete is covered to protect it from the sun, wind or rain during transport. The time to transport should not exceed more than 20 minutes when mixed concrete is to be carried.
6. *Chutes.* When concrete is to be placed below general ground level, the mixer can be placed on the upper level and concrete discharged to the lower level through a small chute of corrugated iron or timber, leading to barrows, carts or other transport facility or directly to the final position where it has to be deposited.
7. *Pumping of Concrete.* Pumping of concrete is economical and successful where the size of the aggregate does not exceed 50 mm and the concrete is of medium workability. The control of aggregate grading and the batching of mix have to be done effectively. The advantages of this type of work are that no mode for transporting concrete within the working radius of about 600 m from the pump is created. About 2 to 25 cu m of concrete per hour can be delivered. The system is very flexible as the pipe lines can be easily changed. Handling of concrete between the mixer and point of incorporation is avoided which makes a better quality of concrete. The plant consists of a pipe line and a pump. The type of pump will vary slightly with the consistency of the concrete and length of the pipe line. For a continuous discharge, it is essential for the mixing plant to be of an adequate capacity. These pumps are operated by electric power or other means. The pipes are of about 150 mm internal diameter for large capacity pumps and about 120 mm diameter for smaller machines. Concrete can be delivered up to a height of about 40 m vertically or 600 m horizontally. The power required to operate a pump at full range capacity is about 35 h.p. This type of transportation is rarely resorted to for building work.

7.23 PLACING CONCRETE

The concrete should be placed in its final position before the cement reaches its initial set and once placed, it should not be disturbed. The following precautions should be taken while placing concrete:

(*i*) Before the concrete is actually placed in position, the inside of the forms should be inspected to see that the forms are water-tight and have been cleaned and oiled. The facing on the formwork likely to come into contact with the concrete should have been well wetted to ensure that moisture is not removed from the concrete mix by the formwork. It should also be ensured that the reinforcement has been properly fixed and there is no possibility of its getting displaced during concreting.

(*ii*) Concrete should be deposited as nearly as practicable in its final position to avoid segregation due to rehandling or flowing. The concreting should be carried on at such a rate that the concrete is at all times plastic and flows readily into the spaces between the bars.

(*iii*) In all types of work, concrete should be placed towards the face of that already placed since this method of placing will assist compaction and reduce the risk of segregation.

(*iv*) Concrete should be placed in even layers, each of which should be compacted before the next one is placed and before the previous one has set. As a general rule, layers should be from 150 mm to 250 mm thick in the case of reinforced work and upto 450 mm thick in the case of vibrated mass concrete.

(*v*) When concreting is once started, it should be carried on as a continuous operation until the placing of the panel or section is completed. The top surfaces should generally be level. When construction joints are necessary, they should be properly made.

(*vi*) When placing concrete under water, it is essential that the water be still and not flowing. The concrete should be dumped as a batch in its final position, so as to avoid leaching and the segregation that would occur if it were allowed to fall freely through several metres of water.

7.23.1 Placing Concrete Under Water

The methods to be adopted for placing concrete under water can be one of the following:

(*a*) Tremie: The concrete is placed through vertical pipes, the lower end of which is always inserted sufficiently deep into the concrete which has been placed previously but has not set. The concrete emerging from the pipe pushes the material that has already been placed to the sides and upward and thus does not come into direct contact with water.

(*b*) Direct placement with pumps: This is similar to tremie, but the vertical end piece of the pipe line should not move to the side during pumping.

(*c*) Drop bottom bucket: The top of the bucket is covered with a canvas flap. The bottom doors open freely downward and outward when trapped. The bottom doors are opened only when bucket rests on the surface upon which concrete is to be deposited.

(*d*) Bags: Bags, of atleast 0.028 m^3 capacity, of jute or other course cloth are filled almost two-third full of concrete. The spare end is turned under. The bags are placed in stretcher and header course so that the whole mass is interlocked.

(*e*) Gronting: A series of round cages made from 50 mm mesh of 6 mm steel and extending over the full height to be concreted are prepared and laid vertically over the area to be concreted so that the distance between centres of the cages and also to the faces of the concrete should not exceed one metre. Stone aggregates of not less than 50 mm and not more than 200 mm size are deposited out side the steel cages over the full area and height. A stable 1 : 2 cement-sand grout with a water cement ratio of not loss then 0.6 and not more than 0.8 is sent down under pressure of about 0.2 N/mm^2.

7.24 COMPACTION OF CONCRETE

As concrete is being placed it should be compacted thoroughly and uniformly by means of hand tools, vibrators or finishing machines, to ensure a dense structure without any air bubbles, close bond with reinforcement and smooth surfaces. The concrete should be worked well around all reinforcement, embedded fixtures, and into the corners of forms. It should, however, not be worked more than necessary as any disturbance of the concrete in place tends to cause it to segregate, the water and the fine particles moving towards the surface. Compaction may be done either by hand or by mechanical vibration.

Hand Compaction. Ordinary hand methods of consolidating concrete consist of ramming, tamping and spading with suitable tools. Spading is used at or near vertical form faces in order to secure a smooth surface, but overworking will draw out an excess of cement paste which will later craze, crack or scale. For compacting; dry concrete the surface is rammed with a heavy flat-bottomed rammer until a thin film of mortar or paste appears at the surface, showing that the voids of the aggregate have been filled.

Mechanical Vibration. Mechanical vibration as a means of consolidating concrete has many advantages over hand compaction methods. It permits the use of stiff mixes of high strength. The action of vibration is to set the particles of fresh concrete into motion, reducing friction between them, and affecting a temporary liquefaction of the concrete which enables it to settle easily into place. Mechanical vibration of concrete is carried out using vibrators. The intensity and duration of vibration should be just sufficient to cause complete settlement and compaction without any stratification of the successive layers or separation of ingredients. Knowing when the concrete is fully vibrated is a matter of experience. There are, however, a number of indications that help to know this:

(*a*) Air bubbles often come to the surface as the concrete is being vibrated; when these stop, it is generally a sign that compaction is completed.

(*b*) The sound is another indication. On complete compaction the tone becomes constant, especially with external vibrators.

(*c*) On nearing complete compaction the entire batch gives a uniform appearance and the surface starts to glisten. Minute film of cement paste becomes discernible between the concrete and the forms and around the reinforcement.

Over vibration causing segregation, unnecessary bleeding or formation of laitance should be avoided.

7.25 CONSTRUCTION JOINTS IN CONCRETE

As it is not generally possible to place concrete continuously, construction joints are formed following temporary stoppages of the cencreting. Such joints are a possible source of weakness and should be formed with great care. In addition to construction joints, expansion joints have also to be provided in concrete structures to allow for the expansion and contraction of the concrete due to temperature changes but they are beyond the scope of this book. Therefore, only construction joints are dealt with.

Position of Construction Joints. As a general rule an arrangement of construction joints should be settled before concreting begins. The horizontal joint should be arranged with levels to blend with the general architectural appearance. The vertical joints should be situated at positions where the length of the joint is shortest or where a symmetrical pattern can be arranged. In columns, beams, slabs and walls the following general rules should be observed in positioning construction joints:

(*a*) Columns should be filled to a level, preferably a few mm below the junction of a beam or alternatively to the bottom of any haunching.

(*b*) Joints in beams and slabs should be at the points of minimum shear. Thus for beams, the joints should be at the centre of span or within the middle third. Slabs spanning in one direction and of short span should have the joints at middle span if they are at right angles to the direction of spanning. Joints in the direction in which the slabs are designed to span should be within the middle third. Joint in slabs spanning in two directions should be within the middle third of either span.

(*c*) If for any reason a joint has to be made between the slab and the beam, it is then necessary to provide some form of key and to add shear reinforcement to meet any weakness at the joint.

(*d*) Horizontal joints in walls are usually made at such positions as the top of a plinth or the top or bottom of a window opening.

Forming construction joints. All vertical joints should be formed against a stop board as raking joints often result in honeycombing. All horizontal joints should be level. Battens may be nailed to form work to ensure a horizontal line and if desired may be used to form a groove joint. If the thickness of concrete permits, it is desirable to provide a key of the tongue and groove type at all construction joints. The width of the groove should be about one-third the thickness of the concrete. For tank wall and similar work it is preferable to form all joints with a strip of sheet metal about 150 mm wide inserted half-way into the concrete at the top of the lift to form a continuous water barrier.

When resuming work at construction joints on a concrete surface which has hardened, such surface should be roughened. It should then bs swept clean, thoroughly wetted, and covered with a 15 mm layer of mortar composed of cement and sand in the same ratio as the cement and sand in the concrete mix. This layer should be freshly mixed and placed immediately before placing the fresh concrete. Where the work has to be resumed on a concrete surface which has not fully hardened, all laitance should be removed by scrubbing. The surface should then be thoroughly wetted and all free water should be removed. The surface should then be coated with neat cement grout. The first layer of concrete to be placed against this surface should not exceed 150 mm in thickness and should be well rammed against the old work.

7.26 CURING OF CONCRETE

Curing is the process of preventing the loss of moisture from the concrete while maintaining a satisfactory temperature region.

The chemical action which accompanies the setting of concrete is dependent on the presence of water. Although there is normally an adequate quantity for full hydration at the time of mixing, it is necessary to ensure that the water is either retained or replenished to enable the chemical action to continue until the concrete is fully hardened. Methods of curing are therefore designed to maintain the concrete in a continuously moist condition over a period of several days or even weeks either by preventing evaporation and the absorption of water by the forms or subgrade or by repeatedly wetting the surface.

Methods of Curing. Concrete can be cured either by water or sealing compounds. Water curing produces the best concrete. It makes the concrete stronger, more impermeable ard more resistant to abrasion. It is also, the most usual method of curing and is effected, either

(*a*) by spraying the concrete with water, or

(*b*) by covering the concrete with a layer of sacking, canvas,. hessian or similar absorbent material which is kept constantly wet.

In the case of the floor slabs and paving, the surface may alternatively be

(*a*) covered with clean wet sand.

(*b*) covered with waterproof paper, which will keep the moisture in the concrete if it is laid in close contact with the surface, or

(c) flooded with water, this is known as ponding, and is done by building small earth or clay bunds around the area to be cured.

The use of sealing compounds and protective coverings for curing is a more specialised method. These compounds are generally proprietary and instructions for application are issued by the manufacturers.

A process of steam curing is widely used these days to cure precast pipes and other units. It accelerates the rate of gain of strength, reduces the time required for curing and enables early removal of formwork, thereby increasing the rate of production with a resulting saving in cost.

Period of Curing. Curing should begin as soon as the concrete is sufficiently hard by covering it or by the application of a water spray. Concrete made with ordinary or blast furnace slag cement requires moist curing for at least seven days under normal temperature conditions. Slow hardening cements such as low heat Portland cement require 14 days or more of moist curing. Concrete made with high alumina cement develops its strength more rapidly, generating a good deal of heat in the setting and hardening process, and this water evaporates more easily from it than from Portland cement concrete. It is therefore absolutely essential to keep it wet on all faces for twenty four hours from the time it begins to harden. It must be kept completely saturated for the first eight hours and thoroughly wet for further sixteen hours.

PROPERTIES OF CONCRETE

7.27 STRENGTH OF CONCRETE

The strength of concrete is usually measured by its resistance to axial compression as this type of force is always present in all kinds of concrete structures. The tensile and shear strengths of concrete are proportional to its compressive strength which, therefore, forms a rational basis for purposes of comparison. The tensile strength of concrete is about 8 to 12% of its compressive strength. Its shear strength, measured as inclined tension, is about 8 tc 10% of its compressive strength.

7.27.1 Factors Affecting the Strength of Concrete

The strength, of concrete varies considerably due to variations in the following factors:

(1) *Quality of Material and Grading of the Aggregates.* The cement and aggregates used have a great influence on the strength properties. The cement used should not have deteriorated by storage or by contact with moisture. The surface texture of the aggregates and their grading also deserve consideration. Unsound and badly graded aggregates cause large variations in the strength of concrete.

(2) *Water/Cement Ratio.* This is the ratio of water to cement by weight. It is the cluief factor affecting the strength of concrete. Other factors being constant, the maximum strength is obtained for a particular water cement ratio. An increase or decrease ot the ratio reduces the strength.

(3) *Cement Content.* The higher the cement content of concrete the greater is its strength provided the water-cement ratio is kept constant. However, a poor quality of aggregate or an increase in the water-cement ratio eliminates the advantages of a higher cement content.

(4) *Age of Concrete.* Concrete gains strength with age, the rate of gain decreasing with time. The compressive strength after 3 days is about 40% of that after 28 days, and the strength after 7 days is about 65% of that after 28 days.

(5) *Methods of Mixing, Placing, Compacting, and Curing.* Adequate mixing, careful placing and proper consolidation are necessary to increase the strength of concrete. Curing is a very important factor in increasing the strength of concrete as the strength depends on the degree of completion of the hydration process. Inadequate curing is a common defect in concrete construction in India. Besides, the temperature during placing and curing also affects the strength. The higher it is, the greater the strength. With temperatures below 5°C, strength is lowered unless special precautions are taken during the placing and curing of concrete.

7.27.2 Strength Requirement And Permissible Stresses

The groups and strength requirement for different grades of concrete are given in Table 7.8

Table 7.8: Proportions and Strength Requirements for different grades of Concrete

Group	*Grade Designation*	*Specified Characteristics Compressive Strength of 150 mm Cube at 28 days in N/mm²*
Ordinary concrete	M 10	10
	M 15	15
	M 20	20
Standard concrete	M 25	25
	M 30	30
	M 35	35
	M 40	40
	M 45	45
	M 50	50
	M 55	55
High strength concrete	M 60	60
	M 65	65
	M 70	70
	M 75	75
	M 80	80

7.28 WORKABILITY OF CONCRETE

Workability of concrete, as generally used, is a relative term which indicates the ease with which concrete can be placed in the forms and compacted. The workability of concrete is normally measured by the slump test. The degrees of workability for various requirements are given in Table 7.17 under article 7.34.3.3 The workability of concrete is influenced by the water content, the size, shape and grading of the aggregates; the ratio of coarse and fine aggregates, and by the use of any admixtures.

7.29 DURABILITY OF CONCRETE

Durability of concrete is its capacity to resist the forces of disintegration owing to natural causes such as weathering, action of water containing chemicals, temperature changes including freezing and thawing, alterations in moisture content and chemical changes. Durability of concrete is governed by the quality of the materials used and their proportioning and the methods adopted for mixing, placing, compacting and curing of concrete. Sound cement, durable aggregates, an optimum water-cement ratio, a dense grading of aggregates, careful mixing and optimum placing, thorough compaction, and adequate curing increase durability.

7.30 ELASTICITY OF CONCRETE

Concrete is not a truly elastic material as the stress in concrete is not directly proportional to the strain. Therefore the modulus of elasticity of concrete is not constant for all stresses. It is observed from loading tests that although initially the stresses are proportional to the strains, as the stresses increase, and long before failure, the strains increase disproportionately. The longer the period over which the loading tests are spread, the greater the departure from the linear relationship of stress and strain. This is associated with creep, a phenomenon discussed in article 7.31.

The modulus of elasticity of concrete is influenced by the strength, age and moisture content of the concrete and the typs of aggregate used. For design purposes the modulus of elasticity of concrete of a given strength in assumed to be constant.

7.31 CREEP OF CONCRETE

The time dependent extra strain which concrete undergoes in addition to the instantaneous strain caused by the instantaneous application of a stress is known as creep. This creep is permanent and does not disappear on removal of the stress. The amount of creep is influenced by the magnitude and duration of the stress, the age of concrete at the application of the stress, the quantity of mixing water, the strength of the concrete, and the properties of the aggregates and the cement used. It has been found that the strain due to creep is about three times the strain that occurs when the load is first applied. At normal working stresses creep is roughly proportional to the applied stresses. The effect of creep is to produce a large though gradual redistribution of stress between the steel and the concrete in a reinforced concrete structure. This tends to relieve the concrete stress and add to the steel stresses.

7.32 SHRINKAGE OF CONCRETE

When concrete sets and hardens in air, it shrinks partly due to the chemical action of the colloids produced by the water reacting on the cement, and partly due to the physical drying out of the mass. The former action is irreversible, but the latter is reversible. The amount of shrinkage strain varies with many factors and may range from 0.02 to 0.10. Other things being equal, shrinkage of concrete is almost directly proportional to the amount of water employed in the mix. Hence if minimum shrinkage is desired the water-cement ratio and the proportion of cement paste should be kept to a minimum. Also shrinkage is reduced by the use of harder and denser aggregates of low absorption and high modulus of elasticity, and cements high in tricalcium silicate and low in alkalies.

The advantage of shrinkage of concrete is that in shrinking, the concrete grips the reinforcement tightly, thus preventing slipping. This shrinkage is the main cause of the bond between the concrete and the reinforcement. But for it, bond would depend solely on some kind of mechanical keying devices. On the other hand shrinkage causes the concrete to crack unless the concrete structure is free to move. It also causes variation in the strains between relatively moist and dry concrete.

7.33 FIRE RESISTANCE OF CONCRETE

Concrete is a good insulator and has a fairly good fire resistance owing to the presence of pores and also water which is present in it in a chemically combined state. This water is removed only when the temperature exceeds about 300°C. However, when the exposed concrete is dehydrated at this temperature, it attains greater insulation capacity and high resistance to heat penetration. The fire resistance of concrete depends mostly on the quality of the aggregate used. Broken brick aggregate makes the concrete more fire-resistant than broken stone aggregate. Broken stone aggregate is better than gravel.

However, the steel in reinforced concrete, being a very good conductor of heat, loses its strength rapidly with rise in temperature. The steel reaches the danger point at about 250°C, and its strength is reduced to half at about 400°C. Therefore to protect the steel an adequate concrete cover should be given to it. This cover generally varies from 13 mm in slabs to 40 mm in columns.

DESIGN OF CONCRETE MIXES

The various methods of proportioning concrete are described below.

The mix proportions are selected to ensure that the workability of fresh concrete is suitable for handling and placing so that after compaction it surrounds all reinforcements and completely fills the formwork. When concrete is hardened, it should have the required strength, durability and surface finish.

In specifying a particular grade of concrete, the following information is normally included:

(*i*) Type of mix *i.e.*, design mix concrete or nominal mix concrete
(*ii*) Grade designation
(*iii*) Type of cement
(*iv*) Maximum nominal size of aggregate
(*v*) Minimum cement content (for design mix concrete)
(*vi*) Maximum water-cement ratio
(*vii*) Workability;
(*viii*) Mix proportion (for nominal mix concrete)
(*ix*) Exposure conditions
(*x*) Maximum temperature of concrete at time of placing
(*xi*) Method of placing

(*xii*) Degree of supervision

In appropriate circumstances, the following additional information may be specified.

(*i*) Type of aggregate

(*ii*) Maximum cement content and

(*iii*) Whether an admixture shall or shall not be used and the type of admixture and the çondition of use.

7.34 MIX DESIGN METHODS

The determination of the proportions of ingregients such as cement, aggregates and water etc. to attain the required strength can be made as follows:

(*a*) By adopting nominal concrete mix, such concrete mix is called 'Nominal mix concrete'.

(*b*) By designing the concrete mix, such concrete mix is called 'Design mix-concrete'.

(*c*) By other methods such as Minimum Voids method and Fineness Modulus method.

These methods are explained below:

7.34.1 Nominal Mix Concrete

Nominal mix concrete is used for concrete of M 20 or lower grade. The proportions of materials in nominal mix concrete are given in Table 7.9.

Table 7.9: Prroportions for nominal mix concrete

Grade of Concrete	*Total Quantity of Dry Aggregate by Mass per 50 kg of Cement.To be Taken as the Sum of the Individual Masses of Fine and Coarse Aggregates. Max*	*Proportion of Fine Aggregate to Coarse Aggregate (by Mass)*	*Quantity of Concrete Water per 50 kg of Cement Max.*
(1)	(2) kg	(3)	(4) litres
M 5	800	Generally 1 : 2	60
M 7.5	625	but subject to	45
M 10	480	an upper limit	34
M 15	350	of 1 : $1\frac{1}{2}$ and a	32
M 20	250	lower limit of	30
		1 : $2\frac{1}{2}$	

Note : The proportions of the fine to coarse aggregates should be adjusted from upper limit to lower limit progressively as the grading of the fine aggregates becomes finer and the maximum size of coarse aggregate becomes larger. Graded coarse aggregate shall be used.

Example. For an average grading of one aggregate the proportions shall be 1 : $1\frac{1}{2}$, 1 : 2 and 1 : $2\frac{1}{2}$ for maximum size of aggregates 10 mm, 20 mm and 40 mm respectively.

The cement content of the mix specified in Table 7.9 for any nominal mix should be proportionately increased if the quantity of water in a mix has to be increased to overcome the difficulties of placement and compaction, so that the water-cement ratio as specified is not exceeded.

7.34.2 Design Mix Concrete

The concrete mix propertioning-Guidelines are covered in IS : 10262-2009. The mix should be designed to produce the grade of concrete having the required workability and a characteristic compressive strength not less than the appropriate values.

Concrete should be of satisfactory quality in both the fresh and handended states. The choice of proportions will be governed by both these conditions. If the plastic concrete is not satisfactory, it cannot be properly compacted and its structural value is gratly reduced. Therefore the mix design is generally carried out for a particular compressive strength of concrete, coupled with adequate workability so that fresh concrete can be properly placed and compacted.

The basic assumption made in the mix design is that the compressive strength of workable concrete is, by and large, governed by the water-cement ratio. Thus another convenient relationship applicable is that for a given type, shape, size and grading of aggregates, the amount of water determines its workability.

However there are various other factors which affect the properties of concrete, for example, the quality and quantity of cement, water and aggregates, batching, transportation, placing compaction,curing etc.

7.34.2.1 Data for mix proportioning

The following basic data are required for design of a concrete mix:

(*i*) Characteristic compressive strength of concrete at 2 days (f_{ck})

(*ii*) Degree of workability desired

(*iii*) Limitations on the water/cement ratio and the minimum cement content to ensure adequate durability.

(*iv*) Type and maximum size of aggregate to be used; and

(*v*) Standard deviation (*s*) of compressive strength of concrete.

7.34.2.2 Target strength for mix proportioning

The concrete mix should be designed for a somewhat higher target mean compressive strength ($\bar{f}_{ck}$) so that not more than the specified proportion of test results fall below the characteristic strength (f_{ck}). The margin over the characteristic strength depends upon the quality control (expressed as standard deviation) and the accepted proportion of results of strength tests below the characteristic strength ($\bar{f}_{ck}$), expressed by

$$(\bar{f}_{ck}) = f_{ck} + 1.65 \cdot s \qquad ...(7.1)$$

where $(\bar{f}_{ck})$ = target mean compressive strength at 28 days in N/mm^2

f_{ck} = characterstic compressive stress at 28 days – N/mm²
s = standard deviation, as per Table 7.10 for the desired grade of concrete in N/mm²

Table 7.10: Assumed standard deviation

Grade of Concrete	Standard deviation
M 10, M 15	3.5
M 20, M 25	4.0
M 30, M 35, M 40, M 45, M 50	5.0

7.34.2.3 Selection of Mix Proportions

(*i*) Water-cement ratio: Since different cements and aggregates of different maximum size, grading, surface texture, shape and other characteristics may produce concretes of different pressive strength for the same free water-cement ratio, the relationship between strength and free water-cement ratio should be established for the materials actually to be used. In case such data is not available, the preliminary free water-cement ratio, by mass, corresponding to the target strength may be selected from Fig. 7.2 As an alternative, the water-cement ratio may be determined from the curves shown in Fig. 7.3 corresponding to the 28 days cement strength to be used for the purpose. The free water-cement ratio selected according to the above two methods is then checked against the limiting water-cement ratio for the requirements of durability and the lower of the two values adopted.

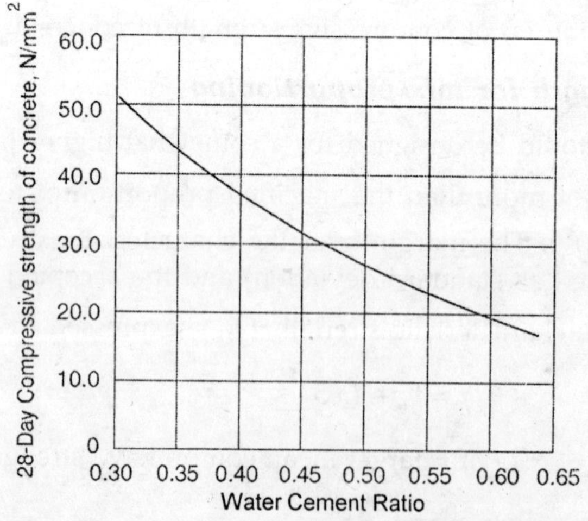

Fig. 7.2. Compressive strength of concrete versus water-cement ratio

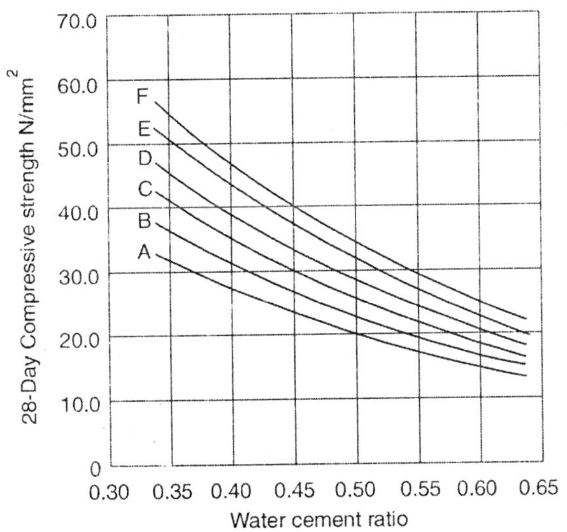

28 Day strength of cement, Tested According to IS: 4031-1988

A = 31.9-36.8 N/mm² (325-375 kg/cm²)
B = 36.8-41.7 N/mm² (375-425 kg/cm²)
C = 41.7.46 N/mm² (425-475 kg/cm²)
D = 46.6-51.5 N/mm² (475.525-kg/cm²)
E = 51.5-56.4 N/mm² (525-575 kg/cm²)
F = 56.4-61.3 N/mm² (575-625 kg/cm²)

Fig. 7.3. Water-cement ratio Vs compressive strength of cement

Table 7.11: Maximum water cement ratio and minimum grade of concrete for different exposure for 20 mm aggregate

Sr.	*Exposure*	*Plain concrete*			*Reinforced concrete*		
		Min cement content kg/m³	*Max free water cement ratio*	*Minimum grade of concrete*	*Minimum cement content*	*Maximum free water cement ratio*	*Minimum grade of concrete*
(1)	(2)	(3)	(4)	(5)	(6)	(7)	(8)
(*i*)	Mild	220	0.60	—	300	0.55	M 20
(*ii*)	Moderate	240	0.60	M 15	300	0.50	M 25
(*iii*)	Severe	250	0.50	M 20	320	0.45	M 30
(*iv*)	Very severe	260	0.45	M 20	340	0.45	M 35
(*v*)	Extreme	280	0.40	M 25	360	0.40	M 40

(*ii*) Estimation of air content. Approximate amount of entrapped air to be expected in normal (non air-entained) concrete is given in Table 7.12.

Table 7.12: Entrapped air

Nominal maximum size of aggregate in mm	*Entrapped air, as percentage of volume of concrete*
10	3.0
20	2.0
40	1.0

***(iii)* Selection of water content and fine aggregate to total aggregate ratio.** For the desired workability the quantity of mixing water per unit volume of concrete and the ratio of fine aggregate to total aggregate by volume are estimated from Table 7.13 and Table 7.14 for the nominal maximum size and type of aggregates.

Table 7.13: Sand and water content/m³ of concrete for grades upto M 35

Nominal maximum size of aggregate mm	*Water content per m³ of concrete kg*	*Sand as percentage of total aggregate by absolute volume*
10	208	40
20	186	35
40	165	30

Table 7.14: Sand and water content per m³ of concrete for grades above M 35.

Nominal maximum size of aggregate mm	*Water content per m³ of concrete kg*	*Sand as percentage of of total aggregate by absolute volume*
10	200	28
20	180	25

The above tables are valid for crushed (angular) coarse aggregates, fine aggregates consisting of natural sand conforming to grading zone II of Table 4 of IS: 383-1970 and workability of concrete corresponding to compacting factor of 0.8. Table 7.13 is based on water cement ratio of 0.6 and Table 7.14 for water-cement ratio of 0.35 (by mass). For other conditions of workability, water-cement ratio, grading of fine aggregate, and for rounded aggregates certain adjustments in the quantity of mixing water and fine to total aggregate ratio should be made, as given in Table 7.15

Table 7.15: Adjustment of values in water content and sand percentage for other conditions.

Change in conditions stipulated	*Adjustment required in*	
	Water content	*Percent sand in total aggregate*
For sand conforming to grading Zone I, Zone III or Zone IV of Table 4 of IS : 383-1970	0	+ 1.5 per cent for zone I – 1.5 per cent for zone III – 3.0 per cent for zone IV
Increase or decrease in the value of compacting factor by 0.1	± 3 per cent	0
Each 0.05 increase or decrease in free water-cement ratio	0	± 1 per cent
For rounded aggregate	– 15 kg/m³	–7 per cent

(iv) Calculation of cement content: The cement content per unit volume of concrete may be calculated from the free water-cement ratio and the quantity of water per unit volume of concrete. The cement content so calculated should be checked against the minimum cement content for the requirements of durability and the greater of the two values adopted as given in Table 7.11.

The minimum cement content given in Table 7.11 corresponds to nominal maximum size of aggregate 20 mm. For other sizes of aggregates, the adjustments are made as given in Table 7.16

Table 7.16: Adjustments to Minimum cement contents for Aggregates other than 20 mm Nominal Maximum size

Sl.No.	Nominal Maximum Aggregate size mm	Adjustments to Minimum Cement contents given in Table. 7.15 (Kg/m³)
(i)	10	+ 40
(ii)	20	0
(iii)	40	- 30

(v) Calculation of aggregate content: The total aggregate content per unit volume concrete can be calculated by the following relations:

$$V = \left[W + \frac{C}{S_c} + \frac{1}{p}\frac{f_a}{S_{fa}}\right] \times \frac{1}{1000} \qquad ...(7.2)$$

and

$$V = \left[W + \frac{C}{S_c} + \frac{1}{1-p}\frac{C_a}{S_{ca}}\right] \times \frac{1}{1000} \qquad ...(7.3)$$

where

V = absolute volume of fresh concrete which equals gross volume (m³) minus the volume of entrapped air

W = mass of water (kg) per m³ concrete

C = mass of cement (kg) per m³ of concrete

S_c = Specific gravity of cement

p = ratio of fine aggregate to total aggregate by absolute volume

f_a, C_a = total masses of fine aggregate and coarse aggregate (kg) per m³ of concrete respectively; and

S_{fa}, S_{ca} = Specific gravities of saturated surface dry fine aggregate and coarse aggregate respectively.

(vi) Combination of different coarse aggregate fractions. Coarse aggregates of different sizes should be combined in suitable proportions so as to result in an overall grading conforming to IS : 383–1970, for the particular nominal maximum size of aggregate.

***(vii)* Calculations of batch masses:** The masses of the various ingredients for concrete of a particular batch size is then calculated.

***(viii)* Trial mixes:** The calculated mix proportions are then checked by means of trial batches. Workability of the trial mix is measured and carefully observed for freedom from segregation and bleeding and its finishing properties. Other trial mixes are prepared by adjusting water content or varying the free water-cement ratio by ± 10 per cent. The mix coresponding to the desired characteristics is then selected.

Problem 7.1

Design a concrete mix of *M* 30 grade from the following data

1. Design conditions

(*a*) Characteristics compressive strength required in the field at 28 days	30 N/mm^2
(*b*) Maximum size of aggregate	20 mm (angular)
(*c*) Degree of workability	0.70 compacting factor
(*d*) Degree of quality control	Good
(*e*) Type of exposure	Severe

2. Test data for material

(*a*) Cement used:	Ordinary Porland cement
(*b*) Specific gravity of cement	3.20
(c) Specific gravity	
(*i*) Coarse aggregate	2.50
(*ii*) Fine aggregate	2.50
(*d*) Water absorption	
(*i*) Coarse aggregate	0.5 per cent
(*ii*) Fine aggregate	1.0 per cent

(*i*) Coarse aggregate

IS Sieve Size mm	*Coarse aggregate Fraction (Per cent passing)*		*Percentage of different fractions*			*Remarks*
	I	*II*	*I 50 percent*	*II 50 per cent*	*Combined 100 per cent*	
20	100	100	50	50	100	Conforming
10	0	57	0	28.5	28.5	to Table 2
4.75		7.5		3.75	3.75	of IS:
2.36		0		0	0	383-1970

(ii) Fine aggregate

IS Sieve Size	*Per cent passing*	*Remarks*
4.75 mm	100	Conforming to grading
2.36 mm	100	zone III of Table 4 of
1.18 mm	93	IS : 383-1970
600 micron	60	
300 micron	12	
150 micron	2	

Solution:

(i) Target strength for mix proportioning

Applying equation 7.1 and using Table 7.10 for standard daviation, the target mean strength for the specified cube strength of 30 N/mm^2

$$= 30 + 5 \times 1.65 = 38.25 \text{ N/mm}^2$$

(ii) Selection of water-cement ratio

From Fig. 7.2 the free water-cement ratio required for the target mean strength 38.25 N/mm^2 is 0.375. This is lower than the maximum value of 0.45 prescribed for exposure for M 30 reinforced concrete as per Table 7.11.

(iii) Selection of water and sand content

From table 7.13 for 20 mm nominal maximum size aggregate and sand conforming to grading zone II, water content per m^3 of concrete = 186 kg and sand content as percentage of total aggregate by absolute volume = 35 per cent.

For change in values in water-cement ratio, compacting factor and sand conforming to zone III the following adjustment is required in accordance with Table 7.15

Change in conditions	*Adjustments required in*	
	Water content per cent	*Percentage sand in total aggregate*
For decrease in water-cement ratio by (0.6 – 0.375) *i.e.*, 0.225	0	–4.5
For decrease in compacing factor (0.8 – 0.7) *i.e.*, 0.10	–3	0
For sand conforming to zone III	0	–1.5
Total	–3%	–6%

Therefore, required sand content as percentage of total aggregate by absolute volume

$$= 35 - 6 = 29 \text{ per cent}$$

$$\text{Required water content} = 186 - \frac{186 \times 3}{100} = 186 - 5.58 = 180.42\ l/\text{m}^3$$

***(iv)* Determination of cement content**

Water-cement ratio = 0.375

Water = 180.42 litres

$$\text{Cement} = \frac{180.42}{0.375}$$

$$= 481.12 \text{ kg/m}^3$$

This cement content is adequate for severe exposure condition, according Table 7.11.

***(v)* Determination of coarse and fine aggregate content**

From table 7.12 for the specified maximum size of aggregate of 20 mm, the amount of entrapped air in the wet concrete is 2 per cent. Then applying equations 7.2 and 7.3

$$0.98 \text{ m}^3 = \left(180.42 + \frac{481.12}{3.20} + \frac{1}{0.29} \cdot \frac{f}{2.50}\right) \times \frac{1}{1000}$$

and
$$0.98 \text{ m}^3 = \left(180.42 + \frac{481.12}{3.20} + \frac{1}{0.71} \cdot \frac{C_a}{2.50}\right) \times \frac{1}{1000}$$

or
$$f_a = 468 \text{ kg/m}^3$$
$$C_a = 1146 \text{ kg/m}^2$$

The mix proportions then become:

	Water	Cement	Fine aggregate	Coarse aggregate
	180.42	481.12	468 kg	1146 kg
or	0.375	1	0.973	2.38

***(vi)* Actual quantities required for mix per bag of cement**

The mix is 0.385 : 1: 0.973 : 2.38 (by mass).

For 50 kg of cement, the quantity of materials are worked out as below:

(*a*) Cement = 50 kg

(*b*) Sand = 48.65 kg

(*c*) Coarse aggregate = 120 kg

$$\begin{bmatrix} \text{Fraction I} = 60 \text{ kg} \\ \text{Fraction II} = 60 \text{ kg} \end{bmatrix}$$

(*d*) Water

(1) For water-cement ratio of 0.375, quantity of water = 18.75 litres

(2) Extra quantity of water to be added for absorption in case of coarse aggregate, = + 0.62 litres

(3) Quantity of water to be deducted for free moisture present in sand, at 1 per cent by mass = – 0.51 litres

(4) Actual quantity of water to be added = 18.75 + 0.62 – 0.51 = 18.86 litres

(*e*) Actual quantity of sand required after allowing for mass of free moisture = 49.16 kg

(*f*) Actual quantity of coarse aggregate required:

(1) Fraction I = 60 – 0.31 = 59.69 kg

(2) Fraction II = 62 – 0.31 = 59.69 kg

Therefore the actual quantities of different constituents required for the mix are:

Water : 18.86 litres

Cement : 50. 00 kg

Sand : 49.16 kg

Coarse aggregate

Fraction I = 59.69 kg

Fraction II = 59.69 kg

7.34.3 Other Methods of Mix Design

7.34.3.1 Minimum Voids Methods

This method aims at evolving a mix which has the minimum percentage of voids so as to obtain a dense concrete. This means that the fine aggregates should be in sufficient quantity to fill the voids in the coarse aggregates. The voids of each aggregate having been determined earlier, the cement is added in sufficient quantity to fill the voids in the fine aggregates. Sufficient water is added to make the mix workable.

7.34.3.2 Fineness modulus method

It has been found that the agggregates can be suitably defined in terms of grading by a term called "Fineness Modulus" which is a numerical index of fineness. It gives an idea of the mean size of particles in the entire body of aggregates. It is obtained by adding the percentage by weight of materials retained on each of the 10 Indian Standard sieves ranging from 76.2 mm to I.S. No. 50 micron sieve and dividing the sum by 100. As a result of considerable experience, certain values of fineness modulus for fine and coarse aggregates and mixed aggregates have been found to give good workability with a minimum qnantity of cement. This value varies with the size of the particles in the body of aggregates. If the finesness modulus of a given sample does not correspond to these rules, the mixture will need more cement for a given water/cement ratio and workability. Any increase in the value of fineness modulus of mixed aggregates from this values will results in a harsh mix while a decrease in the value will give an uneconomical mix for fixed water/cement ratios.

After having mixed the fine and coarse aggregates so as to obtain the desired fineness modulus of combined aggregates, the proportioning of the concrete can be done by actual field trials.

The actual quantity of cement required per unit volume of coarse aggregates and for a desired slump must be checked with the same trial mixes.

7.34.3.3 Methods of arbitrary volumetric proportions

In spite of certain drawbacks this is the most commonly used method of proportioning concrete and the concrete proportioned is thus is known as ordinary concrete. The mix consists of fixed proportions by volume of cement, fine aggregate and coarse aggregate depending upon the type of construction for which the concrete is required. The ratio of fine aggregates to coarse aggregate is generally 1 : 2 because most coarse aggregate have approximately 50% of voids. However, other ratios are also used. The nominal concrete mixes are usually specified and the types of work for which each mix is suitable is given in Table 7.17.

Table 7.17: Mixes for Various Types of Construction

Mix	*Types of Construction*
1 : 6 : 12	Mass concrete foundations.
1 : 5 : 10	
1 : 4 : 8	Mass concrete in super structure, *e.g.* in piers, abutments, massive reinforced concrete members etc.
1 : 3 : 6	
1 : 2 : 4	All normal R.C.C. work including beams, slabs, columns, walls,, small span arches, etc.
$1 : 1\frac{1}{2} : 3$	Impermeable construction, namely for liquid retaining structures, *e.g.* dam. Also used for piles and sometimes for columns..
1 : 1 : 2	Long span arches and high load columns.

The main drawbacks of the method of arbitrary proportions are the following :

(*a*) The lack of proper control over the total quantity of water that goes into a batch of concrete results in a concrete of varying strength and densities.

(6) The fixed specified proportions of cement, fine aggregate and coarse aggregate do not permit taking advantage of favourable types of grading of aggregates to secure a workable mix at the least possible cost of materials and without any harm to the strength of the concrete.

Some other methods of proportioning the ingredients of concrete which are more rational and ensure better quality control are described below.

7.35 SIMPLE VOIDS METHOD

The principle involved in this method of proportioning is that the voids in the mixture of coarse and fine aggregates are to be filled in by the cement paste. Therefore for this method the coarse and fine aggregates have to be determined first.

The volume of fine aggregate required for 1 cu m of coarse aggregate is the total volume of voids in this quantity of coarse aggregate. The volume of cement paste required for 1 cu m of coarse aggregate is the total volume of voids in the fine aggregate used with 1 cu m of coarse aggregate, plus an usual 15% extra, to allow far the additional voids created by the wedging action of the fine aggregates and the cement particles.

7.36 FINENESS MODULUS METHOD

For this method of proportioning, it is essential to know the minimum cube crushing strength (three times the working strength) required of the concrete and also the workability.

In addition, the grading of the fine and coarse aggregates, their weights, their bulking percentages and their water content must be determined.

TESTING OF CONCRETE

The purpose of testing concrete is primarily to ensure that it conforms to the specifications. However, testing of concrete is also necessary to obtain data for designing the concrete mixes. In addition, sometimes it is necessary to test concrete to find the proportions of its various constituents and also to judge its quality by measuring certain of its physical properties.

Following tests are usually carried out on concrete:

1. Tests for consistency of concrete, *i.e.* (*a*) slump test, and (*b*) compaction factor test.
2. Test for compressive strength of concrete.
3. Test for transverse strength of concrete.

For high quality concrete, analysis of freshly mixed concrete, cement content of hardened concrete and modulus of elasticity of concrete are also carried out, for which reference to I.S. Specifications may be made.

The details of the above three tests are given below :

7.37 SLUMP TEST

This gives a rough idea of the workability of concrete. The apparatus consists of a mould of the shape of a frustrum of a cone.

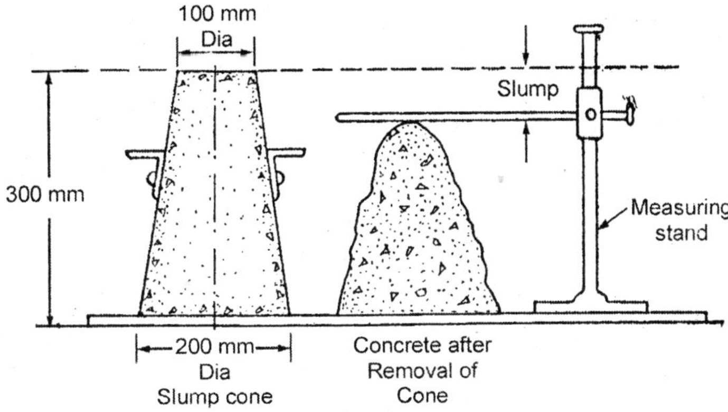

Fig. 7.4. Slump test.

The internal dimensions of the cone are :

Height = 30 mm

Diameter at the top = 100 mm

Diameter at bottom = 200 mm

The mould is open both at the top and bottom and has its axis vertical. It is fitted with handles to enable it to be lifted easily. The internal surface is usually smooth.

The mould is placed on a flat non-absorbent surface. It is filled with concrete in layers of 75 mm, each layer being tamped 25 times with 60 mm diameter and 600 mm long rod having a bullet point at the lower end. The top layer is struck off with a trowel.

The mould is then lifted vertically immediately after filling. The concrete subsides or slumps and the height of the specimen measured when there is no further subsiding. The reduction in height in mm gives the slump values.

The values of slump recommended for various purposes given in Table 7.17.

Table 7.17: Workability of concrete

Placing Conditions	*Degree of Workability*	*Slump mm*
Blinding concrete; Shallow sections; Pavement using pavers		
Mass concrete; Lightly reinforced sections in slabs, beams, walls, columns; Floors; Hand placed pavements; Canal lining; Step footings	Low	25-75
Heavily reinforced sections in slabs, beams, walls, columns	Medium	50-100
Slipform work; Pumped concrete	Medium	77.5-100
Trench fill; in-situ piling	High	100-150
Tremie concrete	Very High	By flow

7.38 COMPACTION FACTOR TEST

For high quality concrete, this method is preferred to slump test for the determination of the consistency of concrete. It consists of two conical hopper shaped moulds provided with doors at the bottom. The doors have catches, which when released, allow the material from the mould to fall down.

The upper mould is filled with concrete to be tested. The door is opened, thereby the concrete falls into the lower mould. Then it is allowed to fall into the cylinder placed immediately below the lower mould, by releasing the catch. The concrete in excess to the top of the cylinder is removed by a trowel and the surface levelled. The weight of concrete in the cylinder is determined, which is the weight of partially compacted concrete of volume equal to that of the cylinder.

The cylinder is emptied and filled with concrete in layers and rammed thoroughly to get maximum density. Again the weight of concrete in the cylinder is determined. The compacted factor is given by

$$\text{C.F.} = \frac{\text{Weight or partially compacted concrete}}{\text{Weight of fully compacted concrete}}$$

A compacting factor of 0.8 represents a concrete mix of low, 0.9 of medium and 0.95 of high workability.

7.39 COMPRESSIVE STRENGTH OF CONCRETE

For this, prisms (150 × 150 × 150 mm) are prepared in steel moulds. The specimens are air dried for 24 hours and then cured in water for 28 days in case Portland cement is used and for 7 days when rapid hardening cement is mixed.

The test is carried out in the compression testing machine immediately after curing. The metal bearing plates are put in contact with the ends of the test piece. The size of the bearing blocks is the same or slightly greater than that of the test specimen. The load is applied axially at the rate of approx. 15 N/mm^2/minute, without any shock. The load is increased, till failure occurs. The total load at the time of failure divided by the area of the test specimen, gives the compressive strength. An average of three specimens should be taken as the representative value, provided the difference between the average and the individual values does not exceed 15% of the average strength.

The values of compressive strength for different grades of concrete are given in table 7.8.

7.40 TRANSVERSE STRENGTH OF CONCRETE

For this test, specimens of rectangular beams with square cross-sections are prepared. The dimensions of specimens depend upon the size of the aggregate used in concrete and are as follows :

Max. size of aggregate	*Size of Specimen*
20 mm	100 × 100 × 40 mm
20–40 mm	150 × 150 × 66 mm
greater than 40 mm	

side = 3 × max. size of aggregate

length = 4 × side

The test specimens are cured as described in the above test.

The test specimen is supported over two parallel steel rollers of 39 mm diameter, with their centres at a distance equal to three times the depth of test specimen (*see* Fig. 34 on page 116). Load is applied to the sides of the specimen as it is cast. A third steel roller 39 mm in diameter is placed on the specimen mid way between supports and the load is centrally applied through it. The rate of load application is adjusted so that maximum load is applied in about 5 minutes. The modulus of rupture is calculated by the formula

$$R = \frac{6W}{a^2}$$

Where R = modulus of rupture in N/mm^2

a = depth of specimen in mm

W = total load in N at which the specimen fails.

The tensile stress in concrete is calculated from the formula

$$\text{Tensile stress} = \frac{F_1}{A_C + m\,A_u}; \text{ where}$$

F_1 = Total tension on the member minus pretesion is steel, if any
A_C = Cross sectional area of concrete excluding finishes
m = modular ratio; and
A_u = Cross sectional area of rainforcing steel in tension

The maximum permissible stresses in tension are as recommended in Table 7.18.

Table 7.18 : Max. permissible strsses in tension

Concrete grade	*Maximum permissible Stresses in tension (N/mm²)*
M10	1.2
M15	2.0
M20	2.8
M25	3.2
M30	3.6
M35	4.0
M40	4.4
M45	4.8
M50	5.2

7.41 NON-DESTINCTIVE TESTS

Non-destructive tests are carried out for estimating the properties of concrete in the structure. The tests normally done include ultrasonic pulse velocity and rebound hammer, probe penetration, pull-out and maturity. Non destructive tests provide alternatives to core tests for estimating the strength of concrete in a structure or can supplement the data obtained from a limited number of cores. These tests are based on determining a concrete property which has some relationship to strength.

For details, refer to IS 13311-Part I and II.

QUESTIONS

1. Give the components of concrete. State the advantages of concrete construction.
2. What functions are performed by different ingredients of concrete? Give the grading limits of aggregates.
3. What modifications can be achieved in the properties of concrete by the addition of admixtures? Give the categories of admixtures.
4. Why is reinforcement provided in concrete? What are the various types of steel reinforcement?
5. List the various types of concrete. Give details of lime concrete.
6. Compare cement concrete and lime concrete. What properties are required in a good proportioning of concrete mix.

7. Write Short notes an:
   (*a*) Plum concrete
   (*b*) Light weight concrete
   (*c*) Air-entrained concrete
   (*d*) No-fines concrete
8. What do you understand by prestressed concrete? What are its advantages over reinforced concrete? Describe briefly the methods of prestressing.
9. Explain the following types of concrete
   (*a*) Precast concrete
   (*b*) Vacuum concrete
   (*c*) Water Proof concrete
10. List the essential requirements of formwrok for concrete. What is meant by stripping of formwork? Give details of various types of formwork used.
11. Why is it necessary to provide concrete cover over reinforcement? How much thickness of cover will you provide for different types of reinforcements. What information is provided in a bar bending schedule.
12. How is the processes of batching and mixing of concrete carried out? What are the various types of mixes usually adopted? Draw a sketch of a tilting drum mixer.
13. Explain briefly the following :
    (*a*) Transporting of concrete
    (*b*) Placing of concrete
14. What is the advantage of compaction of concrete? Briefly describe the various methods of compacting concrete.
15. What considerations govern the positioning of construction joints in concrete? How are these formed?
16. Why is it necessary to cure concrete? What are the different methods of curing? For how much period, the concrete should be cured after laying?
17. Enumerate the factors affecting the strength of concrete. Give the specified compressive strength of concrete in kg/cm^2 for the following grades:
    M 20
    M 30
    M 40
18. Write short notes on:
    (*a*) Workability and durability of concrete.
    (*b*) Elasticity and creep of concrete.
    (*c*) Shrinkage and fire resistance of concrete.
19. What information should be included in specifying a particular grade of concrete? Name the different methods of concrete design.
20. Outline briefly the concrete mix design method as specifed in Indian Code.
21. Describe briefly the following concrete mix design methods
    (*a*) Nominal mix method; and
    (*b*) Minimum voids method.

22. What test is carried out on concrete to determine its workability? Give details.
23. Briefly describe the following tests on concrete :
    (*a*) Compaction factor test
    (*b*) Compressive strength test; and
    (*c*) Transverse strength test.

OBJECTIVE TYPE QUESTIONS

1. Concrete is strong in :
(*a*) Compression (*b*) Tension
(*c*) Shear (*d*) All of those

2. Which one of the following statements is wrong:
(*a*) Concrete is very strong (*b*) Concrete is very durable
(*c*) Concrete is very elastic (*d*) Concrete is very economical

3. The maximum size of coarse aggregate for mass concrete is specified as :
(*a*) 140 mm (*b*) 160 mm
(*c*) 180 mm (*d*) 200 mm

4. The maximum size of fine aggregate for mass concrete is limited to:
(*a*) 2.36 mm (*b*) 4.75 mm
(*c*) 8.0 mm (*d*) 10.0 mm

5. Which of the following statement is wrong:
(*a*) Admixtures improve workability of concrete
(*b*) Admixtures reduces bleeding in concrete
(*c*) Admixtures decrreases setting time
(*d*) Admixtures reduces heat evolution

6. Which of the following is not used as reinforcement in concrete :
(*a*) Mild steel (*b*) Stainless steel
(*c*) Medium tensile steel bass (*d*) hard-drawn steel wires

7. Which of the following is not a pozzolonic material
(*a*) Calcium chloride (*b*) fly-ash
(*c*) Surkhi (*d*) Volcanic ash

8. Unpolished aluminium poweder is used as a
(*a*) Air-entraining agent (*b*) Gas forming agent
(*c*) Retarder (*d*) Pozzolona

9. For mass concrete, ... lime is seldom used :
(*a*) Hydraulic lime (*b*) Semi-hydraulic lime
(*c*) fat lime (*d*) All the above

10. The stripping time for formwork in case of bottoms of slabs upto 4.5 m span should not be less than :
(*a*) 7 days (*b*) 14 days
(*c*) 21 days (*d*) 28 days

11. The suggested workability of concrete (slump) for heavily reinforced sections in slabs, beams etc. is

(*a*) 100-150 mm (*b*) 75 to 125 mm
(*c*) 75 to 100 mm (*d*) 50 to 100 mm

12. The maximum water content per cubic metre of concrete for 40 mm maximum size angular coarse aggregate with 25 to 50 mm slump range is limited to:

(*a*) 150 kg (*b*) 165 kg
(*c*) 186 kg (*d*) 208 kg

13. The maximum water-cement ratio recommended for concrete is:

(*a*) 0.6 (*b*) 0.7
(*c*) 0.8 (*d*) 0.9

14. The factor *t* in the equation $\overline{f}_{ck} = f_{ck} + t\,s$ is recommeded as:

(*a*) 0.84 (*b*) 1.28
(*c*) 1.50 (*d*) 1.65

15. Standard deviation suggested to be assumed for M 20 concrete in N/mm^2 is:

(*a*) 3.5 (*b*) 4.0
(*c*) 4.5 (*d*) 5.0

16. The curing time recommended for determining compressive strength using rapid hardering cement is:

(*a*) 7 days (*b*) 14 days
(*c*) 21 days (*d*) 28 days

17. For mass concrete or ordinary Portland cement concrete, the mix usually recommended is

(*a*) M 10 (*b*) M 30
(*c*) M 50 (*d*) M 70

18. For standard concrete, the mix recommended is :

(*a*) M 20 (*b*) M 40
(*c*) M 60 (*d*) M 80

19. For plain concrete of M15 grade, the maximum free water-cement ratio specified is:

(*a*) 0.4 (*b*) 0.45
(*c*) 0.5 (*d*) 0.6

20. The dosage of retarders should normally be restricted to ... percent by wieght of cementitious material:

(*a*) 0.5 (*b*) 1.0
(*c*) 1.5 (*d*) 2.0

21. The percentage of plasticiser to cement should not be more than :

(*a*) 3.0 (*b*) 2.0
(*c*) 1.0 (*d*) 0.5

22. For moderate exposure, the minimum cement content in kg/m^3 for M15 grade with w/c ratio of 0.6 should be

(*a*) 220 (*b*) 240
(*c*) 250 (*d*) 280

23. Air-entrainment in concrete

(*a*) reduces strength (*b*) increases strength
(*c*) does not alter it (*d*) increases its volume

24. Concrete should be cured at a humidity of

(*a*) 60% (*b*) 70%
(*c*) 80% (*d*) 90%

25. Slump test in concrete is conducted to assess its:

(*a*) compressive strength (*b*) transverse strength
(*c*) consistency (*d*) setting time

26. A compacting factor 0.8 represents a concrete mix of ... workability

(*a*) low (*b*) Medium
(*c*) high (*d*) very high

27. A high workability concrete should have a compacting factor of:

(*a*) 0.80 (*b*) 0.85
(*c*) 0.90 (*d*) 0.95

28. The slump for pumped concrete of medium workability is recommended as:

(*a*) 150 to 200 mm (*b*) 100 to 150 mm
(*c*) 75 to 100 mm (*d*) 50 to 100 mm

29. The pH value of water for concrete shall generally be not less than

(*a*) 4 (*b*) 5
(*c*) 6 (*d*) 7

30. The grade for standard concrete varies from:

(*a*) 25 to 55 (*b*) 55 to 75
(*c*) 60 to 80 (*d*) 70 to 90

Answers

1. (*a*)	**2.** (*c*)	**3.** (*b*)	**4.** (*d*)	**5.** (*c*)	**6.** (*b*)
7. (*a*)	**8.** (*b*)	**9.** (*c*)	**10.** (*a*)	**11.** (*d*)	**12.** (*b*)
13. (*c*)	**14.** (*d*)	**15.** (*b*)	**16.** (*a*)	**17.** (*a*)	**18.** (*b*)
19. (*d*)	**20.** (*a*)	**21.** (*c*)	**22.** (*b*)	**23.** (*a*)	**24.** (*d*)
25. (*c*)	**26.** (*a*)	**27.** (*d*)	**28.** (*b*)	**29.** (*c*)	**30.** (*a*)

FURTHER REFERENCES

Bureau of Indian Standards, New Delhi.
IS Code Nos 383, 456, 10262.

TIMBER AND ALLIED PRODUCTS

8.1 USES OF TIMBER

Wood has always been a very good and attractive construction material since olden times. The advantages inherent in wood make it invaluable in the industrial development of any country.

The use of wood in our country is at present expanding in many sectors with such rapidity that already shortages are being felt.

Uses—The uses are classified under the following categories:

(*a*) Constructional purposes, including building construction, houseposts, beams, rafters, cart building, bridges, piles, poles and railway sleepers;

(*b*) Furniture and cabinet making;

(*c*) Light packing cases;

(*d*) Heavy packing cases (for machinery and similar stores);

(*e*) Agricultural implements and tool handles;

(*f*) Turnery atricles and toys; and

(*g*) Vaneers and plywood.

ADVANTAGES OF TIMBER

Wood suitable for engineering purposes is called timber. Timber has a number of advantages, as a material of construction. Some of the important advantages are:

1. It has more strength than other materials, if compared weight to weight.
2. It can be easily converted into any shape and size.
3. Timber connections are easier to make.
4. It can be used both for load-bearing and non-load bearing structures.
5. It is abundantly used for preparing furniture and other decorative articles.
6. The wastage losses are less as even the smallest section can be put to one or the other use.

7. Timber construction is economical if locally available.
8. Timber is a non-conductor of heat and sound. As such a timber house will be warm in winter and cool in summer. For this reason houses in hilly areas are made mostly of timber. It is also used where sound insulation is desired.
9. If properly seasoned and preserved, it is quite durable.
10. The structural members made of timber are light in weight.
11. It has a good re-sale value.

CLASSIFICATION OF TIMBER

There are hundred of species of timber. Some trees belong to more primitive plant types than others, giving rise to different types of timbers. Various classifications of timber are :

8.2 EXOGENOUS AND ENDOGENOUS TREES

Exogenous trees are those which grow outward. Every year a new-layer of material is deposited between the previous year's growth and the outer surface of the tree (bark). Such trees have a series of consecutive rings, called the annular rings. Timber derived from these trees is mostly used for construction work. Examples are teak, deodar etc.

Endogenous trees are those which grow inwards. A new layer is deposited annually on the inside, instead of on the outside. Thus the older formations are on the outside *i.e.* nearer the outer surface. These trees, are not suitable for construction purposes as they are light and flexible. Examples are sal, bamboo etc.

8.3 DECIDUOUS AND EVERGREEN TREES

The deciduous trees shed their leaves annually in winter whereas evergreen trees do not shed their leaves till the new ones have grown. Timber for construction purposes is derived from deciduous trees.

8.4 SOFT WOOD AND HARD WOOD

This is the commercial classification of trees. The trees that produce these two classes of timber are themselves quite distinct. The former are gymnosperms conifers or cone-bearing plants characteristically with needle-shaped leaves and naked seeds, they have straight fibres and have distinct annular rings. Soft wood are strong in resisting direct pull, but are weak against shear or thrust. They are light in colour, which is uniform throughout the section. Examples are fir and pine.

Hard woods are dicotyledonous broad-leaved plants, characteristically with broad leaves and seeds enclosed in a sea case. The annual rings are well defined and narrow. These are very strong and hard. They can resist compression, tension and shear very well. Examples are sal, teak, shishum, oak etc.

Although this division is a convenient one, the names 'softwoods' and 'hardwoods' are misnomers, some of the so called soft woods being actually harder then the hard woods.

8.5 DURABILITY CLASSIFICATION

The timbers are classified for durability according to the average life of these test specimens as follows:

High — Timbers having average life of 120 months and over

Moderate — Timbers having average life of less than 120 months but of 60 months or over

Low — Timbers having average life of less than 60 months.

8.6 TREATEBILITY CLASSIFICATION

The classification is based to represent approximately the degree of resistance offered by the heartwood of a species to the penetration of the preservative fluid under working pressure of 10.5 kgf/cm^2. The treatability of timbers has been classified as follows:

(*a*) Heartwood easily treatable

(*b*) Heartwood treatable, but complete penetration of preservative not always obtained

(*c*) Heartwood only partially treatable

(*d*) Heartwood refractory to treatment

(*e*) Heartwood very refractory to treatment penetration.

8.7 CLASSIFICATION OF STRUCTURAL TIMBER

According to this classification the variation in strength properties of wood, specially modulus of elasticity and fibre stress in bending and tension along the grain that would influence the design, are taken into consideration. The timber is classified under ordinary, standard and super groups, depending upon modulus of elasticity '*E*' and modulus of rupture σ_t as follows :

Super—*E* greater than 126000 kg/sq cm and σ_t equal to or greater than 180 kg/sq cm.

Standard—*E* greater than 98000 kg/sq cm but equal to or less than 126000 kg/sq cm and σ_t equal to or greater than 120 kg/sq. cm.

Ordinary—*E* greater than 66000 kg/sq cm but equal to or less than 98000 kg/sq cm and σ_t equal to or greater than 85 kg/sq cm.

Timber of ordinary group are suitable for spans upto 6 metres, that of standard group for spans exceeding 6 metres but less than 12 metres and that of super group for spans exceeding 12 metres.

8.8 CLASSIFICATION OF TIMBER FOR SEASONING PURPOSES

Timber is classified into the following three classes depending upon their behaviour with respect to cracking and splitting during normal air seasoning :

(*i*) Highly refractory woods are slow and difficult to season free from defects. Examples are heavy structural timber such as Sal and Laurel.

(*ii*) Moderately refractory woods can be seasoned free from defects with a little protection against rapid drying conditions. Examples are the furniture class of timber such as Sisoo and Teak.

(*iii*) Low-refractory woods can be rapidly seasoned free from defects even in the open air and sun. If not rapidly dried, they develop defects such as blue stain and mould on the surface. Examples are packing case timbers such as Semul and Salai.

PARTS OF A TREE

Like most other plants, trees have three main parts, namely

(*i*) Roots,

(*ii*) Stem and

(*iii*) Leaves.

Trees usually have a single main stem, the trunk or bole. The timber user is primarily interested in the trunk or bole. This bole has an outer covering called the bark which protects the wood from extremes of temperature and mechanical injury. The inner layers of the bark conduct the food manufactured in the regions of active growth and to the places where it can be conveniently stored. The bark, being a conductor of food materials is often rich in chemical substances, as tarmin and dyes.

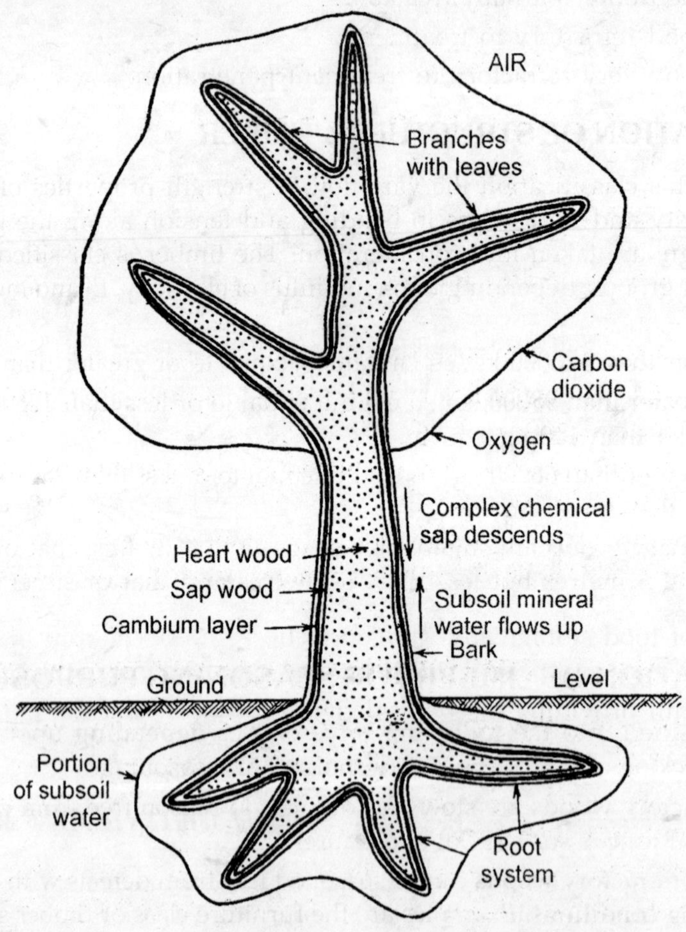

Fig. 8.1. Longitudinal section of a tree showing various parts

Between the bark and the wood is a thin, delicate tissue known as the cambium, which forms a complete sheath, covering the bole and branches. This tissue produces bark towards the outside and wood towards the inside of the tree, and the enlargement in girth of the trunk is brought about entirely by the activity of the cambium sheath. The production of wood and bark tissue occurs only when the cambium is growing ; in temperate region this takes place during the spring and summer months and in tropical climates this occurs practically throughout the year.

Under the bark is the cylinder of wood. In the centre of this cylinder is the pith. This may be upto 12.5 mm in diameter, but in many trees is barely visible. A cross-section of a branch is similar to that of the main stem.

If the cross-section of a bole is examined further in detail, it will be noted that wood of trees, growing under seasonal conditions consists of a series of concentric layers of tissue called medullary rays or annual rings. Each medullary layer comprises the wood produced during a single season. The wood nearest the bark is the youngest. In temperate regions and some tropical countries this provides a fairly accurate means of computing the age of a tree after it is felled, serious interruptions to growth can lead to double or multiple rings. In many tropical regions these may be indistinct and may not be annual, but the new wood is formed in concentric layers.

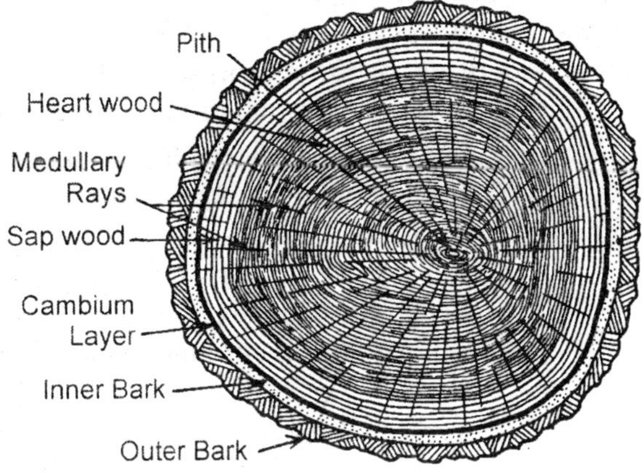

Fig. 8.2. Cross-section of a timber tree.

The work of food storage and sap conduction is performed in most trees only by the outer or youngest growth layers. This is known as the *sap wood.* The central part of the tree is concerned with providing mechanical rigidity to the stem, supports the growth and is known as heart wood.

COMPARISON OF SAP WOOD AND HEART WOOD

It is usual to regard sap-wood as inferior to heart-wood. Generally speaking it is lighter in colour and is less durable and in the log it is wetter than heart- wood. Sap-wood is considered to be immature wood. This is not quite correct. Sap-wood consists of live cells whereas heart-wood is mostly composed of dead cells. In general, commercial logs of average size contain between 25 and 30 per cent of sap-wood, which if discarded would represent an appreciable waste. Important properties of sap-wood and heart-wood are compared below :

1 *Colour.* In some timbers there is no colour difference between sap-wood and heart-wood but in most cases sap-wood is lighter in colour.

2. *Weight.* There is usually no significant difference between the weight of sound sap-wood and sound heart-wood of the same moisture content.

3. *Strength properties.* Mechanical tests indicate that sap-wood of the same moisture content and density as heart-wood, and free of defects, is approximately equivalent in strength properties. In some cases the figures for sap-wood are slightly lower but the differences are not of practical significance.

4. *Durability.* Sap-wood is rich in plant food material that is attractive to certain wood-rotting fungi and insects. Different timbers vary appreciably in this respect. Starch in itself is not sufficient to attract certain insects, the presence of traces of other substances can be removed by prolonged soaking of timber in water. On the other hand some of the substances in the heart-wood are positively toxic to fungi and insects. Sap-wood is, therefore, more readily attacked than heart-wood.

5. *Permeability.* At the time of heart-wood formation, the conducting tissue of the wood usually undergoes modifications and free movement of liquid is interrupted. Due to these and other deposits heart-wood often becomes impermeable to moisture movement. Consequently heart-wood is not so easily impregnated with preservative as sap-wood.

Durability is usually the controlling factor for determining whether sap-wood should be used or not. If adequately treated, sap-wood can be used without any fear in any location including outdoors like fencing posts, gates, railway sleepers, etc. Well treated sap-wood may even outlast heart-wood. In well ventilated, dry internal situations such as joinery work, generally, there should be no objection to the use of sap-wood if the timber is thoroughly seasoned before installation and the site conditions are not expected to alter adversely after installation.

The proportions of sap-wood depend upon the type of tree from which the log is obtained. It varies from 15 to 60 per cent of the wood content of the log. Since sap-wood forms the outer portion of the log, its proportion will be less in a thicker log and more in the logs of smaller thickness.

FELLING OF TIMBER TREES

A standing tree has to be cut to obtain timber, of cutting a tree is known as felling. Important points with the felling of trees are given below :

8.9 AGE OF TREE FOR FELLING

A tree should be in a full growth state or just matured at the time of felling. This is essential to obtain maximum quantity of useful timber. An immature tree will contain large quantities of sap-wood. An over matured tree will develop cracks and other defects. The age for maturity varies with the species.

8.10 PROPER SEASON FOR FELLING

In the spring and late in summer, the wood cells are very active and the proportions of sap and moisture are also large. The starches, sugars, etc. present in the sap-wood cause decay of wood rapidly, with the growth of fungus. During winter and early summer, the cambium layer is inactive and the growth of the tree is dull. Also sap and moisture are present in low percentages. Thus this season is suitable for felling. Moreover the converted sections of timber (logs) are not subjected to the direct rays of the sun thus preventing splitting and growth of fungus.

8.11 PRELIMINARY WORK

Orientation. Before the woodsmen start on the execution of work in a felling area, they should inspect the place well and acquaint themselves with the situation, the grounds, the number and kind of trees marked out, the execution of felling and conditions for removal. This allows them to make the necessary preparation well and to draw up an appropriate plan for the work. If necessary, where things are not quite clear, the wood cutters must get more information.

Marking of Stems. Generally it is advantageous to the woodcutter to put fresh marks on two or even three sides of the trees to be felled, because this helps them keep in view and they do not need to waste time in looking for them.

Determination of Direction of Felling. First thing in each case of felling a tree, is that the wood-cutters should consult together on the direction of its felling so that :

1. The remaining trees stand and the regeneration are damaged as little as possible through the fall and movement of the stem ;
2. The removal of the logs can be carried out in as straight a direction as possible, parallel and not at right angles to the path of removal ;
3. The stem is not damaged in the fall ;
4. The work to be done on the lying stem is made as easy as possible.

Further the form of the stem and its eventual leaning, the crown and its surroundings and the position of the butt-swellings are to be carefully taken into consideration.

Preparation of the Place of Work.

Weeds, shrubs and branches, which would hinder work around the stem to be felled, are to be removed.

8.12 FREEING THE BUTT-END

In order to saw the stump deep near the ground (minimum waste of wood, minimum hindrance to the following work of felling and removal) without damaging the tools ; moss, leaves, earth and stones should be removed with a pick-axe. In a sloping area, most of this work has to be carried out only on the hill side of the stem.

8.13 ROUNDING THE BUTT-END

Rounding of the butt-log has the following advantages :

(*a*) The surface to be cut is reduced.

(*b*) The sawing is facilitated through the removal of fibres, which run at an angle to the cross-cut surface and are difficult to cut through.

(*c*) The desired direction of felling can be adhered to better, as the fall of the stem is not influenced by butt-swellings in an unknown way.

(*d*) There is a definite saving in energy and time when the stem is to be turned for debarking, removing branches etc. and later also while hauling, piling and loading the stems as well as in stacking of cut wood. The stacks or piles are more uniform.

Stems upto a diameter of 300 mm at breast height should as a rule be rounded by one wood-cutter alone. On bigger stems two wood-cutters can work together in so far as they work on opposite sides of the stem and do not hinder each other. They may not strike their axes towards each other as thus they will be exposing themselves to danger.

8.14 REMOVING BARK OF THE BUTT-END

The stems with thick bark, which has to be removed, should be debarked with the axe up to shoulder height immediately after trimming. This work can be more easily carried out on the standing tree than on a lying one.

For hard woods like teak, which are not attacked by fungi, the processes of rounding the butt-end and removing of the bark are carried out 2 to 3 years earlier than the tree is actually felled. This is known as Girdling. The moisture evaporates and further formation of sap is stopped. Thus shrinkage and warping is less during; seasoning.

8.15 CHOPPING OF NOTCH (UNDERCUT)

On the trimmed stem, on the side of the direction of felling a notch or undercut is chopped. This has a level base and a straight inner line and is the so-called tipping edge. This notch makes it possible for the tree to fall over the tipping edge and to break from the stump (part of a tree still in the ground) It further gives the points for the guidance of the saw when sawing.

In order to be able to keep better to the desired direction of fall, the base of the undercut should be 20 to 30 mm lower than the proposed saw cut on the opposite side. The tipping edge must be horizontal, straight and at right angles to the direction of fall. This procedure also prevents the tearing off of fibres of chips from the butt-log.

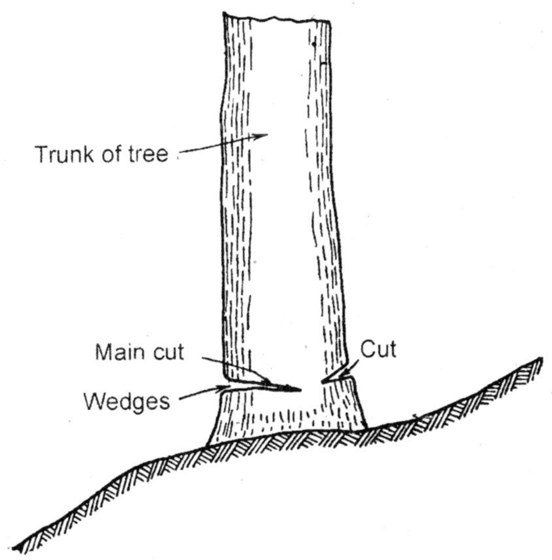

Fig.8.3. Felling of a tree.

The undercut should be chopped as flat as possible and not too big, so that the minimum possible wood is wasted and work is minimised

8.16 SAWING

The saw is to be placed 20 to 30 mm above and on the opposite side of the undercut. After a few strokes, one must check up that the blade of the saw is working at right angle to the axis and at the desired height.

If in the beginning any hindrance compels some deviation from the proposed direction of cutting, it is to be strived to come back to the proposed direction as soon as possible. No tree is to be sawn through to its fall in one operation.

As soon as the saw has penetrated the stem about double breadth of the saw blade, a wedge must be inserted. While sawing further, the sawers must inform each other about the progress of the cut. As a measure of safety, during a pause in cutting, the distance of the saw cut to the tipping edge should be measured and compared. .

In order to be certain that the tree will fall in the desired direction, the sawing should be interrupted, once or twice in case of smaller stems and often in the case of bigger ones, to drive in the wedges and for a check up. If it is observed that the tree has a tendency to deviate to one side from the desired direction of fall, then this side must be sawn a little more so that the unsawn portion of the stem on the other side is broader and more resistant, thus correcting the deviation.

Towards the end of the work, the sawing should be finer, depending on the observations until the unsawn part between the butt-log and the stump, the so-called crest, shows the necessary form in order to act like a hinge and guide the fall of the tree in the desired direction.

Necessity of Wedges.

Wedges are intended to affect the following :

1. In the beginning a wedge is used only to relieve the blade of the saw. For this, one wedge is inserted in the middle of the cut in the direction of fall extended backwards.
2. At the same time this wedge is made use of to ascertain the reaction of the tree when the wedge is struck. Because it is inserted in the middle, it shows the resistance of the unsawn part.
3. Finally the tree with the help of the wedges must be gradually driven over, towards the direction of fall, losing equilibrium. Depending on the lean, the distribution of weight and the thickness of the tree, this driving can necessitate the use of larger number of wedges.

8.17 FELLING

As soon as the tree begins to fall, the woodsmen must step back sideways several steps, so that by the impact of the tree on the ground they are not hit. At the same time they must look up in order to be able to get out of the way of falling branches. Stems which remain suspended in neighbouring areas, which cannot be avoided in thick forests, should under all circumstances be brought down to earth as soon as possible.

If the tree remains hanging inspite of this, then the butt-end should be drawn in a direction opposite to that of the fall of the tree.

CONVERSION OF FELLED TREES

As pointed out earlier, the felled trees should be converted into suitable sections immediately after felling. This is done in the following stages :

8.18 PRELIMINARY WORK

The first step after felling is the removal of branches. This is done by woodsmen. After trimming of branches, the trunk or boles .are measured off and cut to length according to the instructions received. Then the bark is removed. Heavy frost or steep slopes may necessitate a postponement of debarking.

8.19 WATER SOAKING

Bark provides a natural protection to the wood. When it is removed, the log is liable to be attacked easily by fungi and cause decay and rot. To prevent this, the logs are put under water in large ponds called log ponds. These ponds are either natural or artificial. As logs will float to the surface these are held down by means of girders placed across the pond. If this is not possible, the logs are turned on their sides after a few days.

This has an additional advantage of diluting the sap present in the logs.

8.20 STORAGE OF LOGS

The felling of trees can be carried out only in suitable season but the conversion of trees into suitable sections can be carried out throughout the year with the aid of saw mills. Thus storage of logs becomes necessary. There are, in general two methods of storing logs ; the wet method and the dry method. The wet method is explained above. In the case of hard woods like teak etc. wet storing is not essential. In such cases dry method is adopted. The logs are kept on a clean and level floor which is covered by gravel. The logs are kept about 300 mm above the ground level. If necessary, water may be sprinkled to keep the logs in the wet condition.

8.21 SAWING OF TIMBER

The logs of wood from the log ponds or drying yards are converted into suitable sections by sawing. Lumber is a general term applied to cut sections of a log. The conversion is done to facilitate easy handling and according to market requirements. Sawing may be done either manually or by mills. Saw mills are generally used these days. In the process of sawing, the wastage is about 30 to 40 per cent. The methods of sawing timber are :

(*i*) Ordinary sawing.

(*ii*) Tangential sawing.

(*iii*) Radial sawing.

(*iv*) Quarter sawing ; and

(*v*) Combination method.

Ordinary sawing. This is a very easy and quick method of sawing. It is most economical so far as the out-turn is concerned and is mostly adopted in India. The log is moved forward and backward on the platform of a sawing mill, thereby cutting parallel slices or planks.

The planks cut by this method are liable to warp and twist as a result of unequal shrinkage. The central portion of the plank is of "heart wood, where the shrinkage is less but the outer portion is of sap wood which shrinkages more. Thus due to unequal shrinkage, warping occurs.

Tangential sawing. The log is sawn tangential to the annual rings and at right angles to the medullary rays. The sections cut by this method are weak as the radial or medullary rays, which impart strength to the longitudinal fibres, are cut. Tangential sawing is also termed plain sawing or flat grained sawing.

Tangential sawn timber warps more on drying and seasoning in the form of a cup. It cannot be polished evenly. Such timber is not suitable for flooring.

Radial sawing. The log is sawn parallel to the medullary rays and perpendicular to the annual rings. This is a better method of sawing and the timber wears uniformly than the plain sawn timbers. Radial sawn timber shrinks and warps to a less degree and is distorted to the minimum. The annual rings make an angle of 45° and 90° to the broad surface of the board. This wood is suitable for high class work. The main disadvantages are that the wastage is maximum in this method and the cost of sawing is much more as more time and labour are required in turning wood several times.

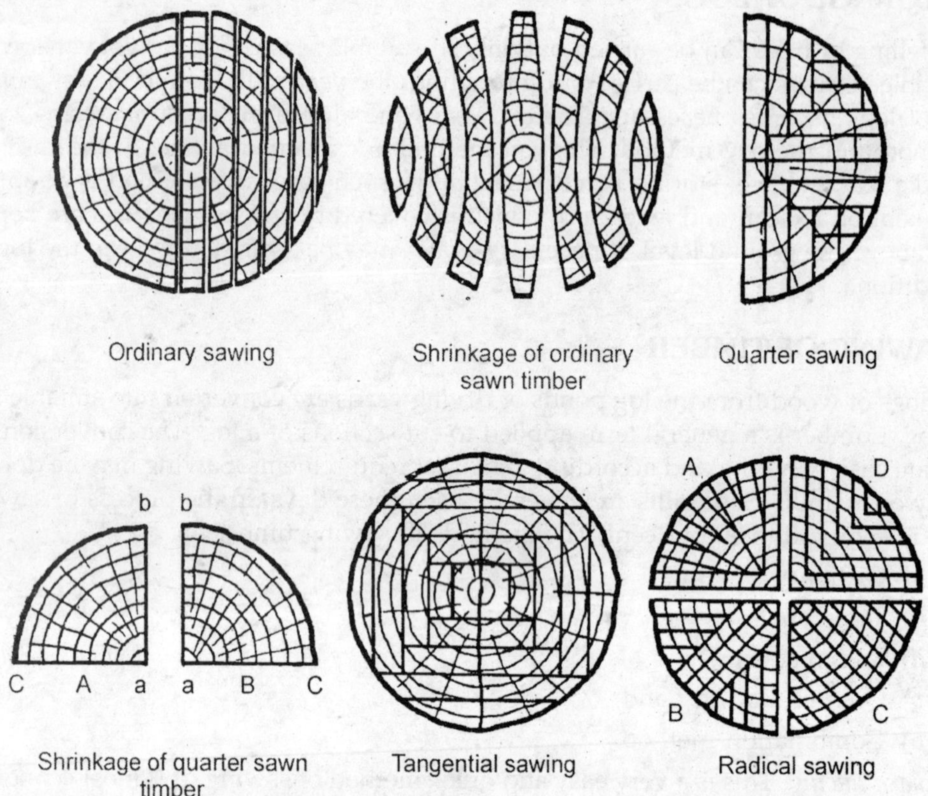

Fig. 8.4. Various methods of sawing timber

Quarter sawing. In this method, the log is cut first into quadrants. Then each quadrant may be sawn by either plain sawing, tangential sawing or radial sawing.

Combination method. Combination of two or more of the above described methods of sawing may be used to cut the log.

ZONES AND IMPORTANT VARIETIES OF TIMBER

8.22 ZONES

The country has been divided into five zones comprising the following areas:

(*i*) **North Zone:** Jammu and Kashmir, Punjab, Himachal Pradesh, Delhi, Uttar Pradesh and Rajasthan.

(*ii*) **East Zone:** Assam, Manipur, Tripura, West Bengal, Bihar, Odisha, Sikkim, Bhutan, Andaman, Arunachal and Meghalaya and Nagaland.

(*iii*) **Center Zone:** Madhya Pradesh, Vidharbha areas of Maharashtra State and the North East part of Andhra Pradesh (Godavari delta area).

(*iv*) **West Zone:** Maharashtra State (except Vidharbbha areas), Gujarat and North West part of Karnataka.

(*v*) **South Zone:** Tamil Nadu, Andhra Pradesh (except the Godavari delta area), Kerala and Karnataka (Except North West part).

The availability of timber is classified under three categories as indicated below:

X—Most common, 1415 m^3 (1000 tonnes) and more per year.

Y—Common, 355 m^3 (250 tonnes) to 1415 m^3 (1000 tonnes) per year

Z—Less common, below 355 m^3 (250 tonnes) per year.

8.23 PROPERTIES OF IMPORTANT TIMBERS IN THE COUNTRY

Table 8.1: Important Timbers

S. No.	*Name*	*Locality where found*	*Weight kg/cm³*	*Description of appearance of tree and quality of wood*	*Uses*
1.	Deodar	Himachal Pradesh, Uttarakhand, Punjab	545	It grows at altitudes of 1200 to 3,000 m. A very lofty straight trunk with only small branches. The wood is therefore very straight in grain, and though light coloured it is handsome, and has the grain well marked and looks very well when smoothened and varnished. It is of great strength and durability, through easily worked. The sap wood is white and the heart wood yellow which becomes brown on exposure. It can be easily seasoned and retains the shape well.	Can be used for any purpose. Used for railway sleepers, railway carriages, structural work, furniture etc.
2.	Chir	U.P., Punjab Haryana and the foot of Himalayas.	575	A similar tree to deodar but inferior in all ways. A soft light and cheap wood.	Useful where great strength is not required. Used for house building, boat building etc. Also used for railway sleepers but after giving preservative treatment.
3.	Fir	U.P., Punjab Haryana and Himachal Pradesh, West Bengal	450	Very similar to chir but much inferior. It is not very durable. Very light and soft.	Packing cases, picture frames, mouldings, light and rough furniture.
4.	Teak or Sagwan	Assam, Maniput, Tripura, Central and	650	A very large tree but more branched than deodar with very large light coloured leaves, the wood is close, straight grained and of a rich brown yellow colour some	Suited for any work from ship building to furniture making. Used in house construction,

		South India and Odisha, West Bengal, Mostly imported from Burma.		what hard and rather heavy, but works excellently and to a smooth polished surface and is of great durability, one of the most valuable timber trees in the world, This is easy to saw and keeps it shape well. Very durable even if used under water ; also resists fire action.	structural works, best quality furniture, flooring etc.
5.	Sal	Punjab, U.P., Bihar, Odisha, West Bengal, Assam and Tamil Nadu	865	A straight gorwing upright tree, with large dark leaves. The wood has a straight close grain and dark brown colour, is hard and very heavy and though not taking a fine or polished surface, works well and from its great strength and durability is a most valuable wood for all but ornamental purposes.	Building construction, railway sleepers and structural work.
6.	Tun	Assam, West Bengal, U.P., and Tamil Nadu	515	A moderate sized spreading tree, a sport trunk and large branches, abundant foliage or long pointed leaves. The wood cannot be cut off to any great size so is only available for furniture and such like work for which it is most excellent as the colour is rich red brown, and the grain is handsome and close, so works to fine polished surface.	Furniture making, musical instruments and small boats.
7.	Shisham	Assam, West Bengal, Odisha, U.P. and Punjab	770	Similar tree in size and branching to tun but with round leaves. The wood is of rich dark brown colour, with well marked grain, but does not work so smooth as tun.Still it works to a fine surface, and is very handsome. It is also a very strong wood.	Railway carriages, high class furniture, house construction and boat making etc.
8.	Mango	All over India	690	Rather straighter in trunk and larger than either tun or sissu. The wood is white and coarse grained, and has no great strength, but still it is a useful wood for rough work.	Plywood, packing cases, toys, tonga bodies etc.
9.	Kail	U.P., Jammu Kashmir, Punjab and foot of Himalayas.	515	It resembles chir to some extent but the colour is pink. This is easy to season. It suffers from a disease known as Termites pipe, a disease similar to T.B. in human being. It can be detected by dipping a freshly sawn piece into water. If diseased, it would turn reddish.	Construction work, light furniture, packing cases, drawing boards, plane-table, railway sleeper, burning charcoal etc.
10.	Kikar	All over Upper India	595	A branching tree with dark bark and light feathery foliage wood. Dark red colour, hard and strong, seldom obtainable in any large size.	Used for any rough work. Used mainly for spokes, tent egs, plough spares etc.

11.	Aini	Maharashtra and Tamil Nadu	595	This compares favourably with teak and is an all purpose wood.	Ploughs, boat making, furniture, structural work etc.
12.	Babul	U.P., Maharashtra, Tamil Nadu and M.P.		Babul is a thorny tree of small leaves and yellow flowers. It grows well even in poor dry soil without water. It is hardly more than 12 m in length and 3 m in girth. It is closed grain, hard pale red colour and crooked.	Cart wheels and bodies, ploughs, tools' handles, structural work etc.
13.	Benteak	Mysore, Travancore, Nilgris	675	Reddish brown in colour quite strong and suitable for general purposes.	Ship making, house construction, furniture, carts, wagon, flooring pavings etc.
14.	Champ	U.P. Mysore, West Bengal	430	A fine timber tree with handsome foliage and flowers. It is of rich brown colour. In Dehradun it reaches 5 m in girth. Slabs 2 m in breadth can be obtained. This wood takes a beautiful polish. Yields good posts and planks.	Furniture especially table, building etc.
15.	Haldu or Hodd	Maharashtra, Tamil Nadu, M.P. and West Bengal	675	This is also a very large tree with a soft, closed, even grained wood. It is light and more easily worked and very susceptible to alternations of temperatures. Splits when nailed.	Best suited for cabinet making, also used for agriculture implements, planks and posts etc.
16.	Rose-wood	C.P., Maharashtra Tamil Nadu and Coorg and in small quantities in Punjab and U.P., West Bengal, Odisha	880	A very strong, hard and tough wood. This is similar to shisham wood.	Building work, tool handles, high class furniture and ornamental work.

STRUCTURE AND IDENTIFICATION OF TIMBERS

8.24 STRUCTURE

Wood is composed of cells which vary in size and shape and in the direction in which they run. Most of the cells are arranged vertically to the axis of the tree, but there are some which also run horizontally. This external difference in the appearance of different timbers is due to the different patterns of cells-aggregates of which it is composed. It is also known that different species of trees produce different kinds of timbers. Some woods are hard and others are soft, some are heavy while others are light. All such peculiarities of a timber are ultimately traceable to the cells and the cells-aggregates of which it is composed.

To examine critically the minute detail of cells of a timber we require a well equipped laboratory. In the field this difficulty is got over by using a hand lens or magnifying glass and a sharp knife. Although by the use of these two simple tools one cannot study the minute details of different types of cells, they are sufficient to enable the observer to detect large cells and the aggregate cell-patterns found in different timbers.

8.25 IDENTIFICATION

Too much importance must not be given to features like colour and weight, as mistakes can be made thereby. Colour in wood is a vary variable feature. It varies, not only in different species, but also in the same species. Colour, however, has in some cases diagnostic value. The weight of wood varies widely. Moreover, a piece of green timber is many times heavier than the same piece when it is dry. It is, therefore, not possible to identify timbers entirely by their weight or colour.

Timber can be best identified by examination of the cell structure visible on the end surface. Longitudinal surfaces may occasionally be useful for identification, but it is the end surface which shows the most conservative features.

The following are some of the more important characteristics. used in the identification of timber :

Growth Rings

Most logs show concentric rings on their cross- section. These-rings indicate growth marks and the layer of wood between two consecutive marks is known as a growth ring. Growth rings are distinct when the variation in the size and the shape of the cells formed at different times of the year is prominent. In India all timbers do not show distinct growth rings. This is due to the fact that some trees. have, more or less, the same intensity of growth throughout the year, and one year's growth merges into the next year's growth without any perceptible changes.

Non-porous, Ring-porous and Diffuse-porous Woods

Vessels or pores are tubular cells with wide openings at both ends. They are comparatively large in size and can be seen as minute holes on the end surface of species of wood. Sometimes they can be clearly seen without a lens, but with a lens they are always clearly visible. When pores are present, they can also be detected on the longitudinal surface of a piece of wood as shallow grooves which are often called vessel lines.

Based on the presence or absence of pores, Indian timbers can be divided into two main groups, namely porous and non-porous woods. As a rule, non-porous timbers do not show any holes on their end surfaces. Some of them, however may have resin canals which although they look like pores on the end surface, are not actually pores. Resin canals are only a few in number in comparison with the regularly distributed pores in a porous wood. When the end surface shows a few scattered holes, one can be almost certain that it is a non-porous timber containing some resin canals. On the other hand, the holes in. a porous wood are more or less numerous. One can also detect a. porous or non-porous timber on the longitudinal surface.

The porous woods can be sub-divided into two groups, namely rings and diffused porous. Ring-porous woods formed during the early part of the growing season are larger than the pores formed during the latter part of the growing season, and they constitute a distinct band along the early part of the ring. In diffuse-porous woods, the early and late pores show no appreciable difference in their size and they are more or less evenly distributed throughout the ring.

Fibres

Fibres are narrow, elongated, thick walled cells with closed tapering ends, individually they are too minute to be visible with a hand lens, but collectively they form different patterns which are of value in identification of timber.

Soft Tissues or Parenchyma

Soft tissues or parenchyma cells are short, thin walled cells. Individually they are not distinguishable with a hand lens, but collectively their distribution and arrangement are of importance in the classification and identification of timbers.

Medullary Rays or Wood Rays or Rays

Medullary rays or wood rays, or rays are groups of horizontally arranged cells running radially from the bark towards the centre of the tree. On the end surface of a piece of wood, they appear as straight lines, sometimes not visible to the naked eye. On the tangential surface also, they are not always visible to the eye but, with a hand lens, or after moistening the surface with water, one can easily see their somewhat dark, spindle-shaped structure. On the radial surface they are not always conspicuous, except in the case of unusual height, when they appear clearly as small plates and produce the effect known as 'silver grain'.

Ripple Marks

In some timbers, rays and occasionally other vertical cells are so arranged as to bring about, on the tangential surface of the wood, the effect of fine, equidistant, wavy lines at right angles to the grain. These are called ripple marks. Ripple marks may not be always visible to the eye but they are distinct with a hand lens.

Resin Canals and Gum Ducts

Non-porous timbers have cavities containing resin, known as resin canals. These run vertically along the grain and are responsible for the dark streaks that are often found on the board faces of pines and spruce. The distribution and size of resin canals are important characteristics of some species, and are often made use of in the identification of timbers.

Amongst the porous woods, also, there are sometimes cavities, containing gum and known as gum ducts. These gum ducts look very much like pores on the end surface but they often contain gum in concentric bands at irregular intervals, and the colour of the gummy deposit is white. The distribution of gum ducts and the colour of the gum is, therefore, sometimes of value in the classification and identification of timbers.

MEASUREMENT OF WOOD

Approximate cubical contents of logs may be determined as follows :

1. *For round logs.* The cubic contents may be obtained by measuring the girth or circumference at three places where it is smallest, near the ends and near the centre. The bark should be removed at these places before measuring. Then

$$\text{Volume (in cubic m)} = \left(\frac{g_1 + g_2 + g_3}{3}\right)^2 \times \frac{l}{4\pi}$$

where g_1, g_2, g_3 = Girth measurement in metres

and l = Length of log in metres.

2. *For square logs.* The cubic contents is found by measuring the girth at the middle of the log. Then

$$\text{Volume} = \left(\frac{g}{4}\right)^2 \times l$$

where g = Girth at the middle

MOISTURE IN WOOD

Apart from the chemical compounds of which the actual wood tissue is formed, the constituent which is of greatest importance is water. In the growing tree or immediately after it has been felled, the amount of water present may be of greater proportion by weight than the solid material itself.

This is roughly 20 per cent based on oven-dry weight of wood substance (*i.e.* roughly 200 per cent). Moisture is present in two forms :

(*i*) Large amount of free water present within the cell cavities.

(*ii*) The cell walls, of which wood is composed, are saturated.

Free water or water filling wholly or partly the cell cavities have little effect on the properties of wood, except for the weight. The weight is increased and thus the handling and transporting charges are increased.

The water saturating the cell walls has a marked effect on the properties of timber. If the water content is reduced to 10 to 12 per cent, the strength is doubled and further reduction still increases the strength.

8.26 METHOD OF DETERMINATION OF MOISTURE CONTENT OF TIMBER

The moisture content of timber is expressed on the basis of the oven dry weight. A cross-section cut, about 20 mm wide (150 to 250 mm away from the end) from a piece of timber typical of the consignment, is weighed, dried in an oven at a temperature of 100°-105°C until there is no further loss in weight, and the moisture content per cent is then obtained from the equation:

$$\text{Percentage moisture content} = \frac{\text{Initial weight} - \text{Final weight}}{\text{Final weight}} \times 100$$

Care should be taken to prevent changes in moisture content between the cutting of the section and the first weighing, and between removal from the oven and subsequent weighing.

Electronic instruments, based on resistance of dielectric measurements and suitably calibrated for moisture content for specific species of timber can also be used for routine determinations.

8.27 EQUILIBRIUM MOISTURE CONTENT

The moisture content of timber varies with the humidity of the atmosphere and temperature and air changes of the place in which it is to be used. Equilibrium moisture condition is that stage at which there occurs no interchange of moisture between timber and air or in other words the humidity of the timber and air is the same. At this stage, the moisture content of wood or air is called equilibrium moisture content. The object of seasoning is to reduce the moisture content upto equilibrium moisture content. The following table shows the various moisture contents suitable under different conditions.

Table 8.2: Maximum Permissible Moisture Content of Timber for Use in Building Construction

Use	*Dry*	*Moderately Dry*	*Moderately Moist*	*Moist*
(*i*) Beams and rafters	12	14	17	20
(*ii*) Doors and windows				
(*a*) 50 mm and more in thickness	10	12	14	16
(*b*) less than 50 mm in thickness	8	10	12	14
(*iii*) Flooring strips for general purposes	8	10	10	12
(*v*) Furniture and cabinet making	10	12	14	15

SEASONING OF WOOD

8.28 OBJECTIVE OF SEASONING

The object of seasoning is to reduce the moisture content of wood under more or less controlled conditions. It helps in the drying out of the free water and certain amount of moisture from the cell walls.

A well seasoned piece of wood may contain about 10 to 12% moisture on an average. This would be in equilibrium with the atmospheric humidity in several parts of the country. This degree of seasoning is necessary for proper retention of shape and size of component parts of high class work. For rough work, and for outdoor uses, drying of timber, say upto 15 to 25 per cent moisture content is considered enough, particularly because it makes preservative treatment easy and effective.

Seasoning of wood is the first step in the efficient utilization of timber, especially in this country, where the climatic conditions are very trying for wood work, and where several commercially important species of wood are hard, heavy and liable to develop defects while drying and in use, unless carefully seasoned. It should be emphasized that all timber, whether teak or other species, need proper seasoning in order to give satisfactory service. Seasoning should be regarded as an integral part of timber utilization. With the development of the timber industry and increased use of Indian teak and several other indigenous hardwoods (non-conifers), it has become imperative that adequate attention should be paid to the process of seasoning wood before use.

8.29 DEFECTS IN UNSEASONED TIMBER

The main defects in the unseasoned timber are :

Shrinkage. After all the free water is dried out (at this point the timber is said to be at "fibre saturation point") the cell walls begin to dry. At this stage the timber starts to shrink. Also, if the free water is withdrawn very suddenly, the wall will probably rupture or collapse.

This explains very briefly the well known shrinkage that occurs in drying timber ; it is greatest in the direction of the annual rings ; *i e.* tangentially to the circumference of the timber, in a radial direction the shrinkage is only about half the amount. In the longitudinal direction shrinkage is negligible.

Unequal shrinkage, which can be caused by uneven drying brings about various splits and checks in drying timber.

The tendency of wood to shrink or swell with changes in the moisture content of the atmosphere is a factor inseparable from the material. This characteristic behaviour is usually called "working" or "movement". This is inherent in the material and cannot be eliminated by any particular method of seasoning or storage, although the deleterious results can be minimised. It should, therefore, be noted that however old a piece of timber is, it will shrink or swell under changes or humidity as much as a recently seasoned piece.

Splits and Checks. In seasoning, the moisture in the timber first evaporates from the surface. The chief difficulty to be overcome is the tendency of the upper layers of a piece of wood to dry out more rapidly than the interior. If these layers are allowed to dry much below the fibre saturation point, while the interior is still saturated, stresses are set up, because the shrinkage of the outer layers is restricted. The stresses may attain such magnitude that the tissues in the outer layers of the wood are actually ruptured, and surface splits and checks result.

Twisting of Timber. If the unseasoned timber is cut into sections, the sections get twisted due to uneven drying. This is due to the release of stresses set up during uneven drying.

8.30 ADVANTAGES OF SEASONING

The advantages of seasoning are :

1. The wood becomes lighter in weight. The weight of wood is reduced to nearly half after seasoning. In the case of light timbers It may become one-third.
2. The wood becomes less liable to change its size. Seasoning brings about dimensional stability, so that the finished timber remains unaltered in size, weight and volume even after use in construction.
3. Wood becomes less liable to change its shape, Timber does not get twisted or warped if properly seasoned. Controlled drying of timbers enables them to retain proper shape and avoid bending which deform timber and make them unsuitable for specific purpose.
4. Wood becomes stronger in mechanical properties. Unseasoned timber is unsuitable as a structural element in load bearing construction. If seasoned, strength of timber is almost doubled. Uncontrolled moisture results in cracking and splitting of its fibres which reduces its strength.
5. Wood becomes suitable for chemical impregnation. Presence of moisture in unseasoned timber may make timber an easy prey to decay due to weathering. To avoid this, timbers are impregnated with preservatives. This is possible only if the wood cells are empty or do not contain water.
6. Wood becomes suitable for painting and polishing.
7. Wood becomes suitable for gluing. The two members can be effectively joined with the aid of glue if seasoned timber is used.
8. Seasoning lengthens life of timber used in buildings and furniture.

8.31 METHODS OF SEASONING

The whole art of successful seasoning lies in maintaining a balance between the evaporation of water from the surface of timber and movement of water from the interior of the wood to the surface.

Seasoning is sometimes begun before the tree is felled. A strip of bark and wood is completely cut away encircling the stem. This is called girdling. This is the general practice with teak in Burma, the main purpose being to reduce the weight of the timber so that logs will float. The reduction in moisture content secured by girdling is very small, even over a period of years, but six to twelve months are usually sufficient to ensure that logs of species that just sink when green will float after girdling. Teak in Burma is girdled three years before felling.

There are two main methods of seasoning timber :

1. *Air Seasoning.* Sometimes called natural seasoning.
2. *Kilning.* Often called artificial seasoning.

In commercial practice a combination of the two will often be more satisfactory and economical. The so called "water seasoning" is a misnomer. There can be no loss of moisture so long as timber remains water logged. Prolonged storage in water, however, reduces the content of starch and other substances of the sapwood of some species rendering it less susceptible to insect attack. With a kiln, any desired moisture content may be obtained.

Actually there is no difference between air and kiln seasoning, as many timber (especially the refractory tropical hardwood) are best air seasoned first for some time and then kiln seasoned as it is usually impossible to kiln dry them whilst green and therefore preliminary air seasoning is often advisable on economic grounds.

8.31.1 Air-Seasoning

General Requirements for Air-Seasoning of Timber

1. Ground for stacking must be clean, smooth, dry and free from growth.
2. Bark and wood waste must be removed away from the stack.
3. To form a base for the stack, foundations can be made with the help of bricks or concrete blocks placed at a distance of 400-500 mm the height being about 500 mm. Longitudinal members of good, seasoned timber preferably treated are placed on these piers. Planks, etc. can then be stacked on these members with the help of spacers or piling sticks.
4. A good general spacer can be 25 × 40 mm of well seasoned timber.

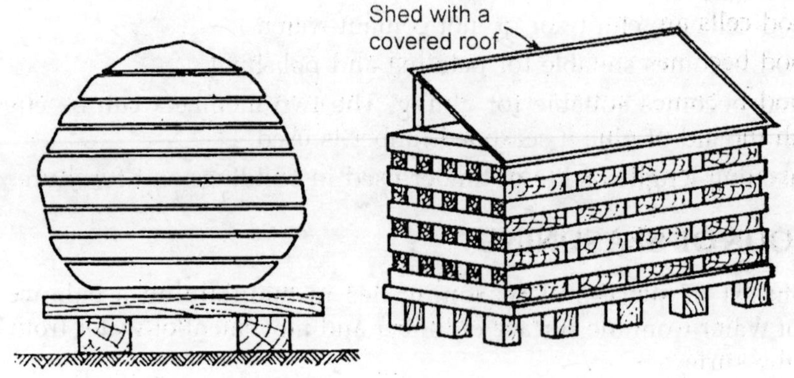

Fig. 8.5. Natural or air seasoning of timber.

5. Spacers should be used for separating the layers of planks and should be of uniform and proper thickness.
6. Spacers should be placed uniformly one above the other.
7. Spacers should be placed 300 to 600 mm apart, the number being what is sufficient to prevent sagging.
8. The width of the stacks should not generally exceed 2 m.
9. As far as possible, timber of different sizes should not be stacked together.
10. At least a space of 300 mm should be left between adjacent stacks.
11. Rotten timber must be immediately eliminated from the rest.
12. Timber should be stacked in such a manner that there is free passage of air to all parts of the stack.

13. Regular inspection helps to reduce decay and prevent its spreading.

Seasoning sheds. To prevent timber from direct exposure to very strong sun and heavy rain, seasoning sheds are constructed with open sides and a covered roof. This is very essential in a tropical country like India where temperature variations during the day are very large. Because of the intense heat during the day, a corrugated iron roof should be avoided. Shingle or rough boarding roof is better.

The precautions to be observed in air-drying in the case of logs, heavy beams or sleepers and planks are given below :

Logs

1. If logs cannot be transported quickly out of the forest, they should be properly stacked.
2. To reduce the possibility of decay, the bark should be removed after felling.
3. Logs should be sawn as quickly as possible and then stacked properly.
4. Painting ends of logs or planks with sealing liquid, or tar or asphalt or shading them with projecting overhanging material like canvas, plywood, etc., helps to reduce end-cracks.
5. In stacking logs, they can themselves be used as spacers.

Heavy Beams or Sleepers

1. Follow general directions, and those given for logs.
2. Stacking and seasoning are easily carried out in the open or under a roof.
3. An easy method of stacking is crossway stacking.
4. Beams or sleepers lying in a slanting position allow rain water to flow off easily.
5. Allow space for air circulation both above and below.

Planks

1. Controlling the circulation of air is the most important factor in the stacking of timber.
2. Uniform spacers should be selected for thickness and length.
3. Spacers should lie neatly one above the other.
4. Planks may be placed, slanting against a support or against each other.
5. Thinner planks should be weighted on top to keep them straight during and after drying.

8.31.2 Kiln Seasoning

Air drying, in general, does not reduce moisture in the timber sufficiently for indoor use in moist climate. Also it is a slow process and takes 2 to 3 years for drying. For these reasons, the accelerated method of kiln drying is often employed.

Kiln seasoning consists of evaporating excess moisture by piling the timber in an enclosed chamber called a kiln, providing maximum control of air circulation, humidity and temperature. These plants are suitable for seasoning timber on a mass scale.

8.31.2.1 Advantages of Kiln Seasoning

1. Drying can be regulated so that shrinkage occurs with a minimum degree. Since temperature controls exist, the rate of drying can be controlled.
2. Lower moisture contents can be reached than with air seasoning. The moisture content can be reduced to any value, generally 5 to 6 per cent by dry weight of timber. This reduction is not possible in the case of air seasoning, in which moisture content cannot be reduced below the equilibrium moisture content.
3. The proccess is very rapid and thus there is a great saving in time.
4. The kiln seasoning has advantages of accuracy and precision.
5. It ensures a dependable supply of good timber in any season of the year. It is a continuous process and is not affected by load.
6. There is uniformity of moisture content in the dried timber.
7. The process not only dries the timber but also sterilises it.
8. The insects are killed and the losses are reduced.

The time required to season the wood in a kiln depends upon a number of factors, of which important ones are :

1. The type of wood to be seasoned, *i.e.,* whether hard or soft wood.
2. Initial moisture content of wood.
3. Efficiency of the kiln.
4. Final moisture content desired.

Usually the time required varies from 1 to 12 weeks.

Types of kilns

There are two types of kilns.

1. Compartment kilns.
2. Progressive or continuous kilns.

8.31.2.2 Compartment kilns

Compartment kilns consist of a number of brick chambers with heating pipes either in the floor or ceiling, fans to keep the air circulating and number of jets to allow the introduction of water or steam. A load of timber is piled inside the chamber and the doors are shut. The piling is done in the same way as in the case of air seasoning. Humid air is blown through the pile and gradually the temperature is raised and humidity decreased until the timber is seasoned. Various instruments record the humidity and temperature during the operation. Test pieces are removed from time to time and the moisture content determined.

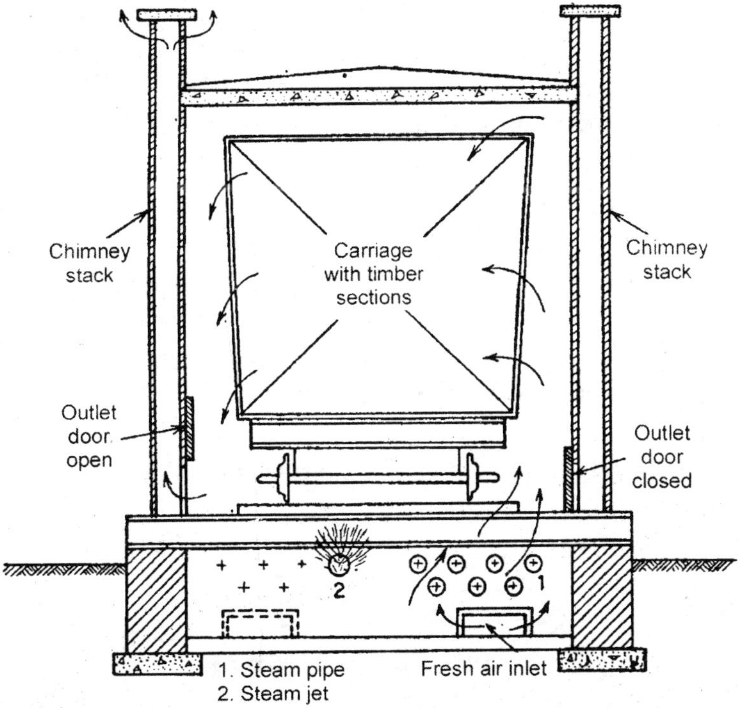

Fig. 8.6. Progressive kiln seasoning of wood.

8.31.2.3 Progressive kiln

This is a continuous kiln and the timber in wagons is charged from one end of the kiln. As it moves slowly in the kiln it gets dried and is discharged from the other end. The hot air is supplied, from the discharging end so that the temperature is less at the charging end and increases towards the discharging end.

Two stage seasoning. A combination method consisting of both air or natural seasoning and kiln or artificial seasoning is preferred in practice on economical grounds. The kilning cost is reduced to less than 50%. The green timber is first converted into the required sections *i.e.* planks, square etc. and air dried to reduce the moisture content to 25 to 30 per cent for 3 to 6 months. These sections are then, kiln seasoned using a short cycle to the desired moisture content.

8.31.3 Other Methods of Seasoning

In addition to the methods described above, sometimes the following methods are adopted in timber seasoning :

1. Electrical method.
2. Chemical method.
3. Steaming or boiling.

Electrical seasoning

This is a very rapid process of seasoning timber. High frequency alternating currents are passed through the timber. The resistance offered by wood to the passage of currents is converted into heat which dries the timber. It may be mentioned that green wood offers less resistance to currents than dry wood. Thus initially the lower temperatures are developed, which increase with the reduction in moisture content. Thus the drying process is gradual.

Chemical seasoning

This is also called salt seasoning. This method is employed in conjunction with the kiln seasoning. Tiie wood to be seasoned is soaked in an aqueous solution of a suitable chemical, usually urea, before it is dried in kilns. The vapour pressure of the chemical salt being less, the surface of the wood draws moisture from the interior, thereby accelerating the process of drying. Thus time is saved in seasoning timber in a kiln.

Steaming or boiling

The air seasoning process is hastened by immersing timber in boiling water or by passing steam around it for some hours. Thus the evaporation will be increased during natural or air drying. The cost is however increased. Also the elasticity and strength of wood is reduced to a certain limit.

DEFECTS IN TIMBER

A defect is an irregularity occurring in or on wood which may lower its strength, durability or utility. Timber being a natural product, is always associated with defects of one form or the other. These defects may occur during the growth of the tree or during conversion and use. The defects are classified as :

1. Defects due to abnormal growth.
2. Defects due to rupture of tissues.
3. Defects during conversion and use.
4. Other defects.

8.32 DEFECTS DUE TO ABNORMAL GROWTH

Knots

Knot is an indication of the branch or limb of a tree which once existed and later has been cut away. The cut away portion is surrounded by subsequent growth, *i.e.* annual rings and the knots are enclosed inside the trunk. Knot is not desirable as it is a source of weakness and makes sawing and plaining more difficult. It may even cause holes in boards after conversion or may result in splits when nails or spikes are driven. Knotty timbers are not used as railway sleepers as it is considered a serious defect, specially near rail seats where spikes are driven. The various types of knots are :

(a) *Dead knot :* A knot not firmly joined throughout to the surrounding wood, sometimes including bark. This is also called unsound knot and is not as hard as the wood itself.

Fig. 8.7. knot

(b) *Decay knot* : *A* knot infected by fungus or insects.

(c) *Loose knot :* A dead knot shrunk away from the surrounding wood.

(d) *Tight knot :* A knot free from decay and held by growth or position in such a way as to retain its place in the piece.

(e) *Live knot :* A live knot is free from decay and cracks and is firmly joined throughout to the surrounding wood. This is also called sound knot.

(f) *Pin knot :* Pin knot is a live knot less than 12.5 mm in diameter.

(g) *Standard knot :* It is a live knot of diameter 12.5 to 40 mm.

(h) *Knot cluster :* Two or more knots grouped together as a unit, with the fibres of the wood defected around the entire unit.

The tensile strength of the timber is reduced more compared to compression and shear strength due to the presence of knots. The magnitude of influence of a knot on the strength of the timber piece depends largely on the nature of the stress to which the timber is subjected and the location of the knot and not on their size. In certain species, when the wood is chosen for a particular architectural effect, knots have a decorative value.

Cross Grains.

Loads acting in direction inclined to the fibres, adversely affect the strength of a wooden piece. Thus any slope of grain due to-natural growth will create stress to loads acting normal to the cross-section of the piece of timber and, therefore, the resistance of the wood fibres to these stresses is low. Such natural inclinations of grains with respect to the axis of a piece of wood are known as cross grains.

Twisted Fibres.

This defect is caused when the tree gets twisted due to the wind blowing in one direction only. The fibres are twisted longitudinally and unequal deposits of tissues are caused inside the trunk. Such timber is structurally useless as many of the fibres will be cut through in squaring. Moreover it will have layers of different age, density, direction of grain etc. Such timber can, however, be used with advantage as masts (long upright poles of timber) in unsawn condition, the twisted fibres add to the strength in such cases.

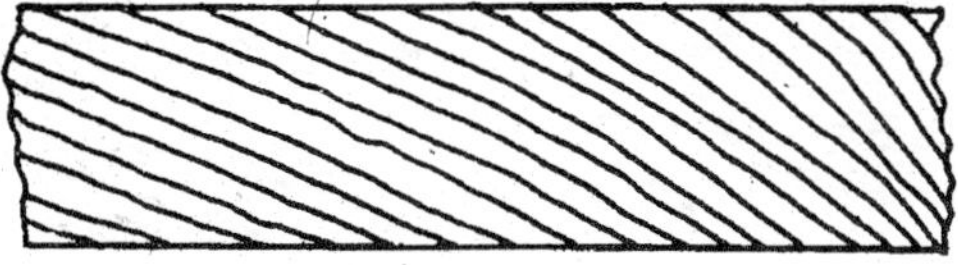

Fig. 8.8. Twisted fibre

Rindgall, Burls or Excrescences.

These are either due to injuries inflicted to the tree when it is young or due to unsuccessful attempts at the formation of branches.

Due to subsequent deposition of annular rings, peculiar curved swellings are caused over the remnants of injured branch.

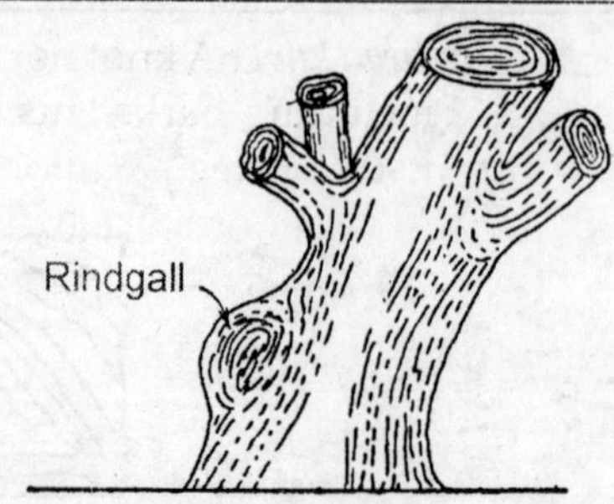

Fig. 8.9. Rindgall.

Upsets.

These defects are visible on that portion of the timber in which the fibres have been injured by shock or crushing during the growth of the tree. The fibres are more or less broken across the log, so that there is every likelihood of a break occurring.

Fig. 8.10. Upset

8.33 DEFECTS DUE TO RUPTURE OF TISSUES

Wood shrinks on drying and this causes rupture of tissues resulting in cracks or splits. The cracks are called shakes and are classified as :

(*i*) Heart and star shakes,

(*ii*) Radial shakes, and

(*iii*) Cup and ring shakes.

Heart and Star Shakes.

A heart shake is a crack in the heart wood. It starts from the centre, *i.e.* the pith and extends in the direction of the medullary rays or at right angles to the annual rings. Thus it causes radial ruptures in the annual rings.

Usually a number of cracks radiate from the pith and due to the shape of the ruptures or cracks, the defect is commonly known as star shake.

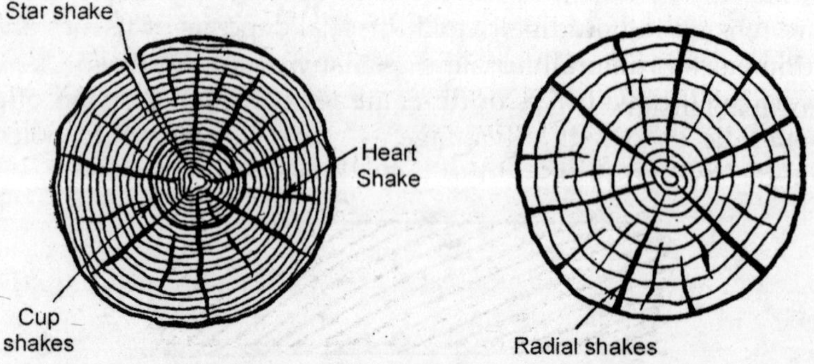

Fig. 8.11. Different types of shakes.

This defect occurs in over-matured trees. The oils, resins, gum and moisture present in the heart wood cells get evaporated and the wood starts drying and cracking from the pith outwards. A living tree, should, therefore, be felled immediately after it has attained the age of maturity to obtain the maximum quantity of good timber.

The other cause of this defect could be the quick drying of the pith, if a tree nearing maturity, is left unused for a long time after felling.

Radial Shake.

A radial shake is a cracking starting from the outside, *i.e.* sap wood and extending towards the centre, *i.e.* the heart wood and pith.

This occurs when the outer tissues of wood dry at a faster rate than the inside ones. This defect is common in seasoned wood and is caused during the process of seasoning due to excessive heat of the sun or cold or frost.

Cup and Ring Shakes.

These are cracks along the annual rings, *i.e.* rupture takes place in a circular direction across the cross-section of a log. If the crack extends, only a part round, it is called cup shake and if the whole way round, it is known as ring shake. These are vertical ruptures between two consecutive annual rings and occur mostly in trees growing on the windward side of the forest.

Fig. 8.12. Different types of shakes.

Wind Cracks

These are shakes on the outside of the timber (*see* Fig. 8.12) resulting from the shrinkage of the exterior surface by atmospheric agencies.

8.34 DEFECTS DURING CONVERSION AND USE

These defects arise due to improper seasoning resulting in unequal shrinkage of the outer and inner tissues of wood. The common defects are :

1. *Bow :* is a curvature of the timber in the direction of its length.
2. *Cup :* is a curvature in the transverse direction of the timber.
3. *Spring :* is a curvature of the timber in its own plane.
4. *Twist :* is a spiral distortion along the length of the timber.
5. *Radial and cup shakes :* have been explained before.
6. *Case hardening :* result due to unequal drying of the exterior and interior surfaces. The exterior surfaces are under compression due to rapid drying while the interior surfaces are in tension. This usually occurs in the bottom sections of heavily loaded stacks kept for seasoning.
7. *Honey combing :* is the separation of the fibres in the interior of the timber resulting from drying stresses.
8. *Split :* is separation of fibres along the grain and extends through the plank from one surface to another.

Fig. 8.13. Defects due to improper seasoning.

The other defects occurring during use of timber are :

Chip Marks. These are the bruises on the dressed surface of the timber made by chips or other small pieces impressed by the rollers or pressure guides of a plaining machine.

Chipped Grain. It is a defect caused by the breaking away of timber below the surface of dressing by the action of the plaining tool or cutter. This is also termed as torn grain.

Wane. The original rounded surface of a tree which remains on the piece of the converted timber.

8.35 OTHER DEFECTS

These are :

1. *Boxed Heart.* The growth in the centre fully enclosed within one piece in the process of conversion.
2. *Dead Wood.* Is the timber obtained from over-matured trees. Its signs are low strength, abnormal pinkish colour and light weight.
3. *Figure.* The ornamental markings on the surface of timber produced by the relative arrangement of the different elements of the timber or by inherent colour matter.
4. *Pitch Pocket.* Well defined lens shaped openings which. contain or have contained resin. This may occur between the growth rings of certain coniferous trees. This is also called resin pocket.

STAINS IN TIMBER

These are discolorations found on the surface of wood, particularly on the light coloured varieties. They can be grouped into the following categories :

1. Those caused by harmless fungi such as the mould fungus and sap staining fungus. Fungi are plants, made up of fine threads (hyphae), invisible to the naked eye unless they are massed or matted together.
2. Those due to wood rotting fungi. Their effect is called dole. This is an incipient decay.
3. Those which occur during the growth of the tree and are caused by soil conditions or frost. An example is mineral stain.
4. Those due to chemical contamination such as iron stain.
5. Those caused by the oxidation of the wood cells. Change of colour may occur by weathering or bleaching or darkening caused by sun-light.

The various types of stains are described below :

Mould Fungi. Grow only on wood with a moisture content greater than 20 per cent. They live on the plant food material, *i.e.* carbohydrates in the cells of the wood. They do not attack or breakdown the cell walls and thus the strength of wood is not affected.

They cause stains in wood of green, grey-blue, pinkish and occasionally white colour. These stains can be removed by light brushing or plaining or rubbing with a sand paper.

Sap Stain Fungi. Sap stain-fungi occur on wood with a moisture content varying from 27 to 30 per cent. The stains are of varying colour, light blue to steel blue in soft woods and grey in hard woods. The fungi do not attack the wood.

Dote. This is a stain either grey or reddish or brown. It is not of much importance unless it is associated with damp or unventilated conditions ; in which case serious attack of the wood may occur.

Mineral Stain. This is grey or brown coloured streaky discoloration and is common in hard woods. This can be removed by bleaching. The strength of wood remains unaffected in most of the cases.

Chemical contamination. This is formed in timbers used for laboratory benches or draining board, but only a few timbers are affected. Acids cause pink or red colours and alkalies green or blue.

Iron Stain is a special type of chemical contamination and is caused by the presence of tannin (various organic compounds in tree barks) in wood ; but only a few timbers contain sufficient tannin to give rise to this kind of stain, such as mahogany, fir etc.

8.36 REMOVAL OF STAINS

Superficial stains can be removed by sanding or plaining ; and a mild form of bleaching agent *e.g.* hydrogen peroxide can be used on finished surfaces. Light coloured stains can be removed by household bleaching agents or ink eradicators. For use on wood, several two-stage hypochlorite bleaching agents are marketed.

DECAY OF TIMBER

Decay is not due to the chemical action of the soil or to the fermentation of the sap, but is the result of the action of certain low forms of plant life called fungi.

Thus the life of a building timber can be almost unlimited if it is kept free from the attack of organisms such as fungi, insects or marine borers. This can be accomplished by keeping timbers out of contact with air *e g.* by completely or permanently immersing timbers in water or burying them deeply. The principal causes of decay are :

1. Presence of sap and moisture.
2. Exposure to alternate wet and dry atmosphere.
3. Moisture accompanied by heat.
4. Lack of ventilation.

Decay accelerates in the following circumstances :

1. When the timber is not seasoned or improper seasoning is done.
2. Using of timber in exposed positions of a building.

3. Using timber underground without a preservative treatment *e.g.* fence posts.
4. Use of timber in structural work such as beams, girders, battens etc. without the application of a paint or preservative or protective covering *e.g.*, slignam.
5. Application of paint to a wet timber will set in interior decay due to fermentation and decomposition of the wood cells and their contents.
6. When wood is attacked by fungi, insects and marine borers.

8.37 FUNGAL ATTACK OF WOOD OR WET AND DRY ROT

Rot is a type of decay in which there occurs decomposition and putrification accompanied by the evolution of gases, mainly hydrogen and carbonic acid gas. Rots are of two types *viz.*, wet rot and dry rot.

In wet rot, the gases evolved during chemical action can escape. Wood tissues, especially those containing sap, are decomposed. Although about 20 percent moisture is essential for decay, yet a higher percentage is required for wet rot. Wet rot can occur even when the tree is standing.

Dry rot occurs only in dead timbers and the principal cause is the lack of ventilation. This helps in the growth of fungi especially in the presence of moisture and warmth. Dry rot sets in first in sap wood. The fungi eat the timber, renders it brittle, lowers the cohesion of its fibres and ultimately reduces wood to powder.

Fungal growth and attack. Fungus cannot support its own growth like ordinary plants by deriving their food from the air by decomposing carbon dioxide. It derives its food from the wood tissues and attaches itself to wood as a parasite and obtains its nourishment second-hand. The growth of fungi starts from minute spores (reproductive cells) or seeds and grow in the form of a soft cotton wool like mass called branching strand (mycelia).

The fungi, causing dry rot, attack cellulose tissues of wood. This fungus is called *Merulius Lechrymans.* It flourishes in damp condition and lack of ventilation at temperatures between 15°—20°C. This germinates as mycelia which can penetrate brickwork or plaster. The timber affected by these fungi, loses moisture, weight and strength ; appears dull brownish in colour and crumbles in fingers. The spores of this fungus are only 0.01 mm in length and can be blown about by a light breeze in the form of a reddish-brown dust. Dry rot can be detected by a musty smell, bending or collapse of wooden sections and the ability to push a knife blade into the wood with little effort. These fungus cannot work in dry wood. There is no such term as 'dry rot' ; the name is derived from the decayed wood which is often dry after it has rotted ; but not while decay is taking place.

8.38 ATTACK OF WOOD BY INSECTS

Wood is also attacked by certain insects, common ones are :

1. *White ants or termites.* These insects pose a serious problem to timber users, especially in a tropical country like India, because they cause damage and loss to timber, even to well seasoned varieties. White ants do not expose themselves to light and build mud tunnels. They live in a colony with a queen ant which alone lays eggs and is larger in size than others. The age of the queen is about ten years and it lays millions of eggs in this period.

The attack of white ants is very severe in soft woods like mango, kail, etc. Hard woods *e.g.* teak, sal etc. contain certain chemical products in their resins which offer resistance to attack of white ants and thus their action is delayed in such timbers. White ants have reduced sound wood structures into hollow shells.

2. *Borer Beetles.* The characteristics and effects of the common types of borer beetles is given in Table 8.3.

8.39 ATTACK OF WOOD BY MARINE BORERS

These flourish in marine or blackish waters only especially in tropical regions. Common types are *teredo, banksia* and *martesia.* These are soft-bodied, worm-like creatures ranging from few centimetre to as much as 3 metre in length. A destructive form is *Teredo navalis.* Marine borers can attack almost any timber. These bore holes in timbers but do not eat the timber. They bore holes to the extent of 20 mm diameter and 50 to 80 mm length.

There is another form of marine borer known as *limnoria,* which eats the wood and honey-combs it, resulting in minute tunnels and ultimate collapse of the structure. It does not attack teak-wood and cannot live in fresh water or under sand.

Table 8.3 Common types of borer beetles

Type of beetle	*Death-watch*	*Powder-post*	*Common furniture*	*House longhorn*	*Ambrosia*
Normal life cycle	3-4 years	1-2 years or more	2 years or more	2—11 years or more	6—12 weeks
Emergence	April—June	May—September	June—August	July—September	All year
Number of eggs laid	Up to 70	30—50	20—40	20—200	———
Points of attack	Decayed timber	Felling site ; sawmills	Old furniture	Roofing timbers	Felling site ; saw mills
Nature of dust	Granular with oval pellets	Fine flour like powder	Finely granular, with cylindrical pellets	Fine powder, with cylindrical pellets	No dust in funnels

8.40 DETECTION OF INCIPIENT DECAY

It is a simple matter to recognize advanced or typical decay. But recognition of early or incipient decay is much more difficult and in some cases is practically impossible without microscopical examination of the wood. Nevertheless, there are discolourations and certain other characteristics produced by such decay that, with a little experience, it can be relied upon ordinarily to denote the presence of incipient decay and to distinguish it from the harmless discolouration. The following criteria are offered as a guide in this respect. No instructions can be followed very successfully without thorough acquaintance with the normal colour variations in the wood of the particular species.

General Characteristics of Incipient Decay

These are

1 . *Abnormal Colour*

(*a*) Any discolouration extending from zones of obvious decay indicates infection.

(*b*) Areas in which the colour of the wood is some shade of brown not in the ordinary range of brown for the normal wood of the species may be infected. Shades of brown caused by decay are variable ; in addition to what might be called "pure" brown, such shades as yellow brown, greenish brown, and greyish brown are common in hard woods. In coniferous woods, distinctly in Douglas-fir the earliest decay discolouration is often a shade of purple. In some cases, the decay fungus soon bleaches the wood to a cream, straw, or yellow colour.

(*c*) Aside from the colour itself, an abnormal mottled appearance of the wood often indicates decay.

(*d*) If a discolouration is limited to certain growth rings it is not so likely to be decay as when it spreads across the rings.

(*e*) Decayed wood commonly has a duller colour than normal wood, and is sometimes described as having a "dead appearance".

(*f*) Some decays, more commonly in hard woods than coniferous woods, produce what are known as zone lines. These are very narrow black, orange, or yellow lines that tend to run somewhat in the direction of the grain. They are often most prevalent at or near the border of the most conspicuously discoloured areas. Sometimes they are associated with very little general discolouration, but their presence is a certain evidence of decay.

(2) *Softness and Brashness.* Abnormally soft or brash wood should be suspected of decay even if no abnormal colour can be detected.

(3) *Roughness of Surface.* In some cases wood containing even incipient stage of decay, is almost always rough.

(4) *Pitting of the Wood.* Many decays produce few to numerous small pits or pockets in the wood instead of any general destruction or instead of being pitted, the infected zones may simply contain small spots that are softer and more intensely discoloured than the intervening wood. In either case the wood should be rejected.

8.41 CONTROL OF DECAY

A prime essential in preventing decay between felling of the tree and use of the lumber is to get the wood air dry as rapidly as possible without excessive warping and to keep it dry. Decay fungi cannot develop in wood that has a moisture content of 20 percent or less. Logs cut in seasons when the temperature is higher than about 5°C should be sawed as soon as is practicable. The freshly sawed lumber should be either kiln dried or dried in accordance with good air-seasoning practice, and thereafter should be protected from rain or other sources of moisture until used. Storage directly on the ground should be particularly avoided.

Logs cannot, of course, always be utilized promptly, nor can lumber always be air dried rapidly. In either instance, a large measure of temporary protection can be inexpensively provided by treating the surfaces of the wood with water solutions of appropriate toxic chemicals. Organic mercurial and phenolic chemicals are marketed for the control of blue stain in both logs and lumber, and the same ones are effective against decay for limited periods. Directions for using them are provided by the distributors. Water storage is often employed for logs.

PRESERVATION OF TIMBER

8.42 DETERIORATION OF WOOD

Although no timber is immune to deterioration and ultimate disintegration if exposed for a sufficiently long period of ordinary atmospheric conditions, the serviceable life of individual pieces of wood may vary considerably, depending on the species concerned, the amount of sapwood present, the use to which the timber is put, and the situation and atmospheric conditions to which it is exposed. For example, sound wood has been recovered from the Egyptian tombs and from piles driven into mud hundreds of years ago. In these, and similar instances, preservation is to be attributed to protection from the atmosphere rather than to inherent durability of the timber used. On the other hand, even the most durable timbers, may last only a few years if placed in a warm, damp and badly ventilated position.

The principal causes of deterioration of wood in service are fungal infection, termite or other insect (marine-borer) attack, mechanical failure and fire. The resistance of a timber to these agencies of destruction may frequently be decreased by the use of a suitable chemical, applied as a preservative. In practice, preservatives are usually applied to non-durable timbers so as to render the treated wood sufficiently resistant to the agents of deterioration to warrant its replacing a naturally durable, but more expensive timber. There are a number of chemicals and proprietary products used, and several methods of application are advocated. The selection of the most-suitable chemical and method of treatment are of the utmost importance and must be based on a thorough understanding of the scope and limitations of preservative treatments. No method confers complete immunity and a treatment suitable for one particular set of conditions may not be useful for others.

It is important to keep in mind the special circumstance of each particular job when selecting a wood preservative. For example, a substance that is readily soluble in water may be excellent for indoor use but worthless for outside work, and conversely, a substance with a pronounced odour may be satisfactory for outdoor work, but totally unsuitable for indoor use. With reservation, even the best preservative only prolongs the life of wood, it does not confer immunity from attack for ever.

8.43 ADVANTAGES OF TREATED TIMBER

Advantages derived in the extensive uses of treated wood are as follows :

1. *The Annual Cost of the Material is Reduced.* If the untreated wood lasts 5 years in service, the treated wood would last at least 15 to 20 years making considerable savings in annual costs.

2. *Cost of Replacement is Reduced.* Using the above example, at least three replacements can be avoided apart from the fact that the basic cost of material during the last 15 years has more than doubled.
3. *Cost of Maintenance is Reduced.* Treated wood generally requires hardly any or no maintenance.
4. *Eases Timber Availability.* Owing to the depletion of valuable timbers the required quantity and grade of the better timbers is impossible to obtain in large quantities. Treated timber fills this gap in a way compromising to the required properties. Valuable timbers can be channelled for either exports or in such places where they can not be replaced.
5. *Special Uses of Treated Wood.* It has been found in certain basic uses such as poles used in electrical supports, railway sleepers, cooling towers, etc., that replacements with substitute materials are either more expensive or unavailable. In such places secondary species properly treated, not only provide with better wood utilisation but also meet an important immediate need.

8.44 DESIRABLE PROPERTIES OF PRESERVATIVES

Properties desirable in an ideal wood preservative are enumerated below, they can be useful as a basis of comparison. The ideal wood preservative should be :

1. Highly poisonous (toxic) to fungi and insects.
2. Readily penetrating into wood.
3. Chemically stable, (*i.e.* not readily volatilised or easily decomposed) and for outdoor use, not easily leached out.
4. Easy to apply and not dangerous or harmful to those applying it or subsequently handling the timber.
5. Non deleterious in effect on the timber treated.
6. Cheap and readily obtainable.
7. Non-corrosive to iron, steel, or other materials, according to circumstances.
8. Fire-resistant, or at least not liable to increase the inflammability of wood.

   The following additional qualities are sometimes, important especially for indoor application.
9. Odourless.
10. Colourless and free from effect on subsequent painting and finishing process. Paint is not a preservative in the true sense. Its action is purely a weather-proofing one, while a preservative is entirely water proofing. Paint prevents the sudden wetting of the surface and protects it from mechanical weathering. For this reason paint must not be allowed to crack and scale off.

8.45 TYPES OF PRESERVATIVES

There are three broad groups of preservatives :

1. Tar-oil type (creosote etc.)
2. Water-soluble type (sodium fluoride, zinc chloride, etc.)
3. Solvent type (mostly proprietary solutions).

Tar-oil type. Derived from coal or wood tar and includes what is, perhaps, the best known preservative, creosote. This type is very toxic to fungi and insects and resistant to leaching, and is very suitable for exterior work. It is however, characterised by an odour which renders it unsuitable for use near foodstuffs, as these are liable to become tainted. Creosote shares the above characteristics, and it may by added that it is also cheap, and does not corrode metals and is easily absorbed by the sapwood of most non-durable timbers. It is difficult to paint over, but special primers may be used to overcome this trouble.

Water soluble type. Consists of chemicals which form a toxic solution. The type is characterised by being generally non-tainting, non-inflammable, and non-straining. Some are corrosive to metals and if used for exterior work, should be forced well into the wood by pressure treatment ; otherwise there will be a tendency to leach out. After drying, all these can be painted over. Some of the chemicals' most generally used are sodium fluoride, zinc chloride, zinc sulphate, mercuric chloride, copper sulphate, copper chromate zinc meta arsenite. etc. Some are used in combination with others to reduce the leachability. Some of them are poisonous to humans, as well as to fungi (for example, arsenic and mercuric salts), and should be used with care.

Where long distant freight charges are involved it is economical to use concentrated chemicals and solids without the solvent. It must not be overlooked that the pressure treatment of timber with water-soluble salts increases the moisture content of such treated timber appreciably, and the timber must be allowed to dry out adequately .before it is used, otherwise some distortion or shrinkage problems may be experienced.

Solvent types. Preservatives of this group consist of toxic substances dissolved in a volatile solvent. They are resistant to leaching can be painted over when dry and have good penetrating qualities. Most of these preservatives are highly inflammable and so require care in handling. Some have definite odours which tend to taint foodstuffs. Such preservatives are generally more expensive than water soluble preparations.

8.46 APPLICATION OF PRESERVATIVES

The three principal methods are (1) brush coating, (2) dipping and (3) pressure treatment. Just getting some definite quantity of a preservative into a timber is no guarantee that the treatment will prove satisfactory. Depth and uniformity of penetration depends on the structure of the wood. Sapwood absorbs preservatives more readily than heart wood, the early wood better than the late wood, and timber of same species more readily than that of others. Because of its better absorptive power, sapwood should be retained. Better penetration is obtained with hot than with cold methods of application. Air-dry timber absorbs preservatives better than green wood. Timber should therefore be seasoned before treatment. This also ensures that checking defect will occur before and not after treatments and the checks themselves facilitate impregnation. All trimming and boarding, etc. of timber must be done before the wood is treated, or, when this is impossible to arrange, surfaces subsequently exposed should be coated heavily with the preservative.

Brush Coating. This is the simplest method of applying a preservative, but it is perhaps the least efficient. Two or more coats should be given and the treatment repeated every two or three years. This gives some measure of protection, particularly with interior woodwork, not exposed to mechanical wear.

Spraying and Dipping. Spraying is usually not so effective as brush coating, but dipping is more effective, because the wood is in contact with an excess of preservative for the duration of the dipping process. The length of the immersion period will control the amount of preservative absorbed. The absorption will be even better if the preservative can be heated.

Open Tank Method. A special form of dipping, in effect a modified pressure treatment, known as the open tank method, is much more effective than brush coating or dipping.

The method requires a tank with some form of heating apparatus, in which the timber to be treated can be completely immersed. The timber is submerged in the cold preservative which is then gradually heated to a temperature of 75° to 90°C. This is held for $\frac{1}{2}$ to 4 hours depending on the timber, the preservative and the absorption required. The preservative is allowed to cool with the timber still immersed.

The simplest form of heating is external firing, but because of the risk of fire, steam-heated coils inside the tank are to be preferred.

Pressure Processes. The most effective method of treating is the use of a pressure process for forcing the preservative into wood. Penetration is obtained to a considerably greater depth than is possible . by any other method.

The usual type of plant consists of a large cylinder, upto 55 cubic-metre capacity, a storage tank, a pressure chamber, and steam heating boiler. The wood is run into the cylinder on trolley, the air is exhausted to form a vacuum and the preservative forced into the cylinder under pressure of about 1.5 N/mm^2 and at a temperature of 85°C. The air is withdrawn from the wood cells, which then get filled up by the preservative.

In the full-cell method the preservative is allowed to remain in the cellular spaces. In the empty-cell method the free preservative is withdrawn by suction and only that absorbed by the wood tissue remains. The full-cell process is too expensive for ordinary purposes and the usual method is to allow for a definite absorption per cubic metre. Different preservatives require some variation in method.

For refractory timbers the absorption of preservative can be increased by incision. This consists of making incision parallel to the fibres to the depth of penetration required. An increase of 30 to 60% absorption may be obtained in some cases.

Boucherie. Process. This is generally used only with water soluble preservatives. Poles, bamboos, posts, etc. are treated immediately after felling, in the green state. The preservative is supplied from a raised tank, under a hydrostatic pressure through a pipe, to the base of the log in a horizontal position. The preservative solution replaces the sap and the treatment is complete when it appears at the top end of the log.

RESISTANCE OF WOOD AGAINST FIRE

Many people regard timber as highly inflammable. This is quite untrue. Some species of timber are quite resistant to fire, *e.g.* teak, panauk etc. and timbers in large sizes are very difficult to burn. This has been proved many times where large beams have withstood exposure to intense flames for fairly long period. Only a short depth of the beam gets charred after such exposure.

Several methods are now employed for making wood fire resistant. One method to exclude air from the wood is by means of concrete, cement asbestos, metal and special paints. More effective than these is a chemical retardant, the most widely used being ammonium, phosphate, with which the timber is treated.

Another method is to use a chemical giving off gas when heated and which will mix with the combustible gases given off by the wood and render them non-inflammable.

WOOD PRODUCTS

The natural resources of the supply of timber are limited. Moreover, land under forests is cleared and used for other purposes. It takes good number of years for a new tree to mature and to supply timber from it. The demand for natural timber especially for construction purposes has far outstripped the production. As a matter of fact, the shortage of timber is not only to be found in India, but it is world-wide. Realising this, the advanced countries have started producing a variety of timber based products using timber waste and secondary species of timber. Examples are plywood, hard boards, soft boards, particle board etc. In India, by now there are a large number of factories which are producing these wood-based panel products and their use in construction is increasing.

8.47 PLYWOOD AND ALLIED PRODUCTS

8.47.1 Veneers

Veneers are thin sheets of wood which are knife cut by various processes. They vary in thickness from 0.4 mm to 6.0 mm. or more. Walnut is a very suitable wood for veneers. In India, the supply of this wood is limited. Other Indian woods used for this purpose are teak, sisso, elm, panauk and Indian rose wood.

Veneers are used in construction where lightness, decoration, durability, non-shrinkage and stiffness are required. They are mostly used for the manufacture of ply-wood, laminated boards, batten boards etc., for which veneers are jointed by means of adhesives such as glue. The defects present in the natural timber can be eliminated in veneers by discarding the portion containing defect.

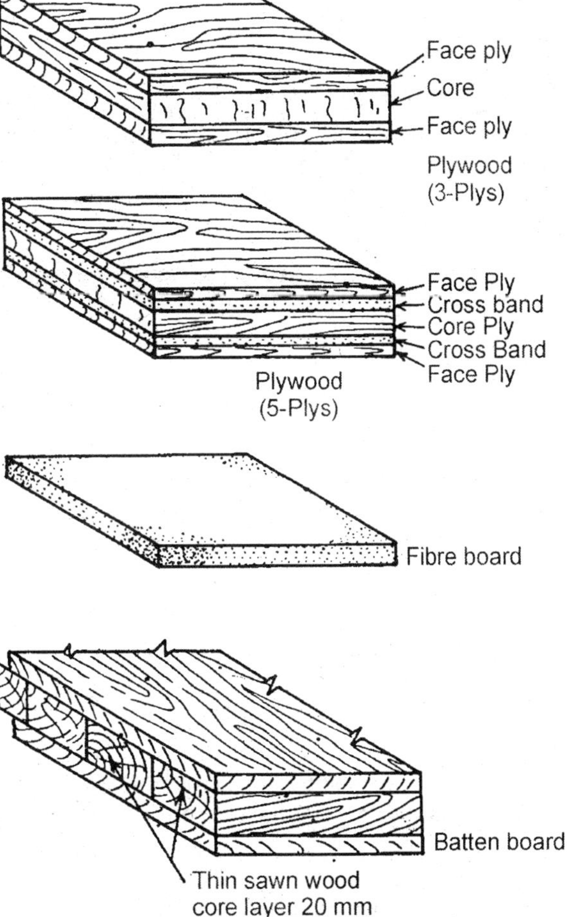

Fig. 8.14. Different wood products.

The four different types of veneers, distinguished by different cutting processes are :

(*i*) Rotary veneers,

(*ii*) Sliced veneers,

(*iii*) Half round veneers,

(*iv*) Sawed veneers.

After cutting the natural timber into thin layers, all the portions containing defects such as knots etc. are removed. They are then seasoned in a kiln to reduce moisture content to the desired amount. The slices are then joined at their edges with glues. This process is known as edging. Thus a continuous roll of veneer is formed, out of which sheets of desired dimensions can be cut.

8.47.2 Plywood for General Purposes

Plywood is made by cementing together thin boards or sheets of wood into panels. There are always an odd number of layers 3, 5, 7, 9 etc. The plies are placed so that the grain of each layer is at right angles to the grain in the adjacent layer.

The chief advantage of plywood over a single board of the same thickness is the more uniform strength of the plywood along the length and width of the plywood and greater resistance to checking and splitting with the change in moisture content.

Usually synthetic resins are used for gluing. Phenolic resins are usually cured in a hot press, which compresses and simultaneously heats the plies between hot plates which maintain a temperature of 90°C to 140°C and a pressure of 1.0 to 1.5 N/mm^2 on the wood. The time of heating may be anything from 2 to 60 minutes depending upon thickness.

When water glues are used the wood absorbs so much water that the finished plywood must be dried carefully. When synthetic resins are used as adhesives, the finished plywood must be exposed to an atmosphere of controlled humidity until the proper amount of moisture has been absorbed.

Plywood for general purposes are of three grades depending upon the adhesives used for bonding the veneers : and it is further classified into six types namely AA, AB, AC, BB, BC and CC based on the quality of the two faces, each face being of three kinds namely A, B and C. After pressing the finished plywood should be reconditioned to a moisture content not less than 8 per cent and not more than 16 per cent.

Dimensions.

The dimensions of the plywood boards are given in the following order :

The first dimension represents the length, that is the dimension parallel to the grain of the faces ; the second the width ; and the third, the thickness.

Size.

Unless otherwise specified, plywood boards are of the sizes given below :

2400 × 1200 ; 2400 × 900 ; 2100 × 1200 ; 2100 × 90 ; 1800 × 1200 ; 1800 × 900 ; 1500 × 1200 ; 1500 × 900 ; 1200 × 1200 ; and 900 × 900

Thickness.

Plywood boards are made of the thickness given below :

Table 8.4 Thickness of Plywood Boards

Board	*Thickness*
3 Ply	3 mm
	4 mm
	5 mm
	6mm
5 Ply	5 mm
	6 mm
	8 mm
	9 mm
7 Ply	9 mm
	13 mm
	16 mm
9 Ply	13 mm
	16 mm
	19 mm
11 Ply	19 mm
	22 mm
	25 mm
Above 11 Ply	As ordered

Uses of Plywood.

Plywood is used for sub-flooring, wall and roof sheathing, webs of arches, and built-up girders of section I. Plywood is also used extensively in prefabricated construction and for lining concrete forms when a smooth-finish of concrete is desired. For interior work plywood is used for walls, panelling, sub-flooring, cabinets, counters, etc. Plywood may be obtained with plain or embossed surfaces or surfaced with hard-wood veneers of almost unlimited variety. Plywood surfaced with plastic has been developed for use in concrete formwork. Since its surface is smooth and requires no painting, it may be reused many times.

8.47.3 Veneeıed Decorative Plywood

Decorative plywood is manufactured either by bonding the various layers of veneers in one operation or by bonding decorative veneers on the synthetic resin bonded plywood of suitable thickness. After pressing the plywood should be reconditioned to a moisture content not more than 15 per cent.

When the decorative veneer is used on one side of the plywood only, a suitable back of required thickness balancing the strength of the decorative face veneers should be provided.

Uses. Decorative plywood is used for furniture making, panelling of all kinds and interior decoration.

8.47.4 Laminated Lumber

Solid timbers in long lengths and of substantial cross-section have always been difficult to obtain, since long straight logs of adequate size, without defects are scarce. Drying such timbers is difficult, requiring many months and usually result in splits, twists, and other deformation.

Laminated lumber fabricated from standard boards, glued together into timber sizes, while somewhat more costly overcomes most of the difficulties enumerated above. Lumber in board form can be kiln dried without any serious deterioration and glued up to any size. End joints adopted to anticipated loadings can be staggered in different layers and permit the use of relatively short boards.

Adhesives are now available stronger than the wood itself, and of maximum durability under severe weather and marine exposure. In laminated lumber construction, selected material can be used in outer layers, where maximum stresses are encountered and materials of lower grade in the lower layers. Laminated lumber assemblies can be bent during the gluing process into curves for bow string trusses, cantilevers, arches and other structural requirements. The versatility of laminated lumber is much greater than the solid timbers.

8.47.5 Blackboards

Blockboard is a board having a core made up of strips of wood, each not exceeding 25 mm in width, laid separately or glued or otherwise joined to form a slab which is glued between two or more outer veneers with the direction of the grain of the core blocks running at right angles to that of the adjacent outer veneers.

Blockboards are of two types *i.e.,* Type 1 and Type 2 ; and each type is further divided into two grades, *i.e.* Grade 1 for exterior use and Grade 2 for interior use.

Commercial Type 1 Blockboard. These are blockboards with faces of commercial timber and are used in ordinary furniture, table tops, partitions and panelling to be painted over.

Decorative Type 2 Blockboard. These are blockboards with ornamental veneers on one side or on both sides and used in high class furniture, panelling, interior decorations, partitions etc.

Dimensions

(*a*) *Thickness.* Blockboards are manufactured in the following thicknesses in mm : 12, 16, 20, 25, 30, 32, 35, 40, 45 and 50.

(*b*) *Sizes.* Blockboards are manufactured in the following sizes.

Length (mm)

2400, 2100, 1800, 1500 and 1200

width (mm)

1200 and 900

Uses. Blockboards are used for furniture, wall panelling, ceilings, partitions etc.

8.47.6 Batten Board

A board having a core made up of strips of wood usually 80 mm wide, each laid separately or glued or otherwise joined together to form a slab which is glued between two or more outer veneers with the direction of the grain of the core battens running at right angles to that of the adjacent outer veneers.

8.47.8 Lamin Board

A board having a core of strips, each not exceeding 7 mm in thickness, glued together face to face to form a slab which in turn is glued between two or more outer veneers, with the direction of the grain of the core strips running at right angles to that of the adjacent outer veneers.

8.47.8 Chip Board

Chip board is manufactured from wood shavings and synthetic resin. Wood shavings are impregnated with synthetic resin and then pressed into boards.

8.48 HARD BOARD

Hard board is manufactured from the following categories of wood raw-materials :

(*i*) Wood waste obtained from saw mill, veneer, furniture making, etc.

(*ii*) Short logs and broken logs obtained during process of logging.

(*iii*) Cut timber which are incapable of bearing the cost of harvesting.

(*iv*) Non-commercial species which are not useful for conversion into sawn wood or pulp.

Manufacture. For the manufacture of hard-board the wood is converted into chips by using high speed chippers. The chips are then softened with steam and are converted in fibres in a rotating disc mill. During this process water repellent agents like paraffin-wax or drying oils are added and if the board is desired to give increased strength resins like phenol formaldehyde are also added at this stage.

For the formation of boards the process of wet felting and air felting are used. In the wet felting, from a low consistency pulp suspension a sheet or mat is formed. In the case of air felting, the mat is formed from comparatively dry fibres with the help of air or mechanical means. The felted fibre mat obtained by both the methods is pre-compressed before hot pressing. The process employed for pressing may be wet pressing or dry pressing. When the moisture content of the mat is high the wet pressing is done. For this wire screen is used to permit escape of steam during pressing. Such boards show screen marks on one side. When the moisture content of the mat is low enough, no wire-screen is used and the board is just formed by dry pressing and is smooth on both sides.

The boards are given either heat treatment or oil tempering to increase their strength and resistance to moisture. The boards are then subjected to humidification treatment for conditioning them to the equilibrium moisture content. After these treatments, boards are trimmed in required size for use. In case of oil tempering the boards, trimming is done before tempering.

Following are the advantages in the use of hard boards over the sawn wood :

(*i*) Being a homogenous panel hard-boards have uniform strength in all directions parallel to the surface.

(*ii*) As it is produced synthetically, hard-boards are dimensionally more suitable than sawn wood.

(*iii*) Natural defects of wood like shakes, knots, etc. are eliminated on the formation of hard-boards.

(*iv*) Hard-boards have better sound insulation and sound damping properties.

(*v*) Hard-boards are available in large sizes up to 5 × 1.25 m. Their use avoids expensive carpentry work.

(*vi*) Hard-boards are factory made products and therefore standard quality hard-boards are available.

(*vii*) Hard-boards can be manufactured such as pre-perforated boards, strained, embossed, wood grained, enamelled, plastic faced and veneered finish.

The different types of hard-boards which are available and suitable for specific use are as follows :

Standard hard-boards for wall and ceiling linings and panels, partition, flush-doors, furniture, etc.

Tempered hard-boards as above and particularly their additional strength and resistance to moisture and abrasion are needed. For example, for exterior uses, concrete formwork, etc.

Perforated hard-boards for interior decoration, ventilation panels, accoustic materials, etc.

Moulded hard-boards for panelling, furniture, shop and bar fittings, displays and exhibition work, etc.

Enamelled hard-boards for walls and ceilings, partitions, splash backs, bath panels, kitchen cabinets, etc.

Plastic faced hard-boards for all types of lining and panelling work.

8.49 PARTICLE BOARDS

Particle board is made out of small size timber and wood waste residue of all types like logs and tops of trees, edgings of saw mills, end cuts, stumps, etc., and other similar woody material.

Manufacture. Wood in all dimensions is converted into small chips by high speed wood chippers. The chips are then dried to a moisture content below 15% in automatic driers and sprayed with synthetic resin adhesives such as formaldehyde or phenol resins. The chips coated with adhesives are spread uniformly by specially spreading machines. Generally, three-layers spreading is adopted. The top and bottom layers of the particle are finer whereas the central layer of particle is coarser. The particles in the form of a mat are pressed in hydraulic press in the presence of heat and moisture. The finished particle-board is conditioned for a week or so before cutting them into sizes or sanding.

The properties of boards are governed by the quality and size of the chips, their moisture content, amount of adhesive employed, the quality and quantity of hydrophobic agents used, pressing conditions, etc. As all these would be properly controlled, a product of uniform quality could be obtained.

Advantages. The particle-boards offer similar advantages as those by hard-boards.

At present, there are a few factories for the production of particle boards in the country with total installed capacity of 23,000 tonnes. The actual production, however, is much less as the use of particle boards is not on a large scale in construction.

BAMBOOS

Bamboo is a very useful material for construction of houses and other structures. From ancient times bamboo has been used for this purpose in our country and in many other parts of the world. Being a forest product, it is readily available in villages at practically no or very little cost.

Generally, bamboo grows in the forest naturally. Also, at some places, it is cultivated.

It is found in almost every part of our country. A variety of species of bamboo are found ; some of which are very suitable and useful for construction.

8.50 DIFFERENT SPECIES OF BAMBOO

Bamboos are found in tropical, sub-tropical and temperate regions. Bamboo forests are found upto the height of 3400 metres above sea level. Very dense bamboo forests are found in South East Asia from India to China and nearby Islands.

Except in very dry and hot areas and high mountainous regions of Himalayas, one or other species of bamboo is found all over. Generally, luxuriant bamboo forests are found in Assam, West Bengal and Western Coast areas. With variation in climate, thick or thin bamboo forests and different lengths of bamboo culms are found in these places.

There are about 109 species of bamboos which are found in the country. Of these, there are two primary species, *Dendrocalamus stridus* and *Bambusa arundinaceae* which are most suitable for construction purposes. These are found in dry as well as in damp regions but not in very dry areas. With other species of bamboos these two species are also found in different other climates.

The local names of bamboos which are suitable for construction and the areas where they are found are given in Table 8.5.

In the forests, bamboo grows naturally. But now at many places it is also being cultivated in farms. Formerly, for fear of fire in the forests, the forests were cleared by cutting or by burning away the bamboos, but now-a-days bamboo forests are being conserved as bamboo finds use as a raw material in a number of important wood-based industries.

From the economic point of view, bamboo is regarded as a very valuable material. There is very great demand for it. So much so that for the building industry, it has become difficult to obtain suitable species of bamboos. Side by side with the conservation of bamboo forests, production of bamboo is being stepped up by adopting modern methods of bamboo farming.

Tables 8.5 Occurrence of Bamboos in India

Local names of Bamboos which have been used in the construction of houses in our country and the areas where they are found.

Local name of bamboos	*Areas of occurrence*
Jilli	Nagaland, Assam
Khnap, Uspar	Assam, West Bengal
Chevani	South and West India

Kata, Katas, Koto, Behar, Bans, Doga	All over India
Balku, Baluka	Assam, Bihar, West Bengal
Pichli, Nan Bans	Lower Himalayas around Jammu
Betana	Assam, West Bengal
Jati, Tulda	Assam, West Bengal
Bansani	Orissa, Bihar and all over India
Lutang	Assam, Bihar
Kokva, Pecha	Assam, West Bengal, Western Himalayas
Vadha Teria	Assam, Sikkim
Medar, Salia, Solia Bans	All over India
Moli, Tarai	West Bengal
Cheva	Western Coast
Bota Natgi bans	Odisha, Assam
Bejal, Toli, Nal	Assam, Eastern Himalayas
Duloa	Assam
Marlang	Assam

8.51 DEFECTS OF BAMBOOS

Bamboos have some defects which are due to the following :

1. (*a*) Due to dampness, bamboo starts decaying and rotting very soon.
   (*b*) Termites and borers attack bamboos and by eating it make it hollow. This shortens the life of bamboo structure and renders them dangerous sometimes. As such, houses constructed with untreated bamboos may not be very durable.
2. When bamboo is buried in the ground, it starts decaying very soon and is also eaten away by termites and insects. Therefore, bamboo should not be used in the construction of foundations for houses.
3. Bamboo catches fire very quickly and is burnt to ashes in no time. Bamboo houses, therefore, need fire protection treatment.
4. With the passage of time, the strength of bamboo decreases. This is responsible for short life of bamboo houses.
5. In case of even a small impact force, bamboos are likely to leak and split up, and the houses are damaged.
6. Bamboos have joints at different places in its length. Therefore, walls made of bamboo do not have even surfaces.
7. Bamboos are not of equal length and thickness. This presents some difficulty in construction work.
8. Bamboos on drying shrink and consequently their volume changes. This has an ill-effect on construction done with bamboos.

8.52 TREATMENT OF BAMBOOS

Some of the defects of bamboos mentioned above can be removed to some extent by proper treatment before their use in construction. The service life of bamboo when treated is increased and the use of treated bamboos in construction results in more durable houses.

The following two compositions have been found to be useful for treatment and can be conveniently prepared in the field.

(*i*) Copper-chrome-arsenic composition (*Neela Thotha, Surkh-Kahi, Kacha Sankhia*)—In 100 litres of water (water required to fill 5 kerosene oil canisters) dissolve one after another 1 kg of arsenic penta oxide (*Kacha Sankhia*), 3 kg of copper sulphate crystals (*Neela Thotha*) and 4 kg of Potassium dichromate (*Surkh Kahi*) and mix well.

Or

(*ii*) Copper-chrome-acetic acid composition (*Neela Thotha-Surkh Kahi-Tezab Sirka*)—In 100 litres of water (water required to fill 5 kerosene oil canisters) add one after another 4 kg of copper sulphate crystals (*Neela-Thotha*), 4 kg of Potassium dichromate (*Surkh Kahi*) and 200 gm of acetic acid (*Tezab Sirka*) and mix well.

Preparation of Bamboos. Depending on the method of treatment, bamboos are prepared for treatment. For treatment by ordinary dipping method bamboos are cleaned by removing their leaves and twigs, and cut into required sizes. At the nodal point, holes are made.

For other methods of treatment, bamboos along with twigs and leaves are used.

For treatment the bamboos should be freshly cut. If for some reason, it is not possible to treat the bamboos immediately after their cutting, they should be kept under water for two or three days. As far as possible delay in treatment after cutting the bamboos should be avoided.

Dipping Method

First Method. The leaves and branches from the bamboos are removed and the bamboos are cut into required lengths. These are then dipped into the solution for treatment. This method of treatment, however, takes considerable time.

Second Method. Freshly cut bomboos with branches and leaves are so dipped in the solution that their leaves remain in contact with air.

From the transpiration of moisture from the leaves and branches, which remain in the air, the chemical soluti on is drawn up into the bamboos culms. It normally takes about two days for the treatment. This time depends upon the length of bamboos, their species and climatic conditions.

From the changing colour of the leaves, the progress of treatment is observed. After a little practice it can be easily seen when to stop the treatment.

Boucherie Process

Freshly cut bamboos with leaves and branches are connected by means of tubes to the drum containing chemical solution. The bamboos are placed at a slight inclination and the drum containing the solution is kept at a higher level so that the solution coluld flow quickly from the reservoir and may pass from one end of the bamboo to the other. Shortly after allowing the solution to flow through at the lower end of the bamboos, some solution starts dripping. At first the dripping contains sap which is contained in the bamboos. As such this should be allowed to flow out and go waste. After some time the preservative solution starts coming out slowly from the lower end of the bamboos. This is observed by watching the colour of the solution coming out. The solution should be collected and this could be reused for treatment after mixing in it the required chemicals for bringing the concentration of the solution to the required strength. When the colour of the solution coming from the lower end of the bamboo is the same as that of the original solution, it should be understood that the treatment has completed.

With the help of an air pump a slight pressure can be created in the drums containing the solution. This increases the rate of flow of chemical solution and thus reduces the time of the completion of the treatment process. In this manner the treatment can be done easily and in very much less time, within a few hours. Ordinarily, cycle or motor car pump could be used. It is necessary that the drums are made air-tight and are fitted with air valves, etc.

8.53 TREATMENT OF HOLLOW BOMBOOS

With the help of steel incisor, the inner surface of the hollow bamboo is scratched so that more of solution could be absorbed by the bamboo. Then hollow bamboos are hung from the branch of a tree vertically and the topmost nodes are filled with solution.

First of all, the sap which is contained in the bamboo starts flowing out from the lower end. This is allowed to go waste. After some time, the chemical solution starts flowing out at the other end slowly. This should be collected and reused after bringing its concentration to the required strength. It should be remembered that the nodes of bamboos should be filled again and again with solution and they should be full of solution always until the treatment is completed.

Pressure Method. Pressure method of treatment of bamboos gives very good result and has been described in detail earlier in this chapter.

Duration of Treatment. Although the time taken for treating bamboos depends on many factors, it has, however, been seen that generally it takes about five days' time to complete the treatment. It is often seen that 48 hours, treatment is enough for ordinary work. When the treatment is done by Boucherie process with the help of air pump, it takes only a few hours time for the treatment.

After treatment, the bamboos should be dried so that the moisture content is about 10 to 15% and thereafter it should be used in construction. As far as possible dried bamboos should be protected before their use in construction from exposure to rain and dew.

QUESTIONS

1. List the categories under which the uses of timber have been classified. What are the five zones into which the country has been divided to classify important timbers commercially available.
2. How are the timbers classified on the following basis :
   (*a*) Durability
   (*b*) Treatability
   (*c*) Exogenous and endogenous trees
   (*d*) Deciduous and evergreen trees
3. Enumerate the advantages of timber as a construction material.

4. Briefly discuss the classification of timber on the following basis :
   (*a*) Comparative strength coefficient
   (*b*) Seasoning purposes
5. Explain the three main parts of a tree with the help of a diagram. Draw the cross-section of a timber tree.
6. Give the comparison of sap wood and heart wood.
7. Explain briefly the important points in connection with the felling of trees.
8. Describe the stages involved in converting the felled trees into suitable sections.
9. What are the various methods of sawing of timber? Explain these with the aid of diagrams.
10. Describe the important characteristics used in the identification of timber.
11. How is the cubic content of log calculated in the case of round and square logs.
12. Give the method of determining the moisture content of timber. What is equilibrium moisture content? Write down the maximum moisture content in timber for use in beam, doors and windows, flooring strips, furniture and cabinet making in dry and moist wood.
13. List the main defects in unseasoned timber. Explain the objectives and advantages of seasoning of wood.
14. Describe briefly the different method of seasoning of wood.
15. Compare air-seasoning and kiln seasoning, bringing out the advantages of kiln seasoning. What are the factors on which the time required to season the wood in a kiln depends?
16. How is wood seasoned in compartment kiln and continuous kiln?
17. Give the classifications of defects in timber. With the help of sketches explain the defects due to abnormal growth.
18. What are 'shakes' and how do you classify them? With the aid of sketches, explain them.
19. Write short notes on :
    (*a*) Defects during conversion and uses
    (*b*) Stains in timber and their removal
20. List the principal causes of decay of wood and under what conditions it accelerates?
21. How the decay of wood occurs due to :
    (*a*) Fungal attack
    (*b*) Attack by insects
    (*c*) Attack by marine borers
22. What are the general characteristics of incipient decay? How can decay he controlled?
23. List the advantages of treated timber. What are the desirable properties of preservatives?
24. Describe three broad groups of preservatives used for timber and explain the methods of their application.

25. What are veneers and their different types? How are these used in the manufacture of plywood? Give the uses of plywood in various forms.
26. From which categories of wood raw-materials, hard boards can he manufactured. Explain the method of manufacure of hard boards and state their advantages over sawn wood.
27. Write short notes on the following :
    (*a*) Types of hard-boards and their specific uses
    (*b*) Particle boards
28. Where are bamboos found in India? Enumerate the defects in bamboos and briefly describe the method of treatment.
29. Give the location, properties and uses of the following types of timbers:
    (*a*) Deodar (*b*) Sal
    (*c*) Shisham (*d*) Teak of India
30. In What zones of India the following timbers are available? Briefly state their qualities and uses:
    (*a*) Chir (*b*) Kail
    (*c*) Rose-wood (*d*) Kikar
    (*e*) Mango

OBJECTIVE TYPE QUESTIONS

1. Timber has ________ strength than other material, if compared weight to weight
   (*a*) less (*b*) equal
   (*c*) more (*d*) None of these
2. Timber can be used as:
   (*a*) load-bearing structure (*b*) Non-load bearing structure
   (*c*) both (*a*) and (*b*) (*d*) Neither (*a*) or (*b*)
3. Timber is a ________ conductor of heat and sound
   (*a*) non (*b*) fair
   (*c*) good (*d*) best
4. The wastage losses in timber are
   (*a*) nil (*b*) less
   (*c*) more (*d*) greatest
5. Exogenous trees are those which grow:
   (*a*) outward (*b*) inward
   (*c*) both inward and outward
6. Deciduous trees shed leaves
   (*a*) quarterly (*b*) half-yearly
   (*c*) yearly (*d*) bi-yearly

7. Deciduous trees shed leaves in
   (*a*) Summer
   (*b*) Winter
   (*c*) Autumn
   (*d*) Rainy season
8. Soft woods are strong against:
   (*a*) shear
   (*b*) thrust
   (*c*) compression
   (*d*) direct pull
9. The modulus of elasticity of standard timber varies as
   (*a*) 600 to 9800 N/mm^2
   (*b*) 9800 to 12600 N/mm^2
   (*c*) 12600 to 15000 N/mm^2
10. The modulus of rupture of ordinary timber is equal to or greater than
   (*a*) 7.0 N/mm^2
   (*b*) 7.5 N/mm^2
   (*c*) 8.0 N/mm^2
   (*d*) 8.5 N/mm^2
11. The modulus of rupture of super timber should be equal or greater than
   (*a*) 17 N/mm^2
   (*b*) 18 N/mm^2
   (*c*) 19 N/mm^2
   (*d*) 20 N/mm^2
12. The food in a tree is conducted by:
   (*a*) bark
   (*b*) cambium layer
   (*c*) sap wood
   (*d*) heart wood
13. The central part of a tree is termed:
   (*a*) sap wood
   (*b*) heart wood
   (*c*) pith
   (*d*) cambium layer
14. Proper season for felling of a tree is
   (*a*) summer
   (*b*) rainy
   (*c*) autumn
   (*d*) winter
15. The wastage of wood and cost of sawing is maximum in
   (*a*) Ordinary sawing
   (*b*) Tangential sawing
   (*c*) Radial sawing
   (*d*) Quarter sawing
16. In the process of sawing, the wastage of wood is about
   (*a*) 20 to 30%
   (*b*) 30 to 40%
   (*c*) 40 to 45%
   (*d*) 45 to 50%
17. For railway sleepers, the wood commonly used is:
   (*a*) deodar
   (*b*) chir
   (*c*) fir
   (*d*) kikar
18. The wood which has the maximum weight in kg/m^3 is:
   (*a*) Kail
   (*b*) Shisham
   (*c*) Deodar
   (*d*) Rose wood
19. The wood having pink colour is:
   (*a*) Shisham
   (*b*) Sal
   (*c*) Fir
   (*d*) Tun

20. The wood which even resists fire tough is:

(*a*) Shisham (*b*) Sal

(*c*) Fir (*d*) Tun

21. The wood which even resists fire action is

(*a*) Shisham (*b*) Rose wood

(*c*) Teak (*d*) Deodar

22. Plywood is normally manufactured from:

(*a*) Kikar (*b*) Babul

(*c*) Benteak (*d*) Mango

23. The moisture content in timber is determined by drying at a temperature of

(*a*) 95 to 100°C (*b*) 100 to 105°C

(*c*) 105 to 110°C (*d*) 110 to 115°C

24. The maximum permissible moisture content in timber for beams and rafters under moist condition is:

(*a*) 12% (*b*) 14%

(*c*) 16% (*d*) 20%

25. The maximum permissible moisture content in timber for furniture and cabinet making in dry condition is

(*a*) 8% (*b*) 10%

(*c*) 12% (*d*) 14%

26. Seasoning of timber causes reduction of weight to nearly:

(*a*) One-third (*b*) One-half

(*c*) Two-third (*d*) Three-fourth

27. Seasoning of timber results in the increase of strength to

(*a*) $1\frac{1}{4}$ times (*b*) $1\frac{1}{2}$ times

(*c*) $1\frac{3}{4}$ times (*d*) double

28. The defect in timber, which is not due to abnormal growth, is:

(*a*) knot (*b*) twisted fibre

(*c*) bow (*d*) rindgall

29. Pin knot is a live knot less than in diameter

(*a*) 12.5 mm (*b*) 15.0 mm

(*c*) 17.5 mm (*d*) 20.0 mm

30. Pick up the defect in timber due to rupture of tissues:

(*a*) cross-grains (*b*) upsets

(*c*) shakes (*d*) twist

31. The defect which occurs due to improper seasoning is:

(*a*) Knot (*b*) Cup

(*c*) Upset (*d*) Star shake

32. Decay fungi cannot develop in a wood which has moisture content less than:

(*a*) 30% (*b*) 25%

(*c*) 22% (*d*) 20%

33. The thickness of a 5 plywood board varies from

(*a*) 5 to 9 mm (*b*) 9 to 16 mm

(*c*) 13 to 19 mm (*d*) 19 to 23 mm

34. In a plywood, the grains of each layer are at ________ degrees to the grains in adjacent layer:

(*a*) 0 (*b*) 45

(*c*) 90 (*d*) 135

35. Plywoods are classifies into ________ types

(*a*) 4 (*b*) 6

(*c*) 8 (*d*) 10

36. Bamboos should be dried to have moisture content of about

(*a*) 0 to 5% (*b*) 5 to 10%

(*c*) 10 to 15% (*d*) 15 to 20%

Answers

1. (*c*)	**2.** (*c*)	**3.** (*a*)	**4.** (*b*)	**5.** (*a*)	**6.** (*c*)
7. (*b*)	**8.** (*d*)	**9.** (*c*)	**10.** (*d*)	**11.** (*b*)	**12.** (*c*)
13. (*c*)	**14.** (*d*)	**15.** (*a*)	**16.** (*b*)	**17.** (*a*)	**18.** (*d*)
19. (*b*)	**20.** (*a*)	**21.** (*c*)	**22.** (*d*)	**23.** (*b*)	**24.** (*d*)
25. (*b*)	**26.** (*b*)	**27.** (*d*)	**28.** (*c*)	**29.** (*a*)	**30.** (*c*)
31. (*b*)	**32.** (*d*)	**33.** (*a*)	**34.** (*c*)	**35.** (*b*)	**36.** (*c*)

FURTHER REFERENCES

1. Timber Identification Manual - WWF - India
2. B.I.S. New Delhi. IS Code No. 1141 (1993)
   Seasoning of Timber - Code of Practice

FERROUS METALS AND STEEL ALLOYS

Metals are grouped into two classes :

(*i*) ferrous, and

(*ii*) non-ferrous.

A ferrous metal has iron as its principal constituent, whereas non-ferrous metals are those whose principal element is not iron.

FORMS OF IRON

Ferrous metals comprise three general classes :

(*i*) Cast iron,

(*ii*) Wrought iron, and

(*iii*) Steel.

All of them are prepared by conversion of iron ores into pig iron and subsequent metallurgical treatments.

9.1 CLASSIFICATION OF IRON PRODUCTS

Iron products are classified as under :

(*i*) *Pig Iron.* It is produced by reducing iron ores in the blast furnace. Carbon content ranges between 2.5 and 4.5 percent. The iron is cast into rough bars termed 'pigs'.

(*ii*) Cast iron is remelted pig iron after being cast or about to be cast in final form. Its composition need not necessarily differ from pig iron. Cast iron is not malleable at any temperature.

(*iii*) Malleable cast iron is a particular form of cast iron, rendered malleable, partly or wholly by annealing.

(*iv*) Wrought iron is a form of iron aggregated from a solidifying mass of pasty particles of highly refined metallic iron, with which, without subsequent fusion, is incorporated a minutely and uniformly distributed quantity of slag. The carbon content is from a trace to 0.25 percent and slag from 0·6 to 3 percent. It is initially malleable but will not harden when rapidly cooled due to low carbon content.

(*v*) *Ingot iron.* It is a type of iron or extremely low carbon steel which has been cast from a molten condition.

(*vi*) *Steel.* It is an alloy of iron and carbon which is cast from a molten mass. It is malleable in some range of temperatures and may not harden upon sudden cooling. The carbon content is usually less than 1.5 percent but not greater than 2 percent. Steel which owes its properties due to carbon is known as 'Carbon Steel' and others which have distinct properties due to elements other than carbon are termed 'alloy steel'.

The chief iron ores, in order of percentage of iron content are :

(*i*) *Magnetite* : (Ferrosoferric oxide) Fe_3O_4—(iron 72.4 percent). It is Steel grey to black in colour, strongly magnetic and very hard. The specific gravity is about 5.

(*ii*) *Hematite* : (Anhydrous oxide) Fe_2O_3 (iron 70 percent). It is red to brown in colour. This is the major ore in India and also in the world. Indian hematite has high percentages of alumina (1—5%) and silica (1—3%). The specific gravity varies from 1.6 to 5 due to porosity.

(*iii*) *Limonite* : (Hydrous ferric oxide) $2Fe_2O_3 + 3H_2O$ (iron 60 percent). This is hydrated hematite, yellow in colour and soft. It converts into red hematite by dehydration at high temperature. By partial dehydration, brown coloured ores, yield brown hematite. The specific gravity ranges from 1.5 to 4.

(*iv*) *Siderite* : (Ferrous Carbonate) $FeCO_3$ (iron 48.3 percent, CO_2 41.4 percent). This has got grey colour, is hard and is often associated with clay or carbonaceous matter which may impart black streaks or colour. The specific gravity is about 3.5.

PIG IRON

The iron ore is reduced to pig iron by heating and with carbon. The oxygen in the ore combines with part of the carbon to form gases, leaving the iron in the uncombined state. Excess of carbon is provided to prevent reoxidising of iron at high temperatures. A flux is added to help in removing the impurities.

9.2 BLAST FURNACE

The above process is carried out in the blast furnace shown in Fig. 9.1.

Construction of Blast Furnace

It consists of a vertical steel shaft lined with fire bricks. The sections of the furnace are :

(*i*) Lower portion known as hearth : is cylinderical about 6 m in diameter and 3 m high. The molten metal is collected in it.

(*ii*) Inverted frustrum of a cone immediately above the hearth called *bosh.* The walls of the bosh are 7 to 7.5 m diameter at the widest point and 2.5 to 3 m high.

(*iii*) Above the bosh extends the part known as *Stack* or *Shaft* converging to a diameter of about 5 m at the throat and 13.5 to 18 m above the bosh. The top of the stack is provided with two core-shaped bells and two charging hoppers.

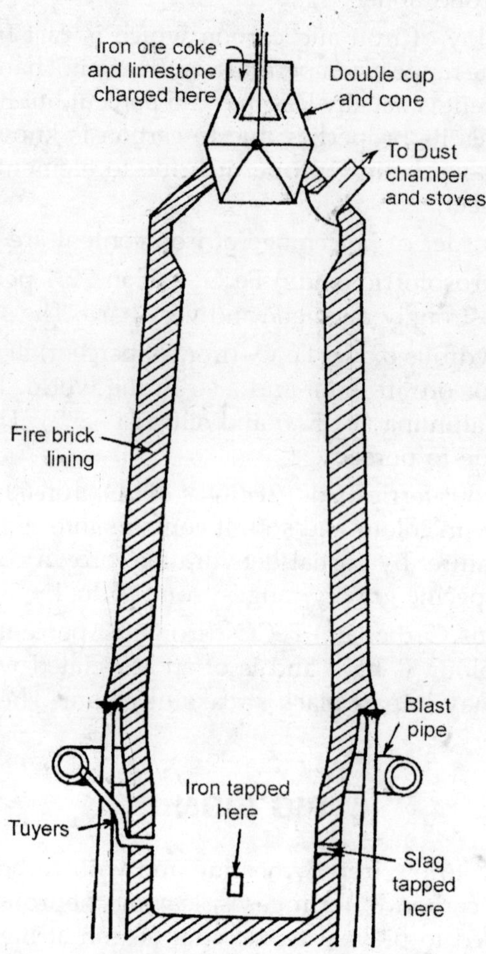

Fig 9.1. Blast furnace

At the bottom of the furnace, arrangements are provided for the admission of hot blast. The gas generated is taken from the top to the dust catcher to remove dust particles.

Charging the Furnace

The charge consists of 60 percent iron ore (or as much as 40 percent scrap iron), 25 percent coke (or other high carbon fuel) and about 15 percent limestone and dolomite. It is fed at the upper hopper. For each ton of iron ; 2 tonnes of ore, 0.9 tonne of coke and 0.4 tonne of limestone are required. The upper bell is lowered, thus allowing the material to fall into the lower hopper. The upper bell is closed and the lower bell opened permitting the material to flow into the stack. The charge gradually works down towards the bottom. Air pre-heated to 500° to 600°C, is forced in near the bottom of the furnace under a pressure of about 0.1 N/mm^2. The oxygen in the air combines with the coke to form carbon monoxide, thereby creating a temperature of about 1600°C at the bottom of the furnace. The blast of air also helps to maintain a steady current of hot gases upward through the furnace. The by-products produced with each tonne of pig iron are 0.6 tonne of slag, 0.1 tonne of flue dust and 5.1 tonnes of blast furnace gas.

Operations of the blast furnace. The blast furnace performs four distinct functions, namely

(*i*) Deoxidation of the ore,

(*ii*) Carbonization of the iron,

(*iii*) Fusion of the charge and melting of iron, and

(*iv*) Separation of the molten iron and slag.

As the iron ore works down the stack, it comes in contact with the ascending gases, containing carbon monoxide. Near the top of the stack, the iron oxide is reduced by carbon monoxide to metallic iron

$$Fe_2O_3 + 3CO = 2Fe + 3CO_2 \uparrow$$

Apparently little reduction is accomplished by the coke directly although it is in contact with the ore. This is because of the fact that the fuel does not burn until it comes in contact with the air near the bottom of the furnace.

It is necessary to add sufficient quantity of coke, so that the ore is saturated with carbon.

This carbonization is very helpful because iron, saturated with carbon, gets sufficiently super heated beyond its melting point and becomes fluid, at the temperatures attainable in the furnace. If this is not done, it will be impossible to melt the free iron.

When fusion takes place, lime (CaO) produced from limestone, part of sulphur, silicon dioxide and other impurities will form slag and expel free iron (fused unoxidised material) from it.

$$CaCO_3 = CaO + CO_2 \uparrow$$

$$CaO + SiO_2 = \underset{\text{(Slag)}}{CaSiO_3}$$

The molten iron will absorb all deoxidised substances like phosphorus, manganese, silica etc. It will also absorb $3\frac{1}{3}$ to 4 percent of carbon. Sulphur will also be taken by iron, if limestone is not present in sufficient quantity and this will make it brittle at high temperatures.

Both slag and molten iron are in liquid state as they come down to the crucible but their densities are different. Slag being lighter floats on the molten metal. Thus pig iron and slag are drawn out separately at about 6 hours intervals through temporary openings punched in the side of the crucible.

Capacity of the furnace. The blast furnaces are designed for capacities ranging from 100 to 1000 tonnes per day. The one shown in Fig. 9.1 will have a daily capacity of about 800 tonnes. It will require the following charge,

Iron ore	—	1500 tonnes
Coke	—	800 tonnes
Limestone	—	400 tonnes
Air	—	3000 tonnes
Water	—	27,000,000 litres.

The slag produced will be of the order of 300 tonnes.

Handling of the product. Usually a steel plant functions in conjunction with the blast furnace. Thus the molten pig iron is taken in ladles into the furnaces or may be stored in tanks until processed.

The other method of handling pig iron is to cast it into 'pigs', weighing about 50 kg ; in a pig moulding machine.

Uses of by-products from the blast furnace. The slag, principal constituents of which are oxides of silicon, aluminium, calcium and magnesium is used in the manufacture of some grades of cement, paint, roofing material, concrete, mineral wool etc. This is also used as ballast for rail road tracks. The hot gases from the blast furnace are used in preheating the air fed at the bottom of the furnace, as they contain carbon monoxide.

CAST IRON

9.3 MANUFACTURE

Cast iron is obtained by remelting pig iron and pouring it into moulds of the desired shape. This helps to reduce impurities and furnishes more uniform product than would be obtained by directly casting molten pig iron from the blast furnace.

Pig iron is remelted either in a cupola or in an air furnace. In certain cases, a small open hearth furnace, similar to that used in making steel, is used to produce a high grade cast iron.

9.3.1 Cupola

Ordinary grey cast iron, used for machine parts, is mostly prepared in cupola. It consists of a stack lined with fire bricks. The sizes of the cupola vary considerably. A 6 m high and 2 m in diameter cupola, which is commonly used, has capacity of about 20 tonnes per hour.

The cupola is charged with 80 percent pig iron and 20 percent coke (or other high carbon fuel). Scrap iron may also be added to the extent of 25 percent of the pig iron content. If scrap iron equals the quantity of metal, an inferior quality of cast iron results. The charge also contains a small quantity of limestone to reduce sulphur content in cast iron and to help in reducing some other impurities.

The cupola operates on the same general principle as the blast furnace ; the difference being that in the cupola no particular chemical change in the iron takes place. Only oxidisation of a little quantity of iron occurs and some sulphur may be absorbed from the fuel.

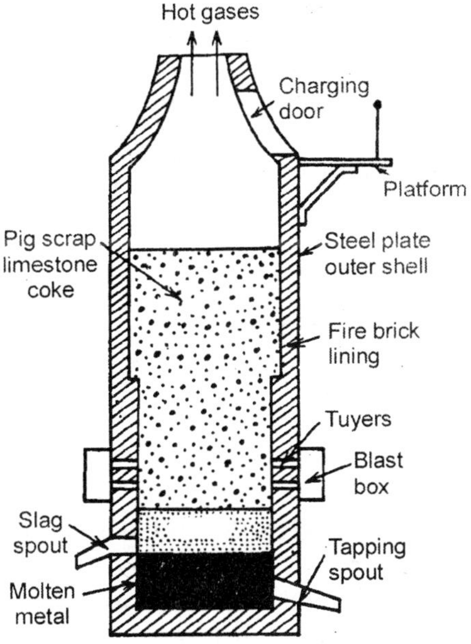

Fig. 9.2. Cupola furnace

9.3.2 Air furnace

It is a type of reverberatory furnace. It consists of a flat, wide chamber known as hearth or bath and an adjacent compartment in which fuel is burnt on a grating. The ceiling of the furnace is lined with fire bricks to help in reflecting heat downward. The hot gases from the burning fuel, melts the metal and the slag comes to the top. A jet of air between the grate and the hearth is introduced to maintain the circulation. The slag forms a protective coating and prevents oxidisation of the metal and absorption; of metal from the fuel. This improves the quality of cast iron produced. Temperature control is also possible. However the amount of fuel required is twice than that required in cupola. Thus the cost of cast iron is increased.

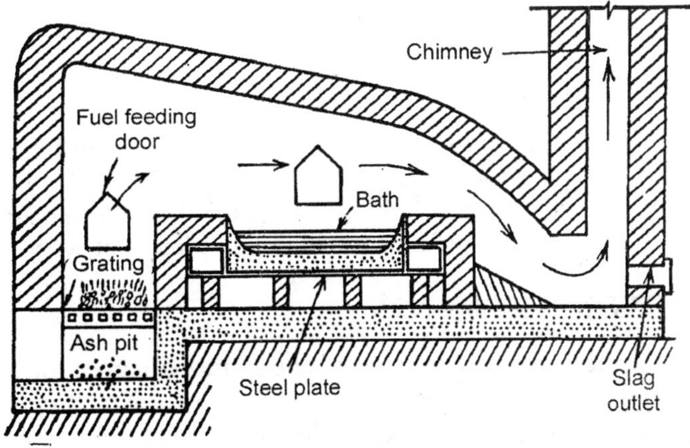

Fig. 9.3. Reverberatory furnace

9.3.3 Duplex Process

In this process, pig iron is given a preliminary treatment in the cupola and is then transferred to an electric furnace having a temperature control. Because of constant temperature, a uniform and superior quality of cast iron results. Cast iron can also be super heated, thereby giving it better casting property.

9.4 TYPES OF CAST IRON

Depending upon the rate of cooling of the molten metal, various types of cast iron result, namely

(i) Grey cast iron. This results from a slow rate of cooling and carbon occurs chiefly in the graphite state. The presence of graphite, makes cast iron weak and brittle and imparts dark colour.

(ii) White cast iron. This is produced by rapid cooling of the molten mass. In this, carbon occurs mainly as the carbide of iron. Because of absence of graphite, the colour is light and as such named as 'white cast iron'. It is strong and hard but also very brittle.

(iii) Chilled cast iron. The mould is lined with metal and cooling is sudden. The surfaces of railroad wagon's wheels have the-wearing surface chilled to make it resistant to wear.

(iv) Malleable cast iron. Rapid cooling makes cast iron brittle and develops high and undesirable initial stresses. By the process of annealing, the metal is made 'malleable' and relieved of the high stresses. Some of the combined carbon is converted into graphite. The graphite formed is not in the form of large flakes but in very small particles, known as temper carbon. Thus malleable cast iron consists of crystals of pure iron interspersed with particles of temper carbon. It is ductile and has high strength and can be used in casting complex shapes.

9.5 PHYSICAL DEFECTS IN CAST IRON

The various physical defects present in cast iron are given below :

Blow holes are gas pockets left in the solid iron. A high sulphur content accelerates the formation of blow holes whereas porous moulding sand helps to reduce them.

Cracks develop if the iron in the moulds does not have an opportunity to shrink properly during cooling. Cracks are also formed at sharp re-entrant angles in a casting because of the formation of crystals with axis at right angles to the surface.

Segregation of impurities. The collection of impurities such as sulphides and phosphides, at a particular place, in the casting create zones of brittleness and stress concentration. Segregation of carbon and silicon results in varying hardness in the cast iron, thereby making it difficult to machine. Coarse grain structure will create planes of weaknesss in the cast iron.

9.6 EFFECTS OF OTHER ELEMENTS

The presence of sulphur, phosphorus, manganese and silicon is unavoidable in cast iron ; as they are always present in the iron ore. Sometimes nickel and chromium are added to improve its properties. The effects of these are described below :

(*i*) *Sulphur* forms sulphide of iron or manganese. It makes the cast iron brittle at high temperatures and increases shrinkage. Sulphur should be restricted to less than 0.1 percent.

Phosphorous. If present in excess of 2 percent, it renders the cast iron brittle and weakens it. However it helps to decrease shrinkage and makes the metal more fluid. Thus this is preferred in sharp castings for ornamental work.

Manganese. This is usually present in quantity less than 2 percent. It combines with sulphur and carbon to form manganese sulphide and manganese carbide. Manganese carbide unites with iron carbide and this decreases the quantity of free carbon, thereby increasing shrinkage and hardening of the cast iron and making it resistant to machining and abrasion.

Silicon. Normally silicon is present from 0.5 to 3 percent in cast iron. Silicon helps in the liberation of free carbon *i.e.* graphite, when it combines with iron. This makes cast iron more soft and weak and good for machining. Silicon decreases shrinkage and increases fluidity. When silicon is present in excess of 6 percent, a harder cast iron results due to the formation of iron silicide. Silicon content of about 15 percent makes the cast iron acid resistant.

Nickel. About 0.5 to 3 percent nickel retard the chemical combination of iron and carbon under rapid cooling, thereby retarding the formation of large graphite flakes. This makes the cast iron stronger, harder and ductile and useful for casting pistons, gears, crankshafts, dies, brake drums etc.

Chromium. About 1 percent chromium is added to increase the percentage of combined carbon for greater strength and hardness. It is added in combination with nickel.

9.7 USES OF CAST IRON

The long established use of cast iron in rainwater pipes, gutters, etc. is almost too well known. It is also used for the manufacture of other articles such as gratings, man-hole covers and balustrades. It is a basic material for the manufacture of steel. In general, cast iron articles should have rounded angles and arrises (sharp edges)

Cast iron pipes

Cast iron pipes are cast either by sand-casting or by centrifugal casting. The former method yields short lengths of pipes whereas the later can produce large lengths. In the centrifugal casting, the motlen metal is filled in a revolving water-cooled steel mould. The centrifugal force due to the revolution hold the metal against the sides of the mould. The method is continuous, no sand is needed and the output is about eight times than the traditional method. The pipes produced by this method are called 'spun' pipes. They are stronger than pipes cast by sand-casting, the increase in strength being 50 to 100 percent ; according to the thickness of the pipe. The finely crystalline structure developed is the cause of this increased strength. Defects such as blow-holes are also eliminated.

Cast iron pipes may get corroded either by electrolysis, direct currents being more damaging or by chemical action of the soil. This is prevented by giving a protective coatings of bitumen on the outside and concrete of thickness varying from 3 to 12 mm, on inside.

WROUGHT IRON

Wrought iron is a pure form of iron and is obtained by eliminating impurities such as carbon, silicon, manganese and phosphorus from pig iron.

9.8 MANUFACTURE

Wrought iron is manufactured by either

(*i*) Puddling process ; or

(*ii*) Aston process.

9.8.1 Puddling Process

A small reverberatory furnace, similar in construction to the air furnace shown in Fig. 9.3 is used in this process. It has a capacity varying from 150 to 700 kg with an average of about 300 kg. It is charged with a good quality of pig iron. Some slag and iron oxide is added to assist in forming a protective coating on the top of the molten metal and prevent it from oxidisation and absorption of impurities. This also helps in the oxidisation. The gases are produced from a fuel, which is generally soft coal. Limestone is also added to act as a flux.

The changes in the composition of pig iron during puddling process involves the elimination or oxidisation of silicon, manganese, phosphorus or carbon. Oxidisation is the essential chemical principal by which refinement of pig iron is brought about. The melting of the charge takes place at a temperature of about 1200°C. The iron oxide (Fe_2O_3) mixes with the molten metal and liberates oxygen which combines with silicon, manganese and part of phosphorous to form oxides or basic slag. Then carbon is reduced to carbon monoxide. A little quantity of iron gets oxidised to iron oxide (FeO). which forms a part of slag.

When the carbon content of the metal is reduced, its melting point increases to 1535°C. Since the temperature in the furnace is less than this temperature, the molten iron gets solidified and removed, in the form of muck balls, each weighing about 100 to 150 kg. The muck balls also contain some molten slag, which is removed by squeezing and rolling them into 'muck bars'. These muck bars are then cut to pieces, stacked, heated and rolled into commercial shapes.

The reactions that take place and the heat generated in each case is given in table 9.1.

Table 9.1: Reactions and Heat Produced in Puddling Process

Reactions	*Calories of heat produced*	
	per kg of metal	*per kg of oxygen*
$Si + O_2 = SiO_2$	6428	5625
$Mn + O = MnO$	1653	5681
$C + 0 = CO$	2430	1823
$Fe + O = FeO$	1173	4106
$2P + 50 = P_2O_5$	5703	4562

9.8.2 Aston Process

In this process, Bessemer converter shown in Fig. 9.4 is used. It consists of a stack lined with fire bricks and capable of being rotated in the vertical plane about a horizontal axis. It has a number of openings at the bottom, called tuyers, through which air is forced in. A converter having a capacity of about 15 tonnes, has dimensions of 2.5 metre diameter and 4.5 metre height.

The converter is kept in the vertical position and charged with molten pig iron. As the air is forced up through the tuyers, oxygen from it combines with the impurities and burns them out. Next the carbon is oxidised to CO and the pure metal, containing some gases only, is formed. To remove these gases, the molten iron is poured into a ladle of molten slag, kept at a temperature well below the freezing point of iron. The sudden reduction in temperature liberates gases with explosive violence. The iron gets shattered into tiny gobbles, coated with slag and settle down in the ladle in a. sponge like mass. The process is known as 'shotting'. The slag is, removed by squeezing and the iron is rolled into commercial shapes.

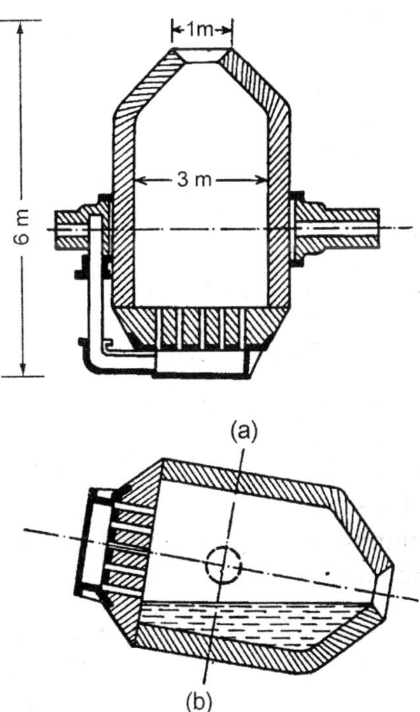

Fig. 9.4. Bessemer converter.
(a) Working position.
(b) Discharging position.

9.9 USES OF WROUGHT IRON

Wrought iron is never cast, it is shaped by hammering, pressing or forging. Because of its high resistance to corrosion, it is used for piping, boiler plates and tubes, ornamental fences and grills, flat and corrugated sheet metal, parts of locomotives and cars, gates and railings.

STEEL

Steel is an alloy of iron and carbon, in which the carbon content is restricted upto 1.7 percent. Steels having only iron and carbon with a small amount of impurities is termed carbon steels and are classified as under, depending upon the percentage of carbon present.

Table 9.2: Carbon Percentage for Various Types of Steel

No.	*Classification*	*Carbon percentage*
1.	Very low carbon steel	0.05 to 0.15
2.	Low carbon steel	0.15 to 0.30
3.	Medium carbon steel	0.3 to 0.6
4.	High carbon steel	0.6 to 0.9
5.	Very high carbon steel	0.9 to 1.7

9.10 MANUFACTURE

Steel is manufactured from pig iron and scraps by removing impurities such as silicon, manganese, carbon, phosphorus etc. by oxidation and adjusting the percentage of carbon to the desired value ; by the addition of ferromanganese or spiegeleisen (iron manganese pig containing carbon) if necessary.

Pig iron can be refined by the following processes :

(*i*) Acid process,

(*ii*) Basic process.

9.10.1 Acid Process

In this process, oxidation of the impurities is brought about in the presence of the oxides of non-metals. First of all iron oxide is formed which react with silicon, manganese and iron to form silicon dioxide, manganese oxide and carbon monoxide or carbon dioxide respectively, which appear in the form of slag on the top of the molten metal. Thus the impurities are minimised and the carbon content can also be adjusted to the required limit. Phosphorus and sulphur are not removed in the acid process.

The reactions that take place are :

$$2Fe + O_2 = 2FeO$$
$$3Fe + 2O_2 = Fe_3O_4$$
$$Si + 2Fe_3O_4 = SiO_2 + 6FeO$$
$$Si + 2FeO = SiO_2 + 2Fe$$
$$Mn + Fe_3O_4 = MnO + 3FeO$$
$$Mn + FeO = MnO + Fe$$
$$4C + Fe_3O_4 = 3Fe + 4CO\uparrow$$
$$C + FeO = Fe + CO\uparrow$$

The slag formation takes place as under :

$$FeO + SiO_2 = FeO.SiO_2$$
$$2FeO + 3SiO_2 = (FeO)_2\,(SiO_2)_3$$
$$MnO + SiO_2 = MnO.SiO_2$$
$$2MnO + 3SiO_2 = (MnO)_2\,(SiO_2)_3$$

The lining of the furnace must be acidic, *i.e.*, made of silica otherwise it will be spoiled by the chemical action.

9.10.2 Basic Process

Limestone is added to act as a flux. Oxygen is supplied either by pig iron and scrap or is forced in as in the case of Bessemer converter. Limestone helps in the formation of more FeO and also absorbs some of the sulphur.

The following reactions take place :

$$CaCO_3 = CaO + CO_2\uparrow$$
$$CO_2 + Fe = FeO + CO\uparrow$$

$$2FeO + Si = SiO_2 + 2Fe$$
$$FeO + C = Fe + CO\uparrow$$
$$Fe_3C + FeO = 4Fe + CO\uparrow$$
$$Mn + FeO = MnO + Fe$$
$$3SiO_2 + 2FeO = (FeO)_2\,(SiO_2)_3$$
$$MnO + SiO_2 = MnO.SiO_2$$
$$2Fe_3P + 8FeO = (FeO)_3\,P_2O_5 + 11Fe$$
$$CaO + FeS = CaS + FeO$$

The lining is done with dolomite [$CaMg(CO_3)_2$] or magnesite ($MgCO_3$).

If required, ferro-manganese or spiegeleisen is added to the ladle otherwise it may reduce phosphorus pentoxide.

The plant used for the manufacture of steel depends upon the carbon content in it. Low carbon steels, which are produced in large quantities and used for structural purposes, are prepared by open hearth process. Bessemer Converter and Duplex processes are also sometimes employed. High carbon steels, which are used for making tools, are produced in the open hearth, cementation, crucible or electric furnace process. Some details of these processes are given below :

9.10.3 Bessemer Process

The constructional details of the Bessemer Converter have been described under article 9.8.2, in connection with the manufacture of wrought iron. The charge consisting of molten pig iron or scrap is brought in iron ladles and poured into the empty Bessemer Converter, which is kept tilted on its side. The converter is then turned upright and air is blown in. The oxidation is affected through these finely divided air currents passing through the molten charge thereby 'burning out' the impurities and converting pig iron into steel. It takes about 14 minutes to produce 25 tonnes of steel in the acid process. The impurities either appear in the form of slag or are blown out as gases.

In India, mostly acid converter is used as the pig iron contains low percentages of phosphorus. The basic process is more costly than the acid as air is to be blown for a longer period, lining is costlier and get worn sooner, flux has to be added and the additional mechanical handling is required for pouring slag.

The steel is then poured into ladles in which ferrosilicon or aluminium is added to remove oxygen by forming SiO_2 or Al_2O_3.

9.10.4 Open Hearth Process

This process is also known as Siemens Martin's process and is similar to the air furnace described under article 9.3.2. The lining given is acidic or basic, depending upon the phosphorus content in the charges. The charge consists of the pig iron and steel scraps (limestone is also added in the basic process). It is heated by gases from the fuel chamber or directly by producer gases. It takes about 10 hours for the oxidation to be complete and produces about 100 to 200 tonnes of steel. The molten metal is poured into ingot moulds. The moulds are then stripped from ingots formed when steel solidifies.

9.10.5 Duplex Process

Sometimes advantage is taken of both the acid and the basic processes. The charge is first treated in the acid Bessemer Converter to remove major portions of manganese and silicon. It is then passed to the basic open-hearth, to which limestone flux is also added, to oxidise phosphorus and carbon.

9.10.6 Electric Furnace Process

Where high grade steel, with controlled properties and carbon content, is required and supply of electricity is cheap, electric furnace process is employed for refining steel. High controlled temperatures are developed by electrical energy. The reactions are similar to those in the open-hearth process. The steel produced by this method is most suitable for high grade castings or forgings. This method is not used in India.

The electric furnaces are of two general types ; namely—

(*i*) the arc type, and

(*ii*) the induction type.

In the former type, a well insulated chamber contains carbon electrodes. The charge consisting of high grade pig or scrap steel and a flux may be below the electrodes or it may contain the electrodes in it. In the first case, the heat produced by passing current through the conduction is reflected down to the charge. In case the electrode is inserted in the charge, the charge acts as one of the electrodes.

In the induction electric furnace, a coil is wound around the insulated chamber ; in which charge is put. High frequency current is passed through the coil, which serves a primary winding of a transformer ; the charge functioning as the secondary. The eddy currents developed in the charge produces sufficient heat to melt it.

High operative cost is the only disadvantage of the electrical furnace process.

9.10.7 Cementation and Crucible Processes

The above three processes, *i.e.*, bessemer, open-hearth and electrical reduce the charge to the form of pure iron. Carbon is added as the molten metal is tapped into the ladles.

Cementation and crucible method are used to make steel from wrought iron and carbon is added while the charge is in the furnace. In the cementation process, the wrought iron (solid state) is heated, in the presence of charcoal or other solid carbon or methane gas, to a temperature less than its melting point. As the charge gets heated, absorbs carbon and steel is produced. The amount of carbon absorbed at any point inside the charge will depend upon—

(*i*) Temperature developed ;

(*ii*) Time of treatment; and

(*iii*) Distance of the point from the surface.

The charge should be very pure as no impurities are removed during the process.

Crucible method consists of melting about 50 kg of wrought iron and powdered charcoal in a crucible made of clay or graphite. The charge melts, it absorbs the carbon to form steel which is then moulded.

9.11 INGOTS

After refining pig iron by any of the methods described above, the molten steel is poured either into ingot moulds or directly into the mould to get finished casting. Usually the former method is employed. Ingots are of various sizes, depending upon the final product to be obtained from them. The metal is usually poured from a ladle, which has an opening at the bottom.

9.12 PHYSICAL DEFECTS IN STEEL

The ingots, as obtained above, develop some physical defects, important of which are described below :

1. *Piping.* When the molten metal is poured into the ingot mould, the metal adjacent to the walls of the mould solidifies first and forms a solid shell. In the next increments of time, more metal solidifies, thereby increasing the thickness of the shell. In this process, shrinkage and hence the decrease in volume of the metal takes place, thereby producing a cavity in the upper portion of the ingot. This cavity is called 'a pipe' and may extend even deep into the ingot. The 'pipe' is sometimes covered by a crust of metal at the top.
2. *Blow Holes and Segregation.* These have been discussed under cast-iron.
3. *Non-metallic Inclusions.* These are solid particles of oxidised metal or slag, which may enter into the ingot mould during pouring operation. They reduce the strength of the metal.
4. *Ingotism.* If the molten metal is cooled slowly in the ingots, the crystals have a tendency to grow larger ; which make the steel weak and brittle. Such a defect is known as ingotism and is rectified by rolling or forging the ingot.
5. *Surface Cracks.* This results from unequal cooling. Surface cracks sometimes penetrate deep into the ingot. This makes the steel weak. These cracks should be cut out by pneumatic chisels or burnt out with an oxygen torch, before the ingot is forged or rolled.
6. *Cold Shuts.* These are inclusions of cold solid metal pieces having been forged or rolled into the hot steel having a poor bond with the remaining metal.

CONVERSION OF STEEL INGOTS INTO MARKET FORMS

The steel ingots prepared by any of the manufacturing processes, described above, are converted into market forms such as rods, plates,, sheets, pipes or tubes and various other sections shown in Fig. 9.7 by any of the following methods :

1. Rolling
2. Forging
3. Pressing
4. Casting

These are described briefly in the subsequent paragraphs.

9.13 ROLLING

The ingots are heated in a furnace and forced through sets of rollers to shape them to the desired section. The principle of rolling is shown in Fig.9.5. Ingots are passed a number

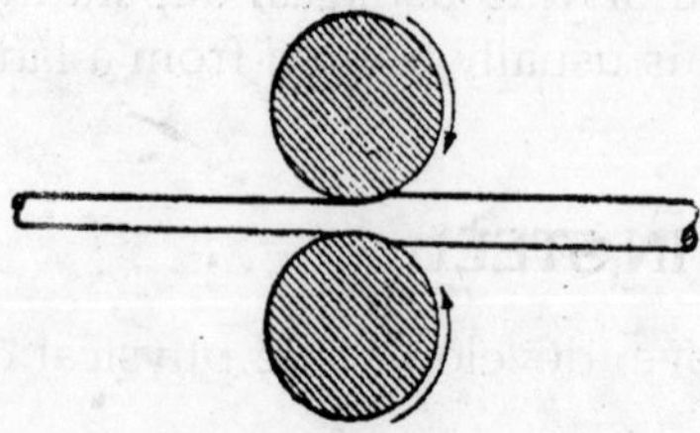

Fig. 9.5. Principle of rolling

of times in the rolling mills before they get converted into the final shape. Fig. 9.6 shows how an ingot is changed into I-Section through sets of rollers. The sets of rollers have grooves, gradually changing from the size and shape of the ingot to those of the finished section.

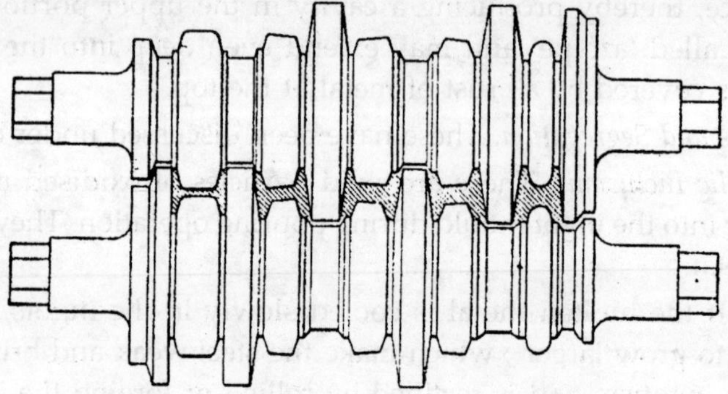

Fig. 9.6. Sets of rollers to roll I-Sections

9.14 FORGING

Though the conversion of ingots by the rolling process is the easiest, yet it cannot turn the ingots into all shapes until special sets of rollers are available to effect gradual change in cross-section. In such cases the method of forging is adopted. The ingot is reduced to the required section by placing it on an anvil and subjecting it to repeated blows by means of a power hammer. By this method, tough, strong and fine-grained bars of any shape can be obtained.

For manufacturing intricate parts such as connecting rods, wrenches, levers etc., the method of drop forging is used. Dies are attached both to the anvil and the hammer head. The metal is fed at regular interval and the article gets forged by the fall of the hammer on to the anvil.

9.15. PRESSING

Hydraulic presses are made use of. These have much deeper action than that of rollers or hammers. The metal gets kneaded slowly to the required section. This method is adopted for making heavy shafts, armour plates, etc.

9.16. CASTING

For very intricate forms, which cannot be rolled or forged or pressed, the method of casting in moulds is adopted. The sections obtained are very tough and strong, but are less ductile. Steel shrinks to a great extent on cooling. Thus in order to prevent cracking, the casting should be removed from the mould as soon as possible after the metal has solidified. Further cooling may be done slowly either under sand or in a heated furnace.

9.17. VARIOUS MARKET FORMS

Steel is available in the market in different forms. Some of the important ones are :

1. *Bars.* Iron bars are made in round, square, or flat forms. The bars are available in diameter or side varying from 6 mm to 76 mm in increments of 2 mm.

2. From 7.6 mm to 5 mm, they are called rods and below 5 mm diameter, they are termed wires.

Flat bars vary in section from 25 mm × 6 mm to 300 × 25 mm and have length of 6.5 m.

2. *Plates.* They vary in thickness from 3 mm to 25 mm.

3. *Sheets.* Sheets have a thickness less than 6 mm and is usually expressed by a number (gauge *i.e.* B.W.G. or S.W.G.)

4. *Corrugated iron sheets.* Corrugated iron sheets have corrugations to enhance their strength and make them suitable as a roof covering material. The corrugated sheets are galvanised to reduce the action of corrosion and are available in different lengths and widths.

5. *Hoop iron.* Hoop iron are thin sheets of 12.5 mm width. They are used as reinforcement in brick masonry or for securing packages for transport.

6. *Steel pipes and tubes.* Depending upon the joints, pipes or tubes are classified as :

(*a*) *Seamless pipes or tubes.* Used for boiler tubes, bicycle tubings and other places where strength and lightness are required.

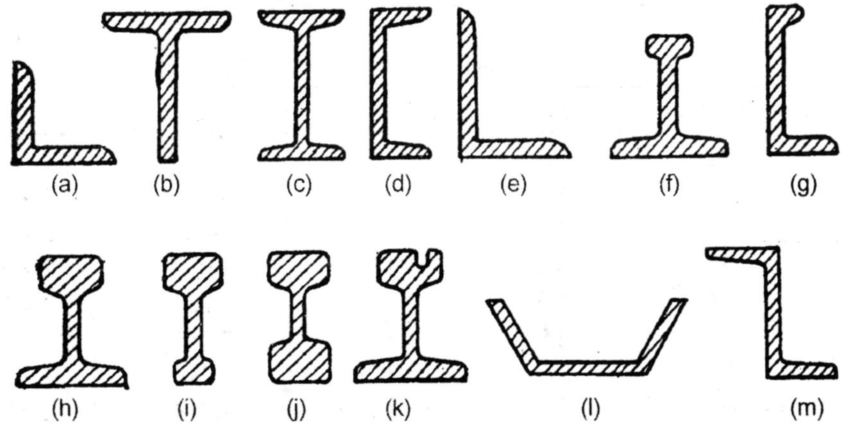

Fig. 9.7. Market forms of steel.

(*a*) Equal angle, (*b*) T-Section, (*c*) I-Section, (*d*) Channel, (*e*) Unequal angle, (*f*) Flat footed rail, (*g*) Bulb angle, (*h*) Flat footed rail, (*i*) Bull headed rail, (*j*) Double headed rail, (*k*) Tram rail, (*l*) Trough section, (*m*) Z-Section.

(b) *Lap welded pipes or tubes.* Used for boiler tubes to resist high pressure.

(c) *Butt welded pipes or tubes.* Used for gas and water pipes, not subjected to very high pressures.

7. *Rolled steel I, L, T, Z sections and channels.* These are shown in Fig. 9.7. They are specified by their overall depth, width and thickness or weight per running metre.

8. *Expanded metal or X.P.M.* These are manufactured from steel sheets and plates. In a steel sheet, slits are cut at regular intervals and in parallel lines. It is then gripped firmly and pulled out. The sheet opens out along the slits, forming diamond shaped meshes.

The finished expanded metal sheets should be free from flaws, joints, welds, broken strands, laminations and other defects. Expanded metal should be protected against corrosion by coating with suitable rust preventive materials.

Expanded metal is available in various sizes of mesh and thicknesses of sheets. The size of mesh is based on the measurements of the shortway mesh (SWM) and long way mesh (LWM) of the diamond shown in Fig. 9.8. The properties and dimensions of expanded metal sheets having (SWM) from 20 mm to 25 mm are as given in Table 9.3.

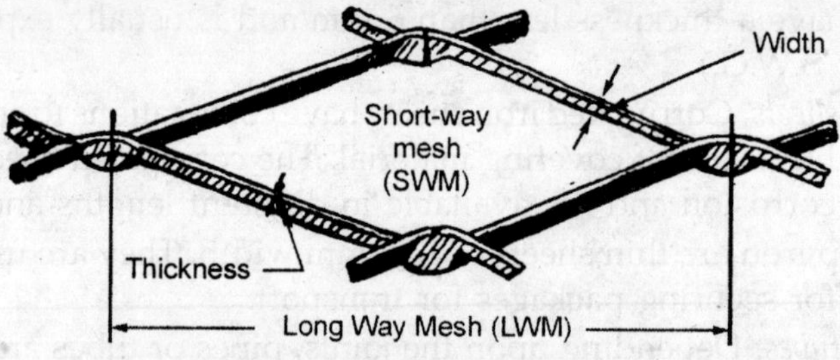

Fig. 9.8. Expanded metal

Table 9.3: Properties and Dimensions of Expanded Metal

Size of mesh		*Dimensions of strands nominal*		*Nominal wt. per sq m*	*Size of sheet normally stocked*
SWM	*LWM*	*Width*	*Thickness*	*kg*	*mm*
25	75	3.5	3.15	6.050	2 × 3
25	75	3	2.14	3.914	2 × 3
25	75	3	1.60	2.627	2 × 3 and 1 × 3
25	75	3	1.25	2.067	2 × 3 and 1 × 3
20	60	3.5	3.15	7.782	2 × 3
20	50	3.5	3.15	7.782	2 × 3
20	60	3	2.24	4.748	2 × 3
20	50	3	2.24	4.748	2 × 3
20	60	3	1.60	3.464	2 × 3 and 1 × 3
20	50	3	1.60	3.464	2 × 3 and 1 × 3
20	60	2.5	1.25	2.285	2 × 3and 1 × 3
20	50	2.5	1.25	2.285	2 × 3 and 1 × 3

Expanded metal is used for reinforcing concrete in bridges, tanks, pipes, roofs, walls, lintels, foundations, etc. This is also used as lath (ground for plaster) in plaster for ceilings, windows, encased steelwork etc.

Table 9.4: Comparison of cast iron, wrought iron and Steel

Sl. No.	Property	Cast Iron	Wrought Iron	Steel: Low or medium carbon or mild	Steel: High carbon or hard
1.	Carbon content	2 to 4.5 percent	0 to 0.25 percent	0.5 to 0.6 percent	0.6 to 1.7 percent
2.	Melting point	1200°C	1500°C	1400°C	1300°C
3.	Specific gravity	7.25	7.70	7.80	7.90
4.	Structure	Coarsely crystalline and granular ; grey, mottled or white	Fibrous ; bluish or silky	Fibrous ; bright	Granular ; fine
5.	Hardness, tempering and hardening	Very hard ; can be hardened by heating and sudden cooling	Tougher and more elastic than cast iron. It cannot be hardened or tempered	Tougher and more elastic than wrought iron. It can be hardened and tempered but with difficulty	Tougher and more elastic than mild steel. Can be hardened and tempered easily
6.	Ductility and malleability	Non-ductile and malleable (brittle)	Highly ductile and malleable	Moderately ductile and malleable	Low ductility and malleability
7.	Welding, forging or rolling property	Cannot be welded or rolled into sheets	Can be easily forged, welded and rolled	Can be easily welded, rolled and forged	Cannot be easily welded, rolled and forged
8.	Shock absorption	Incapable of absorbing shocks	Cannot absorb sudden and excessive shock	Absorbs shock	Absorbs shock
9.	Magnetic property	Cannot be magnetised	Can te temporarily magnetised	Can be permanently magnetised	Can be permanently magnetised
10.	Corrosion	Does not rust easily	Rusts more readily than cast iron	Rusts readily	Rusts readily
11.	Effect of salt water	Becomes soft	Behaves better than cast iron	Not effected	Not effected
12.	Ultimate compresive strength in tonnes/sq cm	7.0	2.0	4.75	12.0 to 25
13.	Ultimate tensile strength in tonnes/sq cm	1.5	3.75	5.5	8 to 11
14.	Uses	Since it does not rust, it finds use in those parts which are	Used as a construction material, especially where parts are	*0 05 to 0.15% carbon.* Boiler plate, rivets, bolts nails, ship plate,	*0.6 to 0.8.% carbon* Hammers, cutlery, small dies, drills,

		exposed to atmosphere such as-sewers, drain pipes, manholes, water pipes, lamp posts, short columns etc. It is also used for brackets, automobile engine cylinders, pistons, flywheels and other components not subjected to heavy shock or tension. Being weak in tension, it is not favoured as a structural member.	to be forged or welded. Wrought iron is never cast, all shaping is accomplished by hammering, pressing or forging. Used for lap and butt welded pipes, flat and corrugated sheet matal, ornamental fences and grills, chains, railway couplings etc.	forge work, sheet steel for tinning and galvanising. *0.15 to 0.25% carbon.* General structural steel, for bridges, buildings etc., rivets, forge work. *1.0 to 1.2% carbon.* Machine parts such as axles, shafts, connecting rod, piston rods, steel castings. *0.4 to 0.6 carbon.* Railroad rails.	cold chisels etc. *0.8 to 1.0% carbon.* Carbon steel springs, lathe tools, dies, strong steel wire and cable. *1.0 in 12% carbon.* Large lathe tools, dies and drills, knives etc. *1.2 to 1.7% car on.* Saws, files, razors, jewelers dies, balls etc.

HEAT TREATMENT

Steel is subjected to heat treatment to develop in it specific condition or structure for use in a particular work. Heat treatment means not only heating but also cooling steel.

9.18 IRON-CARBON EQUILIBRIUM DIAGRAM

For proper understanding of the heat treatment processes, molten steel is considered as an alloy of iron and carbon. Some of the methods which are used in controlling the variations in strength, ductility and other properties can be explained with the aid of iron-carbon system shown in Fig. 9.9. The temperature range is from 0°C to 1535°C, the melting point of pure iron. The range of carbon content is from 0% to 8% which is the maximum for iron-carbon system ; because an alloy containing greater than 8% carbon is too weak and brittle and has little practical significance.

The diagram presents only the condition of slow cooling which occurs normally in air ; and is valid only for combinations of iron and carbon without other alloying metals such as nickel, chromium, silicon, etc. The addition of these metals will change the equilibrium diagram.

Phase Change in Pure Iron. If pure iron, in a molten state, is cooled slowly in air, it undergoes changes to delta (δ), gamma (γ), beta (β) and alpha (α) states. The temperatures at which these phase changes take place are given in table below :

Phase	*Temperature*	*Remarks*
Liquid	Above 1535°C	
δ—Iron	1400— 1835°C	Negligible carbon solubility and has no metallurgical significance.
γ—Iron	900— 1400°C	Carbon solubility 1.8%, known as austenite with dissolved carbon.
β— Iron	768— 900°C	Nil carbon solubility, non-magnetic.
α—Iron	Less than 768°C	Nil carbon solubility, called ferrite.

9.18.1 Iron-carbon Alloys : (Austenite, Cementite, Ferrite and Pearlite)

When an, alloy of iron and carbon is formed, the freezing temperature is lowered. For example, γ-iron has a freezing temperature of 900°C. When 0.85% carbon is added, the freezing temperature is reduced to 730°C. If iron content is more than 0.85%, the curve rises upto 1.8% carbon content. Thus an eutectic point is obtained at 730°C for 0.85 percent of carbon.

The ordinary rolled section such as I-beam, contains about 0.25 percent carbon. If such steel is cooled slowly from 1535°C, it remains liquid upto a temperature of about 1520°C. At this stage crystals of a material containing about 0.1% carbon begin to form.

Thus we have δ-iron crystals in liquid metal.

This results in increase of the carbon content of the remaining liquid, which solidifies further as the temperature drops. At about 1470°C, the entire mass freezes. The mass consists of a solid solution of carbon in iron and is known as *austenite* plus γ-iron. As the temperature drops, the solid solution remains unchanged until the temperature of about 850°C is reached (point B in Fig. 9.9). At this stage, remarkable transformation occurs. The steel, which has cooled to a dull red, begins to glow and liberate heat. An increase in volume also takes place, suggesting that internal adjustment is imminent. Iron changes from γ to β form. Then at a temperature of about 768°C (point C), β-iron form changes to α form. Since the solubility of carbon in α-iron is limited to 0.03 percent, the carbon which was dissolved in γ-form will begin to be rejected from austenite in β and α iron. However, the remaining austenite (with carbon content—1%) will absorb free carbon. The result is that crystals of α-iron, known as *ferrite,* begin to be precipitated from austenite. The selective precipitation of ferrite continues, with the remaining austenite absorbing more carbon, until the temperature reaches 730°C. At this temperature, the carbon and part of the iron combine chemically to form iron carbide, Fe_3C. known as *cementite.* Cementite, instead of forming in separate grains, combines mechanically with some of the ferrite and appear in flat sheets interspersed with similar sheets of ferrite within individual grains. These grains are known as *pearlite,* because of their pearl like appearance in white light. Pearlite, a mixture of ferrite and cementite, is the iron-carbon eutectoid containing 0.85% carbon. From 730°C to room temperature (below point D), steel containing 0.25% carbon, is a mixture of grains of ferrite and pearlite.

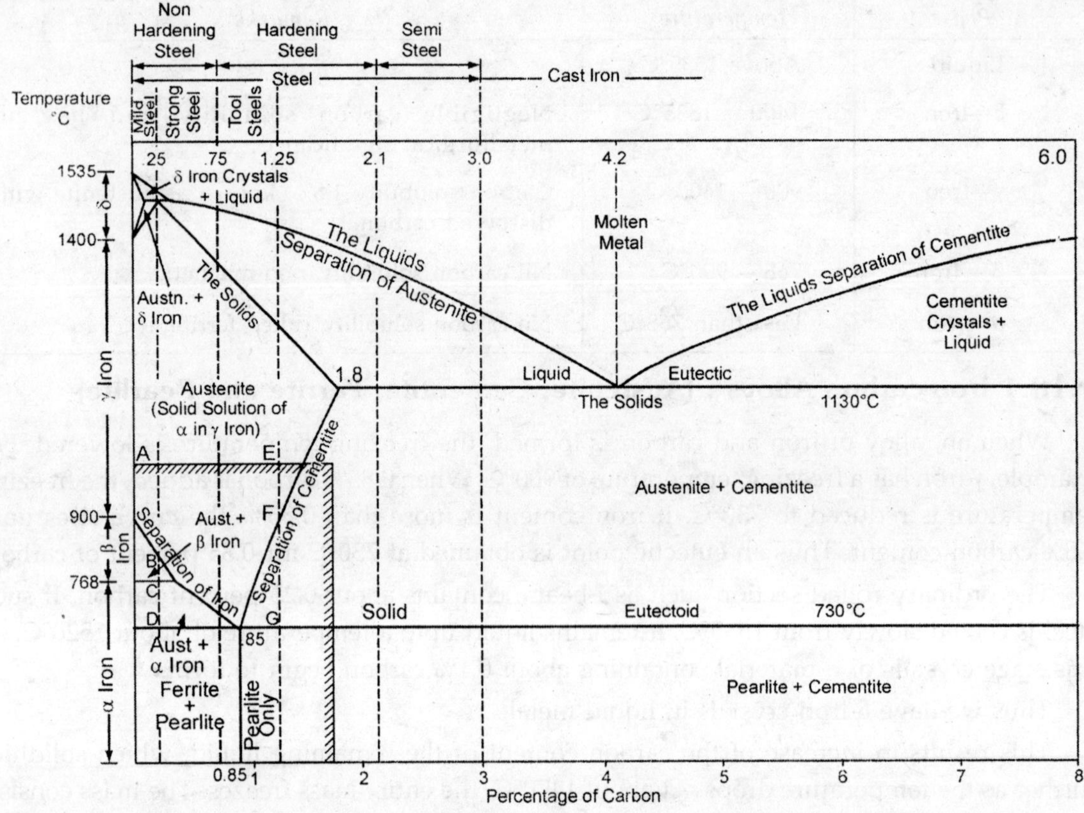

Fig. 9.9. Iron-carbon equilibrium diagram.

Similarly steels with other percentages of carbon can be analysed from molten stage to room temperaturcs. Table 9.5 gives the composition of steel at room temperature, for various carbon percentages.

Table 9.5: Composition of Steel for Various Carbon Percentages

Sl. No.	*Carbon percentage*	*Composition of Steel*	*Type of Steel*	
1.	upto 0.25	Ferrite + pearlite	Mild Setel	Non-hardening Steel
2.	0.25 to 0.75	—Do—	Strong Steel	
3.	0.75 to 0.85	—Do—	Tool Steel	
4.	0.85	Pearlite only	—Do—	Hardening Steel
5.	0.85 to 2.1	Pearlite + Cementite	High Carbon Steel	
6.	2.1 to 3	—Do—	Semi-Steel	

9.19 EFFECT OF RAPID COOLING ON MOLTEN STEEL

The above discussion relates to structural changes in steel, when the cooling takes place slowly and under controlled conditions. Further, it may be seen that the molten solution of iron and carbon or cementite in δ-iron ; changes first to δ-iron crystals plus liquid and then to austenite (solid solution of α in γ-iron). Then depending upon the percentage of carbon, austenite changes to either ferrite + pearlite or pearlite only or pearlite + cementite.

The transformation of austenite to pearlite takes place through the following four stages, on slow cooling.

1. Austenite to martensite
2. Martensite to troostite
3. Troostite to sorbite .
4. Sorbite to pearlite

However, if on the other hand instead of slow cooling, rapid cooling is effected by quenching or other intensified cooling means, the mixture may not have sufficient time or opportunity to pass through all the above transformations. Thus the transformation can be effected at any of the above structural forms ; by adopting a suitable rate of cooling.

A brief description of these structural forms is given below :

Martensite. This results when austenite is cooled rapidly in cold water or oil. γ-iron changes to α-iron and cementite is in dispersion in α-iron. Cementite has no time to change to pearlite. It is intensely hard and has a needle like structure. This type of structure is usually associated with all fully hardened steels.

Troostite. This is formed either by a slow rate of cooling mostly done in an oil tank called primary troostite or during reheating for tempering of the tool steel, known as secondary troostite. This is less hard than martensite, due to greater amount of α-iron in it. As the temperature is raised, troostite increases in amount, uptil a temperature of 1400°C is reached. At this stage, troostite transforms to sorbite.

Sorbite. This is the next transformation stage to that of troostite. This is effected by a still slow rate of cooling. It can also be formed by 'drawing' a fully hardened steel at 675°C. It is harder than pearlite but softer than troostite.

Pearlite. This is the final and stable stage of transformation of austenite. It consists of cementite and ferrite in the proportions 1 : 6. The above three stages are unstable.

9.20 HEAT TREATMENT PROCESSES

The various heat treatment processes are :

1. Hardening
2. Quenching
3. Tempering
4. Drawing
5. Annealing.

In addition, carburizing or case hardening, which are steel conversion rathur than true heat treatment processes, are also described under heat treatment processes for convenience. A brief description of these processes follows :

Hardening, as the name implies, is the process of increasing the hardness of metal by suitable treatment usually involving heating and cooling. Steel is heated to a temperature above 885°C (red heat), in case carbon content is less than 0.85% and 730°C if carbon is greater than 0.85% followed by sudden cooling by quenching. This results in the transformation into intermediate forms called martensite, troostite or sorbite. The degree of hardenability depends on the proportions of these three forms ; which in turn is a function of the rate of cooling and the quenching medium.

Steel containing very low carbon percentage cannot be hardened.

Quenching is the process of rapid cooling from an elevated temperature by contact with liquids, gases or solids. The quenching medium of cold water and brine solution give rapid cooling rates while oil provides moderate rate of cooling. The rate of cooling also-depends upon the size of the piece being quenched.

Tempering is a technique used for the production of a strong carbon steel. It consists in quenching the heated iron-carbon alloy in a medium which is held at a temperature between 200 to 480°C.

The quenching medium such as molten lead, etc. must reduce the temperature of the metal very quickly. After the transformation is complete, the material is removed from the bath and cooled to room temperature.

This makes steel softer and less brittle.

Drawing is the process of reheating hardened or normalized steel to a temperature below the transformation range (400 to 700°C), followed by any desired rate of cooling. This process converts martensite produced by quenching into softer and more ductile product of austenite. The two processes of tempering and drawing differ only in the temperature at which the work is done.

Annealing is a process involving heating and cooling to induce softening. This term is also used to cover treatments intended to :

(*i*) Remove stresses

(*ii*) Alter mechanical or physical properties

(*iii*) Remove gases.

The process of annealing is mostly applied to members which have been previously quenched. This is accomplished by a number of types of annealing processes, some of which are defined below :

(*i*) *Full annealing* in which the alloy is heated to a temperature above the transformation range, maintained at this temperature for a given time, and is then cooled slowly below the transformation range.

(*ii*) *Process annealing* usually applied in sheet and wire industry. The alloy is heated to a temperature close to, but below, the lower limit of transformation range and subsequently cooled.

(*iii*) *Normalizing*, in which alloy is heated to a temperature above the transformation range and subsequently cooled in still air at room temperature.

(*iv*) *Patenting.* This is the process of heat treatment applied to medium or high carbon steel in wire making prior to wire drawing. The alloy is heated above transformation temperature followed by cooling in air or molten lead bath.

(*v*) *Spheroidizing* is the process of heating and cooling steel that produces a rounded or globular form of carbide in the structure.

(*vi*) *Malleabling* is annealing white cast iron in which the combined carbon is wholly or in part transformed to free carbon and in some cases part of the carbon is removed completely.

9.21 METHODS OF HARDENING STEELS

(*i*) *Core-hardening or Case hardening or Carburizing*. This is done by putting the article in a box filled with bone dust, wood charcoal dust, etc., and burning to red heat for about 8 hours. At this high temperature, more carbon is absorbed by iron at the surface of contact. Article is soaked in water when 1.5 mm to 9 mm thick film of eutectoid steel (steel containing about 0.85% carbon) is formed. Steel subjected to such hardening should be low carbon steel.

(*ii*) *Surface hardening*. This is effected by heating the surface rapidly with a blow pipe and then quenching it by spraying water, or by any other suitable medium and thus rendering the surface hard.

9.22 NITRIDING OR NITROGEN CASE HARDENING

This process is carried out for steels alloyed with chromium, molybdenum, aluminium, vanadium etc. The surface of such steels is hardened by nitrite of iron.

Nitriding process is carried out generally in electric furnaces at about 510°C, when the parts are placed in air tight box provided with inlet and outlet tubes through which ammonia gas is circulated. Before the nitriding operation, the parts should be heat treated to impart desired physical properties to the core.

This process is not used for plain, carbon steels, as they do not absorb nitrogen. The process gives an extremely hard, non-brittle surface.

9.23 SORBITIZING

This is the process often employed in the treatment of rails and is known as Sandberg sorbitic process. The metal from the hot rolls, when at about 500°C, is cooled at a definite range by means of blasts of air and atomizing or spraying water.

This process increases the tensile strength of metal considerably, The cooled parts are then passed through a furnace at a controlled temperature of 400°C to 450°C, whereby the internal stresses caused by cooling are relieved.

STEEL ALLOYS

These include nickel, chromium, manganese, nickel-chromium, tungsten and silicon steels. They are used for tools, stainless steel products and for many other special purposes.

9.24 NICKEL STEEL

Nickel steel is made by open hearth, electric or crucible furnaces ; nickel steels are used for structural members where specially high strength is required, contains about 3.25% nickel and 0.25% carbon ; high nickel contents such as 36 to 38% produce the alloy known as invar, which has a very low coefficient of expansion and is used in precise surveying and levelling, and for clock pendulums.

9.25 CHROMIUM STEEL

Chromium steel contains 1 to 2% of chromium added as a hardening agent ; the resulting steel is also more impact resistant than ordinary steels. The elastic limit of chromium steel is high and it has been used for certain parts of points and crossings in railway track. Rustless chromium steels are made either by open-hearth or electric furnace processes; the chromium content is as high as 11 to 15%; the carbon less than 0.45%, silicon less than 0.3% ; manganese less than 0.5%, and sulphur and phosphorus as low as possible. The alloy takes a high polish and is fairly hard but will not resist some abrasives ; it is, however, uneffected by moisture, fresh or brackish water and organic acids from fruits, vinegar or cold nitric acid or hot or cold ammonia ; caustic alkalies, sulphuric or concentrated nitric or hydrochloric acids will attack it. Chromium steels exposed to atmosphere attain a thin continuous film of chromium oxide on the outer surface ; which also takes a high polish and is really a protective coating.

9.26 NICKEL CHROMIUM STEEL

Nickel chromium steel is made in the electric, crucible or open hearth furnace, the later being the more common ; the proportion of chromium to nickel is about 1 to $2\frac{1}{2}$, 1.1 to 1.25% of chromium to 2.5% of nickel being usual. This alloy is difficult to forge or roll. It is ductile and resilient and has a high elastic limit and abrasive resistance. It is used for armour plate and has also been used for structural purposes. High nickel-chromium steel is rust resistant ; it contains about 2% of nickel and about 18% of chromium.

9.27 MANGANESE STEEL

Manganese steel contains about 11 to 14% of manganese, 1 to 1.3% of carbon, 0.3% silicon, and sulphur and phosphorus 0.05 to 0.08%. Common variety is austenite. This alloy is very ductile ; it cannot be machined by ordinary methods but has to be ground to its required shape. It has the special property of hardening under wear and it also resists deformation. Its chief uses are for points and crossings and crusher jaws.

9.28 TUNGSTEN STEEL

Tungsten steel is made by crucible or electric furnace ; up to 9% tungsten can be used and this produces a steel with a high yield point. Its chief use is for tool steel ; such a steel containing 1 to 5% of tungsten can machine 0.3 tonnes/cm^2 steel at a rate of 12 metres per minute.

9.29 SILICON STEEL

Silicon steel is made from an iron to which 4% of silicon is added to increase strength. A 3.5% silicon steel can be bent at right angles without material damage. If silicon exceeds 4%, brittleness sets in. This steel is used chiefly in electrical transformers owing to its high electrical permeability and low hysterisis losses. Mild steels containing 25% of silicon have been used for structural purposes.

9.30 HIGH SPEED STEELS

High speed steels. Chief property required for high speed steel is that it may be run at a red heat without losing its hardness. Most of these high speed steels contain 15 to 20% tungsten, 3 to 5% chromium, 05 to 2% vanadium, 0.6 to 0.8% carbon ; with silicon, sulphur and phosphorus in much smaller quantities. 3 to 5% cobalt with a small quantity of molybdenum is also being alloyed with steel to replace tungsten. High speed steel is also used in parts which must withstand high heat and wear, as is required for the exhaust valves of gas engines etc.

QUESTIONS

1. Give the classification of iron products. What are the chief iron ores?
2. Draw a neat sketch of blast furnace and explain how iron ore is convented into pig iron. List some of the uses to which by-products from the blast furnace can be put.
3. How is cast iron manufactured in (*a*) cupola and (*b*) air-furnace? Draw a sketch of anyone of these.
4. Enumerate the different types of cast iron. What physical defects are present in cast iron?
5. What are the effects of the presence of sulphur, phosphorus, manganese, silicon, nickel and chromium in cast iron? Give the uses of cast iron.
6. Define wrought iron and how it is manufactured? List the uses of wrought iron.
7. What is steel? How pig iron can be refined to form steel? Give the reactions that take place in
   (*a*) Acid process ; and
   (*b*) Basic process.
8. Explain briefly the following methods of manufacture of steel:
   (*a*) Bessemer process
   (*b*) Open hearth process
   (*c*) Electric furnace process ; and
   (*d*) Cementation and crucible processes.
9. What is ingot? Discuss the physical defects present in ingots.
10. Describe the various processes of converting steel into market forms.
11. In what different forms, steel is available in the market.
12. What is an expanded metal (XPM)? Draw a sketch and state its properties and dimensions.
13. Compare the properties and uses of cast iron, wrought iron and steel in a tabular form.
14. Draw iron-carbon equilibrium diagram and explain the phase changes in pure iron.
15. How are iron-carbon alloys formed? Give the various structural changes in steel when it is cooled slowly.
16. What are the effects of rapid cooling on molten steel?

17. Explain the various heat treatment processes of steel.

18. Write short notes on:

(*a*) Methods of hardening steel

(*b*) Nitriding

(*c*) Sorbitizing

19. Enumerate the properties of following steel alloys:

(*a*) Nickel steel (*b*) Chromium steel

(*c*) Nickel-chromium steel (*d*) Manganese steel

(*e*) Tungsten steel (*f*) Silicon steel ; and

(*f*) High speed steels

20. List the use to which the above steel alloys are put.

OBJECTIVE TYPE QUESTIONS

1. Which of the following is not a ferrous metal

(*a*) Cast iron (*b*) Wrought iron

(*c*) Steel (*d*) Copper

2. The carbon content in pig iron varies from

(*a*) 0.5 to 2.5% (*b*) 2.5 to 4.5%

(*c*) 4.5 to 6.5% (*d*) 6.5 to 8.5%

3. The carbon content in wrought iron varies from

(*a*) 0 to 0.25% (*b*) 0.25 to 0.5%

(*c*) 0.5 to 0.75% (*d*) 0.75 to 1.0%

4. The carbon content of steel should not be greater then

(*a*) 1.0% (*b*) 1.5%

(*c*) 2.0% (*d*) 2.5%

5. Which of the following iron ores has the maximum specific gravity

(*a*) Hematite (*b*) Magnetile

(*c*) Limonite (*d*) Siderite

6. Limonite has ______ percent of iron

(*a*) 72.4 (*b*) 70

(*c*) 60 (*d*) 48.3

7. Slag, flue dust and hot gases are the by-products produced with

(*a*) pig iron (*b*) wrought iron

(*c*) steel (*d*) ingot iron

8. Grey, white, chilled and malleable are the types of

(*a*) pig iron (*b*) cast iron

(*c*) wrought iron (*d*) steel

9. Which furnace is used for the manufacture of cast iron
   (*a*) Blast furnace (*b*) Bessemer converter
   (*c*) Cupola furnace (*d*) Electric furnace
10. The carbon percentage of medium carbon steel ranges between:
   (*a*) 0.3 to 0.6 (*b*) 0.6 to 0.9
   (*c*) 0.9 to 1.2 (*d*) 1.2 to 1.5
11. Aston process is used for the manufacture of
   (*a*) Pig iron (*b*) Cast iron
   (*c*) Steel (*d*) Wrought iron
12. Pig iron is refined to obtain
   (*a*) Cast iron (*b*) Wrought iron
   (*c*) Steel (*d*) All these
13. Cold shut is a defect in
   (*a*) Cast iron (*b*) Wrought iron
   (*c*) Steel (*d*) All these
14. Steel wires have a diameter less than
   (*a*) 6 mm (*b*) 5 mm
   (*c*) 4 mm (*d*) 3 mm
15. Expanded metal is manufactured from steel
   (*a*) Sheets (*b*) Bars
   (*c*) Wires (*d*) Tubes
16. Melting point of wrought iron is
   (*a*) 1200°C (*b*) 1300°C
   (*c*) 1400°C (*d*) 1500°C
17. Cast-iron is:
   (*a*) Non-ductile and malleable (*b*) Low ductile and malleable
   (*c*) Medium ductile and malleable (*d*) Highly ductile and malleable
18. The ultimate compressive strength in tonnes/sq cm of wrought iron is
   (*a*) 12 (*b*) 7
   (*c*) 4.75 (*d*) 2
19. The ultimate tensile strength of mild steel in tonnes/sq cm is
   (*a*) 8 (*b*) 5.5
   (*c*) 3.75 (*d*) 1.5
20. β-iron is formed at a temperature of
   (*a*) Less than 768°C (*b*) 768–900°C
   (*c*) 900–1400°C (*d*) 1400 to 1835°C

Answers

1. (*d*)	2. (*b*)	3. (*a*)	4. (*c*)	5. (*b*)	6. (*c*)
7. (*a*)	8. (*b*)	9. (*c*)	10. (*a*)	11. (*d*)	12. (*c*)
13. (*c*)	14. (*b*)	15. (*a*)	16. (*d*)	17. (*a*)	18. (*d*)
19. (*b*)	20. (*b*)				

NON-FERROUS METALS AND ALLOYS

Non-ferrous metals include Copper, Aluminium, Tin, Lead, Nickel, Zinc, Molybdenum, Chromium, Vanadium, Tungsten etc.

COPPER

10.1 OCCURRENCE OF COPPER

The following are the important minerals of copper :

(*i*) Copper pyrites—$Cu_2S.Fe_2S_3$.
(*ii*) Copper glance—Cu_2S.
(*iii*) Malachite —$Cu(OH)_2\ CuCO_3$.
(*iv*) Cuprite or red copper—Cu_2O.
(*v*) Azurite— $Cu(OH)_2\ 2CuCO_3$.

The principal ore from which copper is extracted is copper pyrites.

10.2 EXTRACTION OF COPPER

1. *From Native Copper.* The ore containing native copper is ground and then concentrated by washing away the rocky impurities. Then it is heated on the sloping platform of a furnace, molten copper flows down and is separated from the impurities which do not melt.

2. *From Oxides and Carbonate Ores.* The ore is crushed, ground and concentrated by the *Gravity Process*. It is then calcined to obtain the oxide.

$$CuCO_3 \rightarrow CuO + CO_2$$

The calcined ore is then smelted with carbon and flux in a reverberatory furnace and metallic copper is obtained by the reduction of the oxide.

$$CuO + C \rightarrow CO + Cu$$

$$Cu_2O + C \rightarrow 2Cu + CO$$

3. *From Sulphide Ores.* This is the most important method for the extraction of copper and involves the following steps :

(*i*) *Concentration.* The ore is crushed and ground in ball mills and then concentrated by the Froth floatation process.

(*ii*) *Roasting*. The concentrated ore is then roasted on the hearth of a reverberatory furnace in excsss of hot air. Arsenic and antimony which are present as impurities in the ore are volatilised off as their oxides. Copper pyrites on roasting give a mixture of cuprous and ferrous sulphides. These sulphides are then oxidised to their oxides as shown in the following equations. Copper sulphide is only partly oxidised in this furnace and therefore a good aeal of copper sulphide remains unchanged. Sulphur dioxide produced is sometimes used for manufacturing sulphuric acid.

$$Cu_2S.Fe_2S_3+O_2 \rightarrow Cu^2S+ 2FeS + SO_2$$

$$2FeS + 3O_2 \rightarrow 2FeO + 2SO_2$$

$$2Cu_2S + 3O_2 \rightarrow 2Cu_2O + 2SO_2$$

(*ii*) *Smelting*. To the roasted ore, coke and silicon are added and the mixture smelted in a water jacketed blast furnace. Coke acts as a fuel and silica as a flux. The charge is added at the top of the furnace and air is passed in through tuyers provided at the base. Air is necessary for the combustion of coke and the ore itself. Cuprous oxide formed in tbe above reactions combines with ferrous sulphide, FeS to give cuprous sulphide and ferrous oxide, FeO. The latter combines with silica, SiO_2 to produce the fusible slag of ferrous silicate, $FeSiO_3$ which is removed from the slag hole.

$$Cu_2O + FeS \rightarrow Cu_2S + FeO$$

$$FeO + SiO_2 \rightarrow FeSiO_3 \text{ (Slag).}$$

The molten mass at the bottom of the furnace consists of cuprous sulphide, Cu_2S and a little unreacted ferrous sulphide FeS and is known as *Matte*. The molten *matte* is taken out from the exit below.

(*iv*) *Bessemerisation*. The molten *matte* taken out from the blast furnace is introduced in a Bessemer converter. The metal produced remains below the tuyers and is thus not oxidised by the blast of air. The converter can be tilted backward and forward as desired for adding the charge or emptying the converter. A blast of air and sand is passed through the molten matte through the tuyers. Iron sulphide which remained unoxidised in the blast furnace is oxidised to iron oxide which forms a slag with silica and can be removed. Also cuprous sulphide is partially oxidised to cuprous oxide which reacts with cuprous sulphide to produce metallic copper.

$$2Cu_2S+3O_2 \rightarrow 2Cu_2O + 2SO_a$$

$$Cu_2S + 2Cu_2O \rightarrow 6Cu + SO_2$$

The molten copper is then added to sand moulds. On solidification it has a blistered appearance due to the escape of dissolved sulphur dioxide and is known as *Blister Copper*.

(*v*) *Purification of Blister Copper*. Blister copper is contaminated with a number of impurities like iron, nickel, zinc, silver and gold etc. and is purified by the following methods :

(*a*) *Poling*. The impure metal is melted on the hearth of a reverberatory furnace in the presence of air and stirred with green logs of wood. Sulphur and arsenic are changed into their oxides which go out as gases. The impurities of base metals either volatilize off or form a slag. Any copper oxide present in the molten metal is reduced to copper by the hyaro-carbons produced by the green logs of wood.

(*b*) *Electrolysis.* The impure blister copper is made the anode and a thin slice of pure copper the cathode. Acidified copper sulphate solution is taken as the electrolyte. On passing the current, pure copper travels to the cathode and is deposited there. The impurities either settle down at the bottom of the anode and are known as *anode mud* (Ag, Au) or go in solution (Fe, Ni, Zu).

10.3 PROPERTIES

It is light, tough, strong, ductile and malleable metal with good properties of resistance to corrosion in dry air. It can be forged, rolled or otherwise worked hot or cold and drawn into wires. It has high thermal and electrical conductivity.

10.4 USES

(*i*) Electric wires and cables.

(*ii*) Light gauge copper tubing for hot and cold water supply, gas and sanitation services.

(*iii*) Sheets for roofing, flushings and D.P.C.

(*iv*) Lighting protective devices.

(*v*) As a constituent of brass and other alloys.

(*vi*) As a protective coating for other metals.

(*vii*) Dowels in stone masonry.

ALUMINIUM

10.5 OCCURRENCE

It does not occur native. In the combined state it occurs as

(*i*) *Silicate.* Mica, felspar $KAlSi_3O_8$, tourmaline, clay and slate, $A1_2O_3$ $2SiO_2.2H_2O$.

(*ii*) *Oxide.* Corrundum Al_2O_3 : diaspore $Al_2O_3.H_2O$ and bauxite $Al_2O_3.2H_2O$.

(*iii*) *Fluoride.* Cryolite Na_3AlF_6.

(*iv*) *Basic sulphates.* Alunite $K_2SO_4.Al_2(SO_4)_3$. $4Al(OH)_3$.

The chief ore of aluminium is bauxite and occurs in Kashmir, Bihar and Maharashtra in India.

10.6 EXTRACTION OF ALUMINIUM

These days aluminium is extracted from the bauxite, $Al_2O_3.2H_2O$ by electrolysis. The bauxite that occurs in nature contains mainly two impurities (*a*) Ferric oxide and (*b*) silica, SiO_2. Therefore, the ore is first of all purified to obtain pure alumina, $A1_2O_3$. It is then dissolved in molten cryolite, $AlF_3 3NaF$ and electrolysed when aluminium is obtained at the cathode. The purification of the bauxite is carried out by the following three methods :

(*i*) *Baeyer's Process.* This process is used for the purification of the red bauxite containing ferric oxide as an impurity. The bauxite is thoroughly ground and calcined. It is then heated with a strong caustic soda solution (about 50%) at 150°C in an autoclave for some hours. The alumina present in the bauxite dissolves in caustic soda solution in the form of sodium aluminate whereas the impurity ferric oxide, Fe_2O_3 does not dissolve in the alkali. It is then filtered to get rid of ferric oxide, Fe_2O_3 which remains behind as a residue. To the filtrate water is added and it is agitated in the presence of a little freshly precipitated aluminium hydroxide when sodium meta aluminate is hydrolysed to give aluminium hydroxide, $Al(OH)_3$ which when heated gives pure alumina.

$$\underset{\text{Bauxite}}{Al_2O_3} + 2NaOH \rightarrow \underset{\text{Sodium aluminate}}{2NaAlO_2} + H_2O$$

$$NaAlO_2 + 2H_2O \rightarrow Al(OH)_3 + NaOH$$

$$2Al(OH)_3 \rightarrow \underset{\text{Pure alumina}}{Al_2O_3} + 3H_2O$$

(*ii*) *Hall's Process.* The bauxite is fused with sodium carbonate when sodium aluminate (or sodium meta-aluminate) is formed which is soluble whereas the impurities ferric oxide, Fe_2O_3 and silica, are insoluble and it is filtered, thus ferric oxide and silica are separated.

$$Al_2O_3 + Na_2CO_3 \rightarrow 2NaAlO_2 + CO_2$$

The filtrate is then heated to 50–60°C and a current of carbon dioxide is passed through the solution when aluminium hydroxide, $Al(OH)_3$ is precipitated. It is then heated to give pure alumina, Al_2O_3.

$$\underset{\substack{\text{Sodium}\\ \text{meta-aluminate}}}{2NaAlO_2} + 3H_2 + CO_2 \rightarrow 2Al(OH)_3 + Na_2CO_3$$

$$2Al(OH)_3 \rightarrow \underset{\text{Pure alumina}}{Al_2O_3} + 3H_2O$$

(*iii*) *Serpek's Process.* This process is used for the purification of white bauxite containing silica, SiO_2 as an impurity. The bauxite is thoroughly ground and heated with coke in a current of nitrogen. Aluminium nitride is formed which is hydrolysed by water giving aluminium hydroxide and ammonia. Aluminium hydroxide is then heated to give pure alumina. Silica present in the bauxite is reduced by coke to silicon which is volatilised off.

$$Al_2O_3 + 3C + N_2 \rightarrow 2AlN + 3CO$$

$$AlN + 3H_2O \rightarrow Al(OH)_3 + NH_3$$

$$SiO_2 + 2C \rightarrow Si + 2CO$$

$$2Al(OH)_3 \rightarrow \underset{\text{Pure alumina}}{Al_2O_3} + 3H_2O$$

The advantage of this process is that ammonia is obtained as a by product.

10.7 ELECTROLYSIS OF ALUMINA

This process was simultaneously discovered in 1886 by two chemists Charles Martin Hall, an American and Heroult, a Frenchman. They discovered that alumina, Al_2O_3 gets dissolved readily in molten cryolite and ionized in it. Thus by electrolysis of alumina in fused cryolite, aluminium was obtained.

The apparatus consists of a rectangular iron box lined on the inside with gas carbon which acts as the cathode. A number of carbon rods dipping in the molten mass serve as the anode. The electrolyte is alumina dissolved in fused cryolite, $AlF_3.3NaF$. Cryolite is first pumped in the rectangular tank and it is melted either by external heat or by heat produced by passing the current. The temperature is maintained at 1000°C and to the molten cryolite pure alumina is added. On passing the current, aluminium goes to the cathode and is removed from the tapping hole at the bottom. Oxygen is liberated at the anode and combines with carbon to form carbon monoxide and carbon dioxide which escape in the atmosphere. Some powdered coke is sprinkled on the molten alumina to avoid loss of heat by radiation and to cover the intensely blinding glow produced by the action of oxygen on the carbon anodes. The current passed is 100 amps, at 6.8 voltage. When the amount of alumina in the cryolite decreases the resistance of the bath increases which causes a charge control lamp (connected in parallel) to glow. More alumina is then added to the cryolite and electrolysis is continued and molten aluminium is taken out continuously from the tapping hole. It is found to be 99% pure.

Electrolytic Refining of Aluminium. Aluminium obtained by the above process is 99 percent pure. It is further refined by the *Hoope's Process.* It consists of an iron tank which is lined on the inside (at the bottom) with carbon. Three layers of aluminium with different specific gravities are taken in this tank. The bottom layer consists of impure molten aluminium which serves as the anode. The middle layer consists of a molten mixture of aluminium, barium and sodium fluorides and in the topmost layer there is molten pure aluminium which serves as the cathode. On passing the current, aluminium from the middle layer goes to the top layer and an equivalent amount of aluminium passes from the bottom to the middle layer. Pure aluminium is being constantly removed from the top. Aluminium refined like this is 99.99 percent pure.

10.8 PROPERTIES

Pure aluminium is very soft and ductile and therefore it is alloyed with other metals like copper, magnesium, silicon, manganese etc., which increase its tensile strength and hardness while retaining its characteristics of lightnesss and durability. Aluminium is highly resistant to corrosion and very conductive to heat and electricity.

10.9 USES

(*i*) Window frames and sub-frames.

(*ii*) Glazing bars.

(*iii*) Corrugated aluminium sheets for roofing.

(*iv*) Structural members.

(*v*) Wires, bars, rods.

(*vi*) As fine powder for pigment in paints.

(*vii*) Aluminium foil, and

(*viii*) Posts, panels and balustrades.

TIN

10.10 OCCURRENCE

It occurs in the free state only in traces in Siberia. In the combined state it occurs mainly as *Cassiterite* or *Tinstone* SnO_2 which occurs in Austalia, China, Burma, South Africa and Malaya.

10.11 EXTRACTION OF TIN

Tin is extracted from its ore, tinstone which contains about 5–10 percent of the metal in the form of SnO_2, the rest being mainly silica and wolfram (iron tungstate, $FeWO_4$). It also contains, small amounts of iron pyrites, copper pyrites, certain arsenides and manganese tungstate ($MnWO_4$). The extraction of tin from tinstone involves the following steps :

(*i*) *Concentration.* The ore is crushed and ground in ball mills and then concentrated by the Gravity Process wherein it is washed in a current of water which carries with it the impurities of silicious matter. The heavy tinstone, wolfram, manganese tungstate etc. being heavier settle down and are thus separated from the silicious impurities.

(*ii*) *Roasting.* The ore is now roasted in an inclined revolving tube furnace and a current of hot air is passed in. Arsenic and sulphur present in the ore are removed as their oxides.

(*iii*) *Electro-magnetic Separation.* The concentrated and roasted ore is then passed over magnetic rollers to separate it from the magnetic impurities like iron and manganese tungstates.

(*iv*) *Washing.* The concentrated and roasted ore is then washed with water to dissolve the soluble impurities like copper sulphate and iron sulphate which might have been formed during roasting. Also water washes away ferric oxide and other light impurities.

The washed and concentrated ore contains about 65–80 percent of tinstone and is called *black tin.*

(*v*) *Smelting.* The black tin obtained above is then mixed with one-fifth its weight of well ground anthracite and a little lime and fluorspar as flux and smelted as 1200–1300°C in a reverberatory furnace. Tinstone is reduced to metallic tin.

$$SnO_2 + 2C \rightarrow Sn + 2CO$$

The molten tin is drawn off from the slag hole. It contains about 15% of tin and is therefore, smelted with carbon to obtain the metal from it. Tin prepared by the above method is about 99.5% pure.

(*vi*) *Purification.* (*a*) *Liquation.* The ingots of tin prepared above are heated on the sloping hearth of a reverberatory furnace. Tin melts easily and flows down the slope whereas the impurities remain behind.

(*b*) *Poling.* The molten tin is stirred by green logs of wood which bring about further purification.

(*c*) *Electrolytic refining.* The impure tin is made the anode and a sheet of pure tin is made the cathode. The electrolyte consists of an aqueous solution of hydrofluosilicic acid, H_2SiF_6 containing tin and sulphuric acid. When the current is passed, pure tin travels from the anode and is deposited on the cathode.

10.12 PROPERTIES

(*i*) It is a white, lustrous soft metal melting at 232°C and boiling at 2275°C.

(*ii*) It is malleable and ductile but when heated to 200°C it becomes brittle and can be easily powdered.

(*iii*) When cold, then also it is slightly brittle. If a rod of tin is bent, a peculiar cracking sound is produced (called the 'tin cry'). This is on account of rubbing of crystals against one another.

(*iv*) Tin exists in two main allotropic forms, *white* and *grey*, the former having a density of 7.28 and the latter 5.62. The transition temperature is 18°C.

(*v*) Tin is a very soft and weak metal, but offers excellent resistance to corrosion and acid action.

10.13 USES

(*i*) Protection of copper wires and cables.

(*ii*) Tin plating

(*iii*) As a constituent of solders and

(*iv*) As a constituent of many valuable alloys.

LEAD

10.14 OCCURRENCE

It does not occur free in nature. The chief ore of lead is galena, PbS which occurs in England, Spain, North America and Burma. Lead also occurs in nature as *cerussite* $PbCO_3$ and *anglesite* $PbSO_4$.

10.15 EXTRACTION OF LEAD

Lead is extracted from its ore *galena*, PbS by any one of the two methods given below :

1. *Self Reduction Process or Reverberatory furnace Process.*

This method is used for the extraction of lead from richer ores and involves the following steps :

(*i*) *Concentration.* Galena is crushed, ground and sieved. It is then concentrated by the froth floatation process.

(*ii*) *Roasting.* The concentrated ore is roasted in a reverberatory furnace which is provided with air holes. The air holes are opened and roasting is done at a moderate temperature in excess of air. Galena is partially oxidised to lead oxide and lead sulphate.

$$2PbS + 3O_2 \rightarrow 2PbO + 2SO_2$$

$$PbS + 2O_2 \rightarrow PbSO_4$$

(*iii*) *Reduction.* When nearly half the galena has been oxidised the supply of air is cut off by closing the air holes and the temperature of the furnace is raised. Lead oxide and lead sulphate are both reduced to metallic lead by unchanged galena.

$$2PbO + PbS \rightarrow 3Pb + SO_2$$

$$PbSO_4 + PbS \rightarrow 2Pb + 2SO_2$$

Molten lead is then removed time to time from the tapping hole.

Since the reduction is brought about by galena itself, therefore, it is called the *Self Reduction Process.*

2. *Carbon Reduction Process or Blast furnace Process.*

This method is employed for poorer ores and involves the following steps :

(*i*) *Concentration.* It is crushed, ground and sieved. Then it is concentrated by the froth floatation process.

(*ii*) *Sintering.* The concentrated ore is then mixed with coal and lime and roasted in a current of air at a high temperature. Lead oxide is formed which is sintered into lumps of proper size at the high temperature in a sinter machine.

$$2PbS + 3O_2 \rightleftharpoons 2PbO + 2SO_2$$

Lime absorbs sulphur dioxide and thus the reversible reaction goes to completion. A very small amount of the sulphide remains unoxidised.

(*iii*) *Smelting.* The sintered ore is then mixed with coke and lime and smelted in a small blast furnace. The charge is added from above and a hot blast of air sent from below. The lower part of the furnace is water jacketed.

Coke and carbon monoxide produced by the slow combustion of carbon react with lead oxide to give lead

$$2C + O_2 \rightarrow 2CO$$

$$PbO + C \rightarrow Pb + CO$$

$$PbO + CO \rightarrow Pb + CO_2$$

The unoxidised lead sulphide, PbS is oxidised to lead sulphate, $PbSO_4$ which reacts with lead sulphide to give molten lead.

$$PbS + 2O_2 \rightarrow PbSO_4$$

$$PbSO_4 + PbS \rightarrow 2Pb + 2SO_3$$

Silica present in the ore combines with lime to form calcium silicate, $CaSiO_3$ which is removed as a fusible slag.

$$CaO + SiO_2 \rightarrow CaSiO_3$$

The presence of lime serves a very useful purpose. It does not allow the formation of lead silicate, $PbSiO_3$ which would be wasted as a slag. Even if some lead silicate is formed, it reacts with lime to form lead oxide and calcium silicate which is removed as a slag.

$$PbSiO_3 + CaO \rightarrow CaSiO_3 + PbO$$

$$PbO + C \rightarrow Pb + CO.$$

The molten lead and the fusible slag are taken out from two different outlets.

Purification of Lead. Lead prepared by the above method is contaminated with antimony, bismuth, copper, silver and tin. The presence of these impurities makes the metal hard and the removal of these impurities is, therefore, called *softening of Lead.* It can be brought about by the following methods :

(*i*) Lead is melted on the hearth of a shallow reverberatory furnace and a current of hot air is passed through when all the impurities are oxidised and removed as a scum. Only silver is not removed by this method.

(*ii*) Silver can be removed either by Parke's Process or Pattinson's Process.

(*iii*) It can also be purified by an electrolytic method. Impure lead is made the anode and pure lead sheets act as the cathode. The electrolyte consists of lead silicofloride $PbSiF_6$, hydrofluosilicic acid H_2SiF_6 and a little gelatine. On passing the current pure lead is deposited at the cathode. Sulphuric and hydrochloric acids cannot be used as electrolytes because they form insoluble lead salts.

10.16 PROPERTIES

Lead is a very soft, highly ductile, malleable, plastic, non-corrodible metal with low fusion point and very low strength. The metal is extremely resistant to atmospheric corrosion and is affected by soil and sewage effluents or industrial wastes.

10.17 USES

(*i*) Water service pipes, soil and gas pipes.
(*ii*) Gutters, flushings and cistern linings.
(*iii*) Damp-proof courses of buildings.
(*iv*) Lead sheets for roofing.
(*v*) Cable coverings.
(*vi*) Lead wood and solders for plumbing.
(*vii*) Cast lead for ornamental lead work, and
(*viii*) Lead oxides for paints such as white leads, red lead and litharge.

NICKEL

10.18 EXTRACTION

It is extracted from the following ores :

(*i*) Nickel-ferrous magnetic pyrites containing about 3% nickel.
(*ii*) Hydrated nickel magnesium silicate containing 6 to 8% nickel.

Nickel ores are crushed, roasted and smelted in a blast furnace which produces a crude matter of nickel, iron and copper. Iron is removed by Bessemer's process and copper is removed by treatment with dilute sulphuric acid. Residue is partially reduced by hot producer gas and volatilized at a lower temperature into nickel carbon, which is later passed through a heated tower where pure nickel is deposited in granular form ; this process is called 'Mond process' .

10.19 PROPERTIES

Nickel is a brilliant metal, approaching steel in colour ; it takes a good polish and does not tarnish or corrode in dry air. It is attracted by magnets. Its hardness is like soft steel, but is more malleable and ductile. Addition of carbon makes it hard but brittle.

10.20 USES

(*i*) Pure nickel is used as an electrolate.
(*ii*) For alloying with other metals.
(*iii*) For making coins.

ZINC

10.21 EXTRACTION

It is chiefly obtained from sulphide ores such as zinc blende or black tack (ZnS), which ordinarily contains from 1/3 to 1/2 zinc. These ores are often found associated with galena (lead ores) and pyrites (copper ores). Zine carbonates and zinc silicates are also used as ores for the extraction of the metal. The ore is finely ground and slowly roasted in a reverberatory furnace until nearly all the sulphur is removed. Mixed with ground coal, it is heated to white heat on retorts of fire-clay which removes oxygen. It is collected in liquid form in condensers from where it is poured into moulds. Zinc so cast is called smelter and is quite suitable for industrial purposes. When found mixed with impurities, it is further refined by heating to various temperatures just sufficient to melt the impurities, when seperation of metals (impurities) is effected through difference in their specific gravities. Zinc is also extracted by electrolytic refining.

10.22 PROPERTIES

Zinc is a bluish white metal, very soft, light and highly resistant to corrosion.

10.23 USES

(*i*) Sheets for roofing,
(*ii*) Cutters, flashings, kitchen table tops,
(*iii*) Galvanising iron products,
(*iv*) Zinc oxide extensively used for paints, and
(*v*) As a constituent of brass

MOLYBDENUM

Occurrence

This occurs mostly in the form of molybdenite which is a disulphide of Molybdenum (MoS) and resembles graphite.

Properties

Molybdenum is grey in colour ; it is brittle and has a high melting point and a very high tensile strength.

Uses

For alloying with iron, nickel, managenese etc. specially for making tool steels.

CHROMIUM

Extraction

This occurs as an oxide (Cr_2O_3) associated with earthly matters. It also occurs in the form of chrome iron ore ($FeCr_2O_4$) and is found contaminated with silica, alumina, magnesia etc. It is extracted by heating a mixture of ore Cr_2O_3 and carbon in an electric furnace, which is further refined by smelting it with CaO and thereby eliminating its carbon. The product is heated in a crucible furnace lined with calcium chromate which produces almost pure chromium.

Properties

It is a hard metal of brilliant steel grey colour and can be easily polished ; it resists corrosion and retains its polish.

Uses

For making alloys such as stainless steel and other superior varieties of tool steel.

VANADIUM

This generally occurs combined with lead etc.

It is a hard, very brittle and infusible metal, unaffected either by air or water.

Uses

For alloying with steel for producing high grade steel.

TUNGSTEN

This occurs mostly in the form of wolframite ($2FeWO_4.MnWO_4$).

Properties

It has a very high melting point (2800°C) and can be rendered perfectly ductile by repeated heating.

Uses

(*i*) In filaments of electric lamps,

(*ii*) Being cheaper, it has replaced platinum

(*iii*) For alloying with steels to which it imparts toughness, hardness and resistance to shocks.

ALLOYS OF NON-FERROUS METALS

10.24 BRASS

It is an alloy of copper and zinc. Properties vary considerably according to varying proportion of metals. Most commercial wrought brass contains 65% copper and 35% zinc. Brasses are ductile and malleable at ordinary temperature and can be rolled into sheets, turned into tubes, drawn into wires or cast into moulds. Brass resists corrosion well.

Brass is used in buildings for stair treads, grills, protective sheets, and especially for finished hardware. Although bright yellow when fresh, it quickly becomes dull and requires constant polishing.

10.25 BRONZE

It is an alloy of copper, zinc and tin and contains about 80% copper. Phospher-bronze contains about 1% of phosphorus. Properties of bronze vary with the content of different constituents.

Manganese bronze is an alloy (usually white) of pure copper with 2 to 30% of manganese.

Bronze is largely used in building construction for doors, window sash, frames, grills, balconies, balustrades, screens, hardware and a great variety of ornamental purposes.

10.26 Gun-metal

It is an alloy of copper, tin and zinc to which sometimes lead is also added. This has high tensile strength and elasticity. It is to some extent used for strong castings.

10.27 DURALUMIN

This is an alloy of aluminium with 3.5 to 5.5% copper, 0.5%. Mg and 0.5 to 0.8% Mn. It is light with 2. 85 as its specific gravity. It can easily be heat treated, which renders it highly resistant to corrosion. It is used in surgical instruments, air crafts and automobile parts and for other instruments.

10.28 Y-ALLOY

This is an important alloy which has been developed during the last war. It consists of aluminium alloyed with 4% copper, 2% nickel and 1.5% magnesium. It is a costly alloy which is light but very tough and hard, and possesses great strength at high temperature and is therefore very suitable for use in piston heads, gear boxes, propellers, blades etc.

10.29 MAGNALIUM

This contains about 2 to 6% magnesium alloyed with aluminium. It is lighter than pure aluminium, easy to work, exceptionally strong and ductile. It is used as a deoxidizer in copper smelting operations.

10.30 GERMAN SILVER AND NICKEL SILVER

This is a brass containing about 20% nickel ; properties being similar to those of brass. It is white in colour and resists tarnishing very much better but requires regular cleaning to keep it bright.

It is available in sheets, strips, tubes, wires, rods and castings. It is used for making resistance coils as it has very low electric conductivity. Its common use is for immitating silver.

10.31 MONEL

It is an alloy of nickel with about 35% copper. It is as strong as steel, but it is highly resistant to corrosion and tarnishing. It is a comparatively costly alloy. It is available in the form of sheets, strips,, wires, rods, tubes and castings and is used for pipes, propeller shafts-of ships, valves of pumps etc.

10.32 STAINLESS STEEL

It is known by trade name of Alleghary metal. It consists of 18% chromium, 8% nickel, the remainder being iron. It has an outstanding corrosion resistance together with very great strength and toughness. It does not tarnish and needs very little cleaning. It is available in all the forms in which mild steel is produced. Its use for cutlery and household utensils is becoming very popular. It is also used for various hardware goods which require a permanently stainless and polished hard surface.

10.33 BEARING METALS

These are alloys of lead, tin and antimony with copper in varying proportions. White metal contains 86% tin while magnesia metal contains 80% lead. Bearing metals are used in bearings where they are easily melted, and thus protect the bearings which can be relined to a smooth surface.

10.34 DOW METAL

It contains 85 to 98% magnesium alloyed with 4 to 12% Al, 0.1 to 0.4% Mn, added with a little copper and other metals. It is very light and has a specific gravity of 1.8. It is mostly used in aeroplane and automobile parts as it is soft, strong and ductile. It can be cast, forged, rolled or drawn to wire.

10.35 STELLITE

It contains 45% cobalt, 30% chromium, and 15% tungsten with small amounts of Mn, Silica, Carbon and Iron. It is very hard and is used for the tips of cutting tools and facings of des. It does not tarnish and is, therefore, very suitable for use in surgical instruments.

10.36 SOLDERS

Solders are alloys for the purpose of making joints between pieces of metal. They are largely adopted in engineering construction. Such joints are not mere mechanical connections since the solder itself combines with the metal which it unites and forms a new alloy. Every solder must be more fusible than the metals which it unites.

Solders are of two classes :

10.36.1 Hard Solders

(*a*) *Brazing solder.* It consists of brass alloys (Cu 4.9 to 81%, zinc remainder). It is used to join steel parts by using a gas flame to heat the metal to 870° C. Borax is used as a flux to keep the surfaces clean and free from oxides. This solder gives a brazed joint of considerable strength.

(*b*) *Copper brazing.* It is employed in fabrication of metal products using copper wire. The parts are heated in a special furnace to 1150°C in an atmosphere of hydrogen gas when copper flows into the seams and combines with the steel. Hydrogen prevents the formation of oxides on the steel.

(*c*) *Silver solders.* These are alloys of Cu, Zn and Ag (Cu 14 to 30%, Zn 9 to 20%, Ag 42 to 62%). The melting points of these solders range from 650° to 870°C. These are used for brazing steel etc. especially in cases where the other methods are difficult.

Borax is used as a flux.

(*d*) Al, Sn, Zn alloys are also used as solders. No flux of any kind is used. When the melted solder has run well over the surfaces to be jointed, they are held pressed firmly together until the solder has solidified. As aluminium solders are cathodic to Al, and as in the presence of moisture, an electrolytic action causing the disintegration of joints is developed, no dependence should be placed on a soldered aluminium joint.

10.36.2 Soft Solders

It melts at a low temperature. For plumbers work, a solder of 1 part Sn and 2 parts Pb and for tinner's work, 2 parts Sn and 1 part Pb are used. $ZnCl_2$ or NH_4Cl in the form of a solution, or as paste made of the salt slightly moistened with water, or as paste made by stirring the finely divided salt either in vaseline or fats, with or without resin, is also employed.

QUESTIONS

1. List the important minerals of copper. Explain briefly the extraction of copper from sulphide ores. State the properties and uses of copper.
2. In what combined states, aluminium occurs? Give two methods of its extraction from bauxite.
3. Describe the electrolysis method of obtaining and refining aluminium. State its properties and uses.
4. How is tin extracted? Write its properties and uses.
5. What is the chief ore of lead? Explain the various steps involved in its extraction.
6. What do you understand by the term 'Softening of lead'. By what methods it is carried out.

   Give the uses and properties of lead.

7. Write briefly the extraction, properties and uses of the following:
   (*a*) Nickel (*b*) Zinc
   (*c*) Chromium (*d*) Tungsten
   (*e*) Molybdenum
8. Name atleast eight alloys of non-ferrous metals. Describe any three of these.
9. (*a*) What is the composition of stainless steel? State its properties and uses.
   (*b*) What is solder? Describe the two classes of solders.

OBJECTIVE TYPE QUESTIONS

1. Azurite is a mineral of
   (*a*) Tin (*b*) Aluminium
   (*c*) Copper (*d*) Zinc
2. Concentration, roasting, smelting, bessemerisation and purification are the processes involved in the extraction of ______ metal from sulphide ores:
   (*a*) Copper (*b*) Nickel
   (*c*) Lead (*d*) Chromium
3. Red bauxite and caustic soda solution are heated in Baeyers process at a temperature of
   (*a*) 100°C (*b*) 150°C
   (*c*) 200°C (*d*) 250°C
4. Aluminium obtained by electrolytic refining in Hoopes' process is almost ______ percent pure
   (*a*) 99.25 (*b*) 99.50
   (*c*) 99.75 (*d*) 99.99
5. In the electrolysis of alumina, a temperature of ______ °C is maintaried:
   (*a*) 800 (*b*) 900
   (*c*) 100 (*d*) 1100
6. The boiling point of tin is?
   (*a*) 2250°C (*b*) 2275°C
   (*c*) 2300°C (*d*) 2325°C
7. The density of white tin is:
   (*a*) 7.28 (*b*) 7.38
   (*c*) 7.48 (*d*) 7.58
8. Lead is used for:
   (*a*) Cistern lining (*b*) Damp-proof course
   (*c*) Roofing (*d*) All these
9. The nickel ores contain nickel upto
   (*a*) 8% (*b*) 12%
   (*c*) 16% (*d*) 20%

10. The metal used in making coins is:
(*a*) Zinc
(*b*) Lead
(*c*) Nickel
(*d*) Tungsten

11. Zinc extracted from sulphide ores contain zinc about
(*a*) 0 to 10%
(*b*) 10 to 15%
(*c*) 15 to 30%
(*d*) 30 to 50%

12. Zinc has ______ resistance to corrosion
(*a*) No
(*b*) Low
(*c*) Medium
(*d*) High

13. Chromium is used for making ______ alloy
(*a*) Brass
(*b*) Stainless steel
(*c*) Gun-metal
(*d*) Monel

14. Brass is an alloys of
(*a*) Copper and zinc
(*b*) Copper, tin and zinc
(*c*) Copper and nickel
(*d*) Aluminium-copper-nickel

15. For plumber's work solder consists of tin : lead in the ratio of
(*a*) 1 : 3
(*b*) 3 : 1
(*c*) 1 : 2
(*d*) 2 : 1

Answers

1. (*c*)	**2.** (*a*)	**3.** (*b*)	**4.** (*d*)	**5.** (*c*)	**6.** (*b*)
7. (*a*)	**8.** (*d*)	**9.** (*a*)	**10.** (*c*)	**11.** (*d*)	**12.** (*d*)
13. (*b*)	**14.** (*a*)	**15.** (*c*)			

PLASTICS IN BUILDINGS

Plastic materials have come in the market in all possible forms. Being a synthetic product, a great variety of plastics are being produced to suit varying requirements. The new material is a 'controlled product' in the sense that the properties desired of it could be obtained by introducing variations in the constituent materials and the method of production. In the field of building construction, plastics have been found very useful for a variety of purposes. From fittings and fixtures for elegance and decoration to structural component parts, plastics have found use in buildings.

FORMATION AND MANUFACTURE

11.1 FORMATION

Plastics are commonly taken to mean a group of organic fusible and infusible materials which have in common the property of being plastic at some stage in manufacture and in which state they are moulded in the required shape. They are generally natural or synthetic resins (compounds of carbon, hydrogen and oxygen possibly combined with nitrogen, chlorine or sulphur) weighed or filled with fibres or granular substances according to requirements.

Plastics are broadly classified into two categories:

1. Thermo-setting ; and
2. Thermo-plastic

Plastics made from thermo-setting resins undergo a chemical change when they are moulded. These are prepared from resins in a form that softens on heating and can be moulded but further heating (curing) causes the material to harden after which it cannot be resoftened by heat. The reason for this is that due to the chemical changes undergone at higher temperatures, the material leaving the mould is different in chemical composition from that which enters the mould.

Thermoplastics simply soften when heated to fabrication temperatures. No chemical change is involved and the material leaving the mould is identical chemically with that entering the mould.

Very few resins, particularly those of the thermoplastic kind are moulded in their natural state as the resulting products are very brittle. Softeners, termed plasticisers, are therefore added in the course of manufacture in varying proportions. Plasticisers may be granular or fibrous.

11.2 MANUFACTURE

The various processes of manufacturing plastics are :

1. Moulding
2. Extrusion
3. Casting
4. Other methods.

11.2.1 Moulding

This is a very simple process and can be done by any of the following methods :

(*i*) *Positive compression moulding*. The moulds are made in the form of plates ; which are heated. Exact quantity of the raw materials are placed between the plates and pressure applied. The plates are then separated and the moulded article removed and finished.

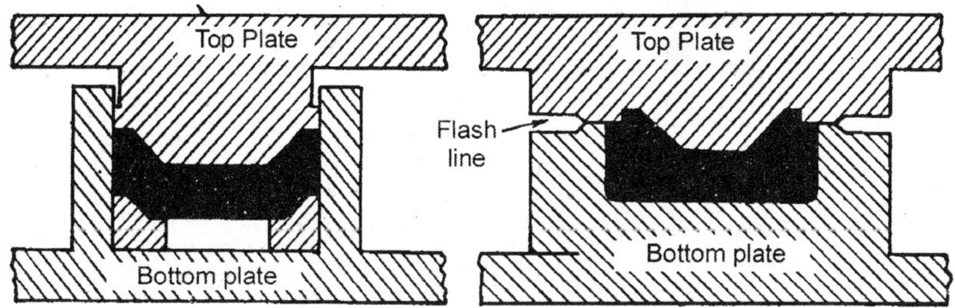

Fig. 11.1 (*a*) Positive and (*b*) Flash compression moulding

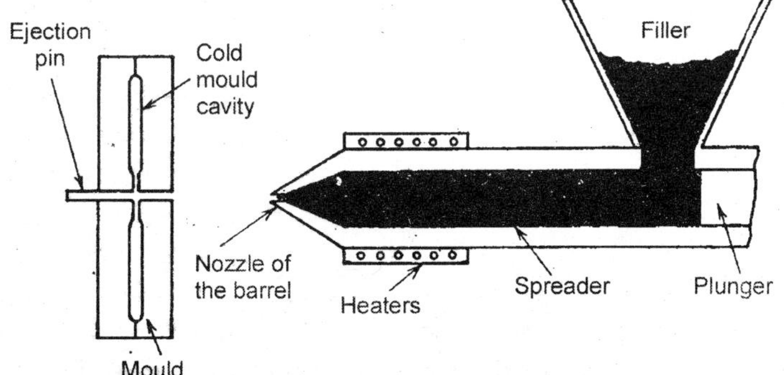

Fig. 11.2. Injection moulding

(*ii*) *Flash compression moulding* is similar to the positive compression moulding, except that excess quantity of raw materials are placed between the plates and the excess quantity squeezing out at the flash line is removed.

(*iii*) *Injection moulding*. The raw material is softened by heating and is injected into cold moulds. This is a much rapid process and is adopted for the manufacture of small articles in the case of thermoplastic materials.

11.2.2 Extrusion

This method is used for plastics to be used for the purpose of making insulated sheets, table tops etc. Thermoplastic tubes and rods are extruded through a die. A screw forces out the material through a nozzle.

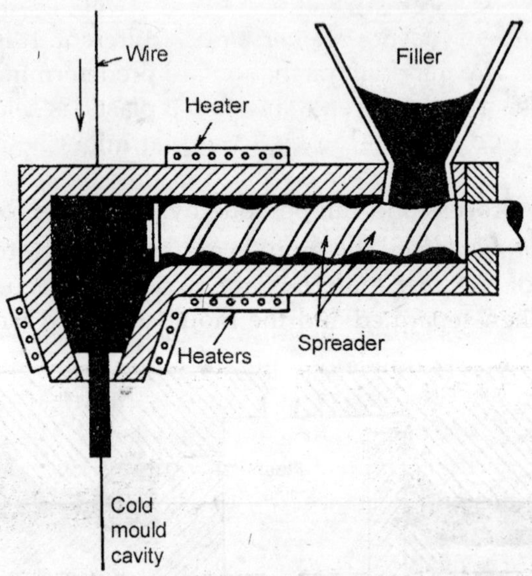

Fig. 11.3. Extrusion moulding.

The wires in cable construction are drawn through a die as shown in Fig. 11.3. As the vinyl resin is extruded, the wire gets covered with the plastic material.

11.2.3 Casting

The casting of plastic materials is similar to that of cast iron casting. The raw materials are heated and cast in moulds. Different colour effects can also be given. Zinc, wooden or plaster of Paris moulds are used. Where they have to withstand high pressures, steel moulds may be used.

11.2.4 Other methods

Others are rolled out to form sheets. Some plastics can be made into tough films while some can be dissolved in solvents or dispersed as emulsions for use in varnishes, paints and adhesives. It is possible to utilise the two stage characteristic of the thermo-setting resins to get laminated sheets. They are also used as bonding agents in the laminated structural timbers. Processes have been evolved for forming the initial products and hardening them in that state to form materials of very low density and high thermal resistance.

APPLICATION IN BUILDINGS

It can thus be seen that plastics are versatile materials which can be adapted to serve a wide range of functions. It is this property that has enabled plastics to start from small beginnings to a major and progressively increasing industry. This does not mean that plastics can be used indiscriminately. The character of the material and the form must be such as to fulfil the function it is intended for and the choice must be made accordingly. Each proposal has to be judged from a consideration of how well it may be expected to satisfy the aesthetic requirements of the architect and his client at a price the latter is prepared to pay.

The properties of different plastics are very much different. It is, therefore, essential that the properties required in the material for the job are predetermined and then the selection of the material is pursued. It will be seen that when a plastic is selected for a given purpose it will rarely be found to possess all the qualities that might be desired and the qualities inadequately represented will have to be balanced against the more important ones present in full measure.

11.3 PLASTIC FITTINGS AND MOULDING

Plastics started as substitutes in the building industry in the war time when there arose a general shortage of traditional building materials. Their use became popular for common electrical and sanitary fittings for buildings ; such as electrical points, switches, holders, lamp shades, insulators, etc., cisterns, and cistern floats, W.C. seats, doors and windows, domestic furniture, etc. Recently, baths and sink units of glass reinforced polyester resin have been used for novelties and elegance.

11.4 PLASTIC FOR FLOORING

Two types of plastic floor finishes have been evolved :

(*i*) Poly-vinyl

(*ii*) Thermoplastic

The poly-vinyl types of floor finishes may be either of poly-vinyl chloride or poly-vinyl acetate.

The first poly-vinyl floor ever laid dates as back as 1933. Later on, the behaviour of material based on poly vinyl chloride (PVC) in service during the war showed them to be resistant to abrasion and unaffected by many chemicals. Polyvinyl chloride resin is a snow-white, brittle resin which softens when heated. Hence, plasticizers are essential before these resins are put to this use. It is this rubber like material that forms the basis of PVC floor finishes. As PVC and its plasticizers are costly a floor finish made wholly of these would be expensive. PVC does not easily stick to other materials, but glues have now been evolved which have given satisfactory results in this respect. The various finishes are classified in three main classes on the basis of their appearance :

1. PVC Sheet and tile
2. Fabric backed PVC sheet and tile
3. Thermoplastic vinyl tiles

11.4.1. PVC Sheet and Tile

PVC and its compounding ingredients are heated and masticated and then formed into a sheet between heated rollers. The sheet is annealed and after cooling, it is rolled up or cut into tiles. Because the basic ingredients are light in colour, bath sheets and tiles can be made in a variety of clean, bright colours and marble effects.

To reduce the cost, only a thin sheet is rolled, the thickness of which is of the order of 1 mm. Such a thin sheet probably has sufficient heat resistance for domestic use but it shows as the irregularities that may be present in the floor on which it is laid. A second method cf reducing the cost is to incorporate fillers, but there is the trouble that PVC cannot carry a large amount of fillers. Moreover, it is not practicable to calender sheets thicker than 3 mm. Another method consists in sticking the thin wearing layers of lightly filled PVC to a backing of lower quality material so that the thickness of the composite product is sufficient to allow laying on normal floors. Commonest of this type are tiles in which the wearing layer of lightly-filled PVC is pressed on to a heavily filled PVC backing.

The process of manufacture of these materials produces a sheet or tile which has got a smooth underside and hence the adhesion problems arise. The adhesives needed to stick them to the sub-floor have been evolved but are at present expensive. A patterned or a roughened backing to assist keying helps to some extent.

The flooring has got very good wearing resistance and resistance to grease and solvents. In badly formulated materials, brittleness may occur after some time.

11.4.2 Fabric backed PVC Sheet and Tile

Two main types are produced ; one is manufactured from a mixture of plasticized PVC, fillers and pigments which is fed while hot on to a hessian backing and passed between heated rollers. The sheets so formed are kept hot for some time to remove strains and are then rolled up. The thickness is generally 2 mm. In the other type PVC plasticisers and pigments are mixed while cold to give a paste which is spread by a roller or knife on to a moving sheet of fabric. The sheet than passes through a hot zone in which the paste is converted to a tough film. Only thin layers can be applied and thick skin can be made by successive treatments. The material is made in sheets upto 3 mm. thick. Hessian or felt is used for backing; the hessian-backed material is intended to be stuck while the felt type is tc be laid loose. These materials are used in the same way as linoleum. The hessian backings allow the use of ordinary linoleum adhesives because the bond is made between the adhesive and the hessian which is well-keyed to the PVC layer.

The materials are suitable for foot traffic but being sensitive to moisture, they should not be used in wet situations.

11.4.3 Thermoplastic Vinyl Tiles

These are manufactured by heating and masticating a mixture of plasticised PVC, asbestos, limestone and pigments. The resulting hot mass is passed through heated rolls from which tiles are cut or punched.

These tiles have very good resistance against oils and greazes and wear. They are suitable for heavy foot traffic and can be used both on wooden and concrete sub-floors. Protection against water for them is also not necessary unless there is a head of water.

All the types of flooring described before are flexible enough to follow the contours of the sub floors which should, therefore, be even, smooth and rigid.

PVC sheets and tiles show a considerable expansion when wetted. If therefore, they are laid on wet concrete, there is a tendency to curl at the wet face and the adhesives in that case are required to restrain this tendency. It is, therefore, essential to provide a damp proof barrier below PVC sheet and tile on concrete laid direct on the ground through which water may rise. Moreover, the sheet should be laid only on dry surface.

Fabric backed types also show an expansion when wetted but there is the added complication that the fabric backing also moves and is liable to decay due to fungi attack. The precautions and restrictions given above apply for these also.

Thermoplastic vinyl tiles are the least sensitive to moisture of all the vinyl materials. Once in position, they are not likely to get dislodged by subsequent wetting of the floor. They cannot, however, be fixed to a wet concrete slab though there is very little effect of moisture on the tiles. They do not serve as a damp proof barrier which may be necessary at very wet sites as water rising to the joints may cause white efflorescence between the tiles.

Before laying, floor should be dried and cleared of all dirt, oil, grease and plaster. The laying technique varies for the different types. For PVC sheets and tiles, the sub floor and the the back of the PVC are both covered with a thin layer of adhesive and when both adhesive films are tacky, the PVC is placed in position and, if necessary, rolled with a light roller to obtain good adhesion. If the tiles are slightly warmed before laying, the improved flexibility makes the operation easier. The adhesive is generally a paste of natural or synthetic rubber with high boiling solvents that dissolve PVC. Hessian backed tiles are laid in the same way as linoleum. Adhesive is spread on the floor and the sheet or tile is pressed on to it after the adhesive has become tacky. The adhesives used are the same as for linoleum and include water paste, spirit gum types, bitumen emulsion and rubber and bitumen solutions. Thermoplastic vinyl tiles are laid in cut-back bitumen adhesives in the same way as thermoplastic tiles described later.

11.4.4 Polyvinyl Acetate Tiles

In the last few years, jointless in-situ floor finishes based on "polyvinyl acetate have been used. These are also known as plastic .emulsion floor finishes.

They are laid by spreading a cold mixture of Poly Vinyl Acetate (PVA) emulsion fil!ers and pigments which, on drying, form a tough layer. The hardening occurs by the removal of the water from the emulsion. For reasonably quick drying and to avoid cracking of the finish, thin layers must be spread and a thick floor finish must be built from several thin layers. The sub floor must be dried and should be primed with PVA emulsion so that its suction is reduced. After the primer has dried (nearly one day), the undercoat, which usually contains 10—20% resin, is applied to a thickness of about 1 mm. This is allowed to dry for one day and then successive 1 mm layers to the number required, each containing more resin, are applied at daily intervals. The top layer should contain nearly 35% resin and the total finish should consist of at least three layers. The temperature should not fall below 10°C and if artificial heating is applied, it should not be used to speed drying too much and should be evenly applied.

The general properties of the finish depend on the fillers used and the proportions of materials in the various layers. But, generally, they are comfortable to walk on, non-slippery, and fairly quiet. Sufficient material is not available to guarantee their wear resistance.

One major disadvantage is that this floor finish also absorbs water, swells and softens in the process but returns to its original state when dried. While wet, it has low wear resistance. This does not mean that it cannot be washed but it only ensures that it should not be used where water saturation may occur. It is possible to improve the water resistance by using a water-proof sealer on the top surface. The value of these methods depends on their remaining in tact under traffic.

11.4.5 Thermoplastic Tiles

Thermoplastic flooring materials are of comparatively recent origin but the expansion of this industry has taken place at a very rapid pace.

These tiles are manufactured by heating and masticating a mixture of thermoplastic-resin binders, fluxes, short fibre asbestos, powdered mineral filler and pigments. The hot mass so produced is pressed through heated rolls from which tiles are cut or punched. The resins used are generally all bye-products of petroleum industry.

The thickness of the tiles is generally 3 mm but 5 mm thickness **is** also sometimes made.

The tiles are brittle at ordinary temperatures and may be broken if they are flexed in the hands. When the tiles are laid on a flat rigid base, they do not flex.

3 mm thick tiles find use in houses but for shops, offices, schools, etc. 5 mm tiles are needed. Ordinary tiles are not resistant to oil or grease but special grades are available for use in kitchens, etc.

11.4.6 Laying of the Floor

The life of the floor depends on the laying. As the tiles themselves cannot support loads, a smooth sub floor is essential. An usual method for providing smooth surface is to finish with a cement sand screed (1 cement : 3 to 4 sand). The thickness of the screed should be approx. 25 mm and after 'striking off' to this thickness, the screed may be allowed to stiffen before finishing smooth with a steel trowel. When the surface is sufficiently hard to walk upon without damage, it should be covered with a layer of wet sand or saw-dust in order to prevent quick drying.

The tiles should not be laid until the base is perfectly dry. It should then be cleaned of all dirt, loose grit, oil grease and plaster, etc. The screed is then covered with an adhesive which is allowed to become tacky before tiles are put in position. If the surface of the base is dusty, a primer is usually supplied and sufficient time is allowed for it to dry. Care must be taken that excessive amount of primers are not used as they are generally slow-drying (to avoid fire risk) and any primer left undried will eventually rise to the top and soften the tiles. The tiles are generally warmed before laying so that they are more flexible and easier to bed to the adhesive. These are then firmly pressed. Edges of tiles where-ever exposed require an edging strip which may be of metal or another plastic.

Although all thermoplastic tiles expand slightly when wetted, it is only in extreme conditions that the tiles may become detached from the base concrete. Such a floor for all practical purposes may be taken as unaffected by water but it does not imply that the floor in itself will form damp-proof barrier and hence whenever there is a water head acting, it would be necessary to lay a damp-proof layer below the tiles.

11.5 PLASTICS FOR WALL PANELLING

The thermoset laminates are available in two categories :

(*i*) The melamine-surfaced phenolic paper laminates which come in sheets ;

(*ii*) The polyester paper laminate which is available in continuous lengths.

The first variety is sold under various patented names viz. Formic, Panelite, Farlite, Richwood, etc. These are tough, have excellent resistance to abraison, heat, mild acids and alkalies, water, oil and solvents. They are made in an unlimited range of coloured patterns, are light in weight and easily worked by carpenter's tools.

Recent development have proved successful in putting the materials at the job with pressure sensitive adhesives. The developments of thicker laminates wherein the material is used to surface a special grade of hardboard has opened up new possibilities for wall applications. Until recently, wall applications have been made by laminating to plywood and using the combination as panelling with wood, plastic or aluminium.

The polyester surfaced paper is lower in cost than the melamine high pressure sheet. Some varieties reach the user with the adhesive already on the back and by the application of a solvent on the adhesive, it can be directly glued on the surface. With others, adhesive is supplied separately for brush application.

11.6 PLASTIC FOR THERMAL INSULATION

Basically, all plastics are poor conductors of heat and in their expanded form they meet the accepted values for thermal insulation. An insulation must also be resistant to the accumulation of water and ice and must not decay or support fungus growth and should be odourless. Foamed polystyrene which is produced as rigid lightweight slab of very low thermal conductivity has negligible water absorption and is unlikely to be attacked by micro-organisms under normal conditions of use. Their properties make them easy to fix with a suitable adhesive and they can also take plaster.

Expanded polystyrene is an excellent roof insulation but the heat distortion temperature is lower than the application temperature of materials used for hot roof applications and the insulation must therefore be protected to prevent thermal distortion. This has been done by applying 6 mm layer of Portland cement mortar or by using one layer of roofing felt. In certain cold storage rooms two layers of polystyrene boards are laid directly on bar joints with Portland cement between the layers and on the top surfaces. This forms a sandwiched construction of the top layers resulting in strength sufficient to meet the building requirements for roof decks. The use of expanded polystyrene as a material in sandwich construction however is still in its infancy.

Foamed isocyanate resin is used as an alternative to foamed polystyrene. This material is particularly useful as it is capable of introducing the foam into existing cavity walls. It is, however, not yet proved that it can be used in this way without increasing the risk of rain penetration through the wall.

Both foamed polystyrene and foamed isocyanate products will burn but because of their very low density their contribution to the fire load will be very small. It is being further investigated whether they could contribute to the spread of fire if used as an internal lining *e.g.* under a roof.

Phenolic resin has been used as a binder and surface coating for glass fibre insulation. The resin makes the material easier to handle and prevents compaction when it is used as a loose fill in cavity walls. No loss in rain penetration is effected.

11.7 PLASTIC VAPOUR AND MOISTURE BARRIERS

Absorptive insulating materials used in the construction of walls and roofs become saturated with moisture because water vapour from the warm interior diffuses through the insulation and condenses inside the cold outer skin of the wall or roof. This reduces thermal resistance and may also promote decay in susceptible materials. Vapour barriers are used to forestall such trouble by preventing the excess of water vapour to the cold surfaces. The impervious types of foamed plastics referred to in the preceding paragraphs under 'thermal insulation' serve the function of both insulation and vapour barrier provided that the joints are effectively sealed.

Polythylene sheet is a tough water-resisting material available in wide rolls and in varying thicknesses. It does not deteriorate with age and is unaffected by extreme temperatures or humidities. As the sheets are available in sufficient widths, there are only a few joints to be sealed and hence this is a potential alternative to bitumen felt or paper as an underlay for roof tiling. It is also used for sheltering materials from rain on building sites, to retain moisture and assist the curing of concrete and as temporary covering for door and window openings. The one drawback of this material is that though it is tough in itself, it tends to tear easily once the edge is broken and therefore large sheets need careful handling in high winds.

Another recently introduced materials, polystyrene film aluminium foil laminate, which has extreme toughness and zero water vapour transmission rate should find application in the building field.

11.8 PLASTIC FOR ROOFING

Corrugated phenolic-resin-bonded paper laminates have been manufactured in dark colours to give opaque roofing materials. These have been in use for a nunber of years and though some change in appearance has taken place yet the appearance is not patchy. The sheets are lighter in weight and have better strength and corrosion resistance than other corrugated roofing material.

For lighting purposes, clear and translucent domes and corrugated sheets in acrylic resins and translucent corrugated sheets in glass-reinforced polyester resins are available. These sheets have profiles matching with other corrugated roofing sheets and thus can be utilised in conjunction with other sheets. The required amount of daylight can be provided by the simple incorporation of an appropriate number of plastic sheets.

The acrylic sheets offer good resistance to weathering. There is, however, the defect that if the sheet has not been properly annealed during manufacture, internal crazing may occur. Glass reinforced polyester resin sheets have been in use for a comparatively lesser time and it is difficult at this stage to predict their behaviour towards weathering and efficiency in retaining their light transmitting properties. They are light in weight and have a high strength in the thickness in which they are supplied.

The fire-hazard depends on the type used and hence it should always be ascertained from the manufacturer whether the materials will be fire-resistant to the required extent or not.

11.9 PLASTIC FOR LOAD BEARING MEMBERS

There has not yet been any reasonable prospect of using plastics as load-bearing members generally in buildings in place of timber, masonry or steel. The softening effects of heat in thermo-plastics and the brittle-fracture of most thermo-setting plastics preclude their use in this way.

Developments have however been made and the glass-fibre-reinforced polyester resins are being used to act as load bearing panels. The method is a sort of a composite construction. Foamed plastics core sandwiched between two sheets of glass-reinforced polyester resin have been tried for single-storey buildings in the United States of America.

If plastics are to be used economically in structural panels in houses, we will probably have to depart from the traditional methods of construction and follow the lead of the recent trends of prefabrication.

Several houses have been built in U.S.A. using 1.25 m × 2.5 m structural wall panels consisting of faces 6 mm thick water proof plywood bonded to a core of expanded polystyrene. The adhesive used is also a plastic ; a type of thermosetting resin.

The expanded plastic core is found to stiffen the structural faces to provide ample and continuous insulation and to act as a vapour seal.

Plastics have been bonded with concrete to give wall panels. The plastic insulation board is placed in the form and then the concrete is poured on the insulation. An intimate bond develops upon setting. A large ware house building has been constructed in U.S.A. which has got precast wall panels with an insulating core of 40 mm thick expanded polystyrene. The thickness of concrete surfacing is 4.5 mm on both faces making a total slab thickness of 130 mm.

As said before, all plastic composite structural panels are also being developed.

All this is still in the initial stages and much more will have to be known about their behaviour under sustained loading over long periods before such products can generally be adopted as practical propositions.

11.10 PLASTIC PIPING

The use of plastic pipe is largely confined to distribution of water, mine drainage, gas distribution and specialised industrial applications. The most widely used plastics for this purpose are acrylo-mitrile copolymer blends, cellulose acetate butyrate, polyethylene, polyvinyl chloride, poly-vinylidene chloride and combinations of glass fibre with thermosetting resins. All these materials have got widely different properties.

11.10.1 Polyethylene Pipes

Polyethylene has been the most popular plastic pipe material in recent years. The reasons for this popularity are its inherent flexibility and a past record of accepted use for wrapping food materials, excellent resistance to breakage from mechanical shock, chemical inertness and light weight. Because polyethylene is flexible, pipe made from it is shipped in coils and because of this property they easily follow the contours of irregular ditches and rough terrain.

The chemical inertness of polyethylene prevents the use of solvent welded fittings. As a consequence, polyethylene pipe sections are usually joined with rigid insert type fittings which are moulded from a rigid plastic such as the copolymer mentioned above or fabricated from a non-corrosive metal. The use of polyethylene pipes has been efficiently recognised for water supplies, over-flow flush and pipe for use in chemical and food industries.

The acrylo-mitrile copolymer blends are rigid plastic materials and pipes made from this are used as standard wall pipes. The pipe is usually threaded, as are the fittings and adapters used in it, but at the same time it can be solvent welded. Both pipe and fittings are produced from the same copolymer blend. It finds a special use in industrial applications to combat corrosive conditions where metal pipe requires frequent replacement. The insertness of the pipe indicates that it is probably entirely satisfactory for sewage disposal and other domestic drain lines, but investigations under actual conditions are yet to be made.

11.10.2 Polyvinyl Chloride Pipes

Polyvinyl chloride pipes and fittings have been used extensively for over 50 years now and are provided for drain lines, and sewage disposal lines. They have usually good ductility, are easily threaded and can be welded using a hot gas stream and PVC rod. Good pipe joints are formed by welding butt-ends and by welding external sleeves or flanges over the end of pipe. One defect however is that these pipes change dimensions with temperature and therefore, provision must be made for the movement of the pipe which may be avoided by burrying or otherwise insulating the pipe. Polyvinylindene chloride is similar to PVC in physical and chemical properties.

In short, behaviour of the plastic pipes is excellent at lower temperatures and specially near about the freezing point of water. This has an advantage over other traditional types of pipes which occasionally burst due to freezing of water. With the temperature rise, the plastic pipes tend to soften Hence, their use is limited to places where the temperature are only moderate. The maximum temperature up to which they can be used is 65°C.

11.11 PLASTIC DUCTS AND CONDUITS

The main difference between a duct and a pipe is that ducts operate at much lower pressures, usually less then 0.15 N/mm^2.

Rectangular and large circular duct work is fabricated from thermoplastic sheet. They are softened by heating and given the shape of the duct. Rectangular sections are usually formed through the use of a bending brake while large circular duct work is rolled on a mandril while warm. The portions of duct may be cemented or thermally welded to form complete sections.

All the thermo-setting plastics may be filled with fibres such as asbestos and cast to form duct work. The product so obtained is slightly heavier than the thermo-plastic ducts. Fibre glass reinforced plastics have also been used in the manufacture of duct systems. It consists of glass cloth or glass mat impregnated with styrene solutions or unsaturated plastics may be used at temperatures upto and above that of boiling water, but they still are very costly.

It may be stated that the plastic duct work has a much lower resistance to liquid and gaseous flow.

No reasonably priced duct work can be recommended for at temperatures nearabout the boiling point of water. It is therefore, essential in designing plastic ducts that they are provided only where the temperatures are expected to be low. It is also important to avoid sudden changes of cross section areas by the use of gradual tapers or adequate fillers. Adequate allowance for expansion must be made and all duct work must be supported properly. This is effected through the use of compressible gaskets or expansion joints. Proper support for duct work is also very important. Horizontal structures should be supported at intervals not greater than 7.5 m. Ring supports should be placed within 150 mm on each side of every flange. Vertical sections should be supported at least every 1.5 m. Plasting duct work should be installed as a free member throughout its entire length.

11.12 SYNTHETIC ADHESIVES

The greatest achievement in plastics has been made in this direction. Water resistant synthetic adhesives have widened the scope for the use of wood in buildings. Synthetic adhesives are used in the production of large laminated members and chip boards.

TYPICAL PLASTICS USED IN BUILDINGS

11.13 THERMOPLASTICS

Type	Uses
Acrylic (polymethylmethacrylate)	Transparent or opal, domed skylights and corrugated roof lights. Has been used to replace window glass where shatter risk is high. Lighting fittings.

Bitumen-based plastics	Mouldings, *e.g.* flushing cisterns.
Coumarone indene resins	These and other resins, used in the manufacture of thermoplastic tiles.
Cellulose acetate	Lighting fittings, door plates, covering for handrails.
Isocyanate (polyurethane)	Foamed in Situ-insulations.
Polythene (low density)	Pipes for cold water services ; film for vapour barriers, weather shields in building operations and for cover for concrete during curing. Mouldings-cistern ball floats.
Polythene (Hight density)	In one type of moulded fitting for joining low-density polythene tubing.
Styrene (polystyrene)	Thin, light wall tiles in bright colours with or without patterns. In slabs of solidified foam as insulating materials. Foamed mouldings for cistern ball floats.
Vinyl (polyvinylchloride (PVC) Polyvinylacetate (PVA)	Wall and floor tiles and coverings, normally with marbled or flecked pattern. Emulsion paints.

11.14 THERMOSETTING

Amino plastics (melamine-(formaldehyde, urea-formaldehyde)	Decorative laminates in light colour for curtain walling units, wall and ceiling linings, table and counter tops. Mouldings in light colours-w.c. seats, etc. Water-resisting adhesives. Wood chip-board manufacture.
Phenolic (phenol-formaldehyde cresol-formaldehyde, resorcinol formaldehyde)	Laminates in dark colours for curtain walling units, 'wall linings. Corrugated roofing sheets. Mouldings in dark colours ; cisterns, w.c. floats, electrical fittings, door, furniture. Water-resisting adhesives. Wood chipboard manufacture.
Polyesters-Usually glass reinforced.	Transluscent, corrugated roof lights corrugated and plain partitions. Baths and sink units. With sand filler, but without glass, as a finish for concrete blocks.
Water-soluble cellulose derivative (sodium carboxy methyl cellulose)	Binders for distempers and crack-filling compositings. Wall-papering adhesives.

QUESTIONS

1. (*a*) Briefly state the two broad categories of plastics.
   (*b*) Explain, with the help of diagrams, the various processes involved in the manufacture of plastics.
2. What is meant by PVC? List the three main classes of floor finished based on their appearance and discuss their merits and demerits.
3. What are the ingredients of thermoplastic vinyl tils and how these are manufactured? Explain the process of laying of the floor.
4. Write short notes on the plastics used for the following purposes:
   (*a*) Wall panelling (*b*) Thermal insulation
   (*c*) Roofing
5. Briefly describe the following:
   (*a*) Plastic vapour and moisture barriers
   (*b*) Plastic piping
   (*c*) Plastic ducts and conduits
6. How plastic can be used for load bearing members?
7. Give the type and uses of typical plastics used in buildings in the form of:
   (*a*) Thermoplastics ; and
   (*b*) Thermosetting
8. Which plastic will you recommend for use
   (*a*) as Window glass (*b*) in Mouldings
   (*c*) for Pipes (*d*) as pertition walls

OBJECTIVE TYPE QUESTIONS

1. Thermoplastics when heated to fabrication temperatures
   (*a*) become hard (*b*) undergoes no chemical change
   (*c*) a compound is formed (*d*) resulting produces are not brittle
2. Plasticisers are added to act as:
   (*a*) softner (*b*) hardner
   (*c*) air-entrainer (*d*) retarder
3. Plastics are used in
   (*a*) Varnishes (*b*) Paints
   (*c*) Adhesives (*d*) All these
4. The thickness of a PVC sheet is limited to
   (*a*) 1 mm (*b*) 2 mm
   (*c*) 3 mm (*d*) 4 mm

5. The thickness of fabric backed PVC sheet is generally
   (*a*) 1 mm (*b*) 2 mm
   (*c*) 3 mm (*d*) 4 mm
6. The plastic used for sky-lights is
   (*a*) Acrylic (*b*) Bitumen-based plastic
   (*c*) Cellulose acetate (*d*) Vinyl
7. Binders for distempers are of type:
   (*a*) Thermoplastic (*b*) Thermosetting
   (*c*) Both (*a*) and (*b*)
8. For emulsion paint, the plastic used is:
   (*a*) Arcyclic (*b*) Polyesters
   (*c*) Vinyl (*d*) Polythene

Answers

1. (*b*)	**2.** (*a*)	**3.** (*d*)	**4.** (*c*)	**5.** (*b*)	**6.** (*a*)
7. (*b*)	**8.** (*c*)				

PAINTS, VARNISHES AND PROTECTION OF METAL

PAINTS AND VARNISHES

Paint is a mixture of solids and liquids and is used to cover surfaces of wood, iron etc. for decorative and preservative purposes. The solids are the pigments or base, and the liquids the medium or vehicle to enable the solid particles to be applied to the surface.

The functions of solids are to provide body or substance, to give colour and obscuring power to the film of paint, and to protect it from too rapid decomposition. The functions of the liquids are to hold the solid particles in suspension and to allow the material to be spread evenly over a surface. They also help to bind the particles together and to give a cohesion and gloss to the painted surface.

Coloured pigments provide the requisite tints to paint and are often known as stainers. They may be natural earth pigments such as-iron oxides, or chemically manufactured colours such as chrome yellow and prussian blue. These pigments are sometimes used in combination with a white pigment called the base, which gives the main bulk and weathering properties.

12.1 INGREDIENTS OF PAINTS

A paint is composed of the following ingredients ;

1. Base
2. Extender or inert filler or adultrant
3. Pigments or stainers
4. Vehicle or medium
5. Thinner
6. Drier

The first three are the solids and the other three are the liquids.

12.1.1 Base

A base is a finely divided solid material, usually having white or light colour. Its functions are :

(*i*) It provides body to the paint.

(*ii*) It determines the character of the paint.

(*iii*) It gives binding power to the paint.

(*iv*) It reflects harmful ultra-violet light.

(*v*) It minimises shrinkage cracks on drying.

(*vi*) It prevents ingress of water to the painted material.

(*vii*) It offers greater resistance to weathening agencies, since it provides a very hard film or coating. Abrasion is minimised.

(*viii*) It imparts strength to the paint.

(*ix*) It prevents the penetration of the paint to the underlying surface.

Commonly used bases are :

(*i*) *White Lead.* It has great covering power and protective qualities and spreads well.

(*ii*) *Red Lead.* It is an excellent preventive of rust, tough and durable and consequently widely used as a priming coat for structural steel.

(*iii*) *Zinc While.* (Zinc oxide and zinc sulphate).

It is a non-poisonous pigment.

(*iv*) *Lithophone.* When zinc sulphide solution is added to barium sulphate solution, a white precipitate is formed, which is extremely white and is used for interior work. It has good covering power.

(*v*) *Titanium White.* It is an oxide of titanium of brilliant whiteness. It is a very fine white pigment with high oil absorption, great elasticity and twice the covering power of white lead.

(*vi*) Blue lead.

(*vii*) Zinc chromate.

(*viii*) Aluminium, bronze and zinc powders.

12.1.2 Extenders

The extenders have little covering or hiding power ; their chief value lies in helping to maintain the other pigments in suspension and to prevent too rapid setting and to improve the brushing quality. The following are the chief extenders used :

(*i*) Magnesium silicate,

(*ii*) Talc,

(*iii*) Diatomite,

(*iv*) Barium sulphate,

(*v*) Calcium carbonate,

(*vi*) Gypsum, and

(*vii*) Ground silica.

12.1.3 Colour Pigments or Stainers

Colour pigments or stainers are very finely divided solid materials, which in addition to serving all the functions of the base, give necessary colour and shade to the paint.

Commonly used coloured pigments are :

(*a*) *Slack Colours :*

(*i*) Lamp black,

(*ii*) Carbon black,

(*iii*) Bone black, and

(*iv*) Graphite.

(*b*) *Earth Colours :*

(*i*) Iron oxide,

(*ii*) Imber, and

(*iii*) Sienna,

(*c*) *Chemicals :*

(*i*) Chrome green,

(*ii*) Hydrated chromium oxide, and

(*iii*) Prussian blue.

(*d*) *Lakes :*

The pigments known as lakes are formed by precipitating coal-tar dyes on a mineral base such as barium sulphate or alumina, thereby colouring it.

12.1.4 Vehicle

The functions of the vehicle are :

1. It holds the ingredients of the paint in liquid suspension.
2. It enables the ingredients of the paint to be spread evenly over the underlying surface to be painted.
3. It gives gloss to the painted surface.
4. Vehicle, such as linseed oil contains acids which react with oxygen and hardens by forming a thin, tough, adhesive, durable and water proofing film.
5. It acts as a binder of base and pigments.

The vehicle used for oil paints is generally linseed oil in raw, boiled or double boiled form. In recent years tung oil, poppy oil and dehydrated castor oil are also being used.

12.1.5 Thinners

Thinners are used to bring the paint to the required consistency. Various types of thinners used are as below :

(*a*) *For Oil Paints :*

(*i*) Turpentine,

(*ii*) White spirit, and

(*iii*) Naphtha.

(*b*) *For Spirit Lacquers :*

Alcohol.

(*c*) *For Cellulose Paint :*

Ethyl amyl acetate.

(*d*) *For Distempers :*

Water.

12.1.6 Driers

These are substances which are incorporated in small amounts in paints or varnishes based on drying oils, bring about an appreciable reduction of their drying time. The function of the driers is mainly catalytic. It should be limited upto 10 percent. They are of two types :

(*i*) *Oil Driers.* They are used in powdered form, *e.g.* Litharge, magnesium dioxide and magnesium borate, and

(*ii*) *Liquid Driers.* They are usually cobalt or vanadium compounds.

12.2 REQUISITES OF A GOOD PAINT

1. It should have a high covering or obscuring capacity. Covering capacity is the area covered by one litre of a paint, when applied on the surface to be painted. It is measured in square metre per litre.

2. It should produce gloss and washable films.

3. It should be fluid enough to be spread evenly in a thin film.

4. It should dry quickly. This is contradictory to requirement at serial number (3). Hence a balance should be struck between its spreading quality and drying characteristic.

5. It should leave no brush marks on drying.

6. It should not show any cracks or shrinkage on drying and also later on.

7. It should provide tough and durable film ; offering adequate resistance to abrasive and weathering agencies.

8. If pigments are added, there should be no fading of colour. In other words, the tinting should be permanent.

9. It should have lasting attractive appearance.

10. It should not be injurious to underlying material or the painter.

12.3 TYPES OF PAINT

12.3.1 Oil Paint

Oil paint consists of a pigment, a vehicle, a thinner and a drier. The pigment divides the bulk of paint and may consist of white or coloured pigments. Sometimes extenders are also added.

12.3.2 Varnish

Varnish is a solution of resin in a drying oil or in a volatile solvent such as alcohol or turpentine. It contains no pigment and hardens into a smooth, hard and glossy coat by oxidation of the oil or evaporation of alcohol.

Both natural resins such as copal, dammer, shellac, rosin etc. and synthetic resins like phenol-aldehydes and urea-formaldehyde are used in varnishes.

Varnish is of two types :

(*a*) *Oil varnish.* It consists of hard resin ; a drying oil, drier, and a thinner.

(*b*) *Spirit varnish.* It consists of shellac dissolved in spirit. A very small amount of plasticizers such as castor oil is sometimes added.

Varnish has the property of bringing out the beauty of the underlying surface. A further use is as a binding agent in certain types of paints.

12.3.3 Enamel

When a pigment is ground in varnish or in a viscous boiled oil instead of linseed oil ; the resultant product has a harder and glossier film and is known as Enamel.

12.3.4 Lacquer

Lacquers have widely increased in use because of their durability, hardness, lustre, smoothness and quick drying properties. They consist of a composition of nitro-cellulose, resin, solvents and softeners. Toughness and resistance to abrasion is produced by nitrocellulose ; adherence, hardness, lustre by resins and elasticity and plasticity by softeners. The solvents are acetates and alcohol and softener is castor oil. They dry with a very smooth surface of high tensile strength which does not scratch and is easily cleaned.

12.3.5 Stains

Stains are used to change or modify the colour of wood and to bring out its grains and texture. The main varieties are :

(*a*) *Oil stains.* The vehicle is oil to which turpentine or benzene may be added. Either pigments or aniline dye may be used to colour the stain.

(*b*) *Water stains.* These stains consist of solution of aniline dyes in water. They dry quickly, but tend to raise the grain of the wood.

(*c*) *Spirit stains.* They consist of dye stuffs dissolved in acetone or methylated spirit. They are clean, dry quickly and penetrate deeply.

(*d*) *Chemical stains.* These contain no colouring matter, but colour change occurs by action of chemicals *e.g.* iron salts, potassium and sodium trichromates etc.

12.3.6 Distempers

Distempers are cheap paints. They are durable and can be applied easily on plastered and white washed surfaces. They are affected by weathering agencies and, therefore, are not recommended for use on the exposed surfaces. They are suitable for interior walls.

Distempers are of two kinds :

(*a*) *Glue Band Distemper.* The medium used is size, a form of powdered glue dissolved in water. Whiting is the pigment and water is the thinner. Glue bound distemper will not stand hard wear and will not last in damp conditions.

(*b*) *Oil Band Distemper.* This distemper has for its medium a solution known as "petrifying liquid". An ingredient in it is a drying oil which has been treated to emulsify other binders used, glue and casein.

The pigments used in distempers according to the quality desired include lithopllone and barytes, coloured with staining colours which are affected by alkalies. Water is used as a thinner.

12.3.7 Glazes

Transparent liquids called glazes are sometimes used over cold water paints to produce a blending of colour and a soft sheet.

12.3.8 Casein Paint

It consists of a mixture of finely ground casein and slaked lime with water.

12.3.9 Stucco, Brick and Masonry Paints

These surfaces contain alkali. Washing of the surface of fresh masonry with zinc sulphate solution is helpful in reducing alkalinity. Recently chlorinated rubber or polyvinyl acetate emulsion paints have become popular because of excellent alkali resistance of the vehicle.

12.3.10 Paints based on Synthetic Resins

The best known and the most used types of resins for paints are the alkyd type. Outstanding properties of paints based on alkyd resins are excellent exterior durability, fast drying, retention of gloss and colour, and superior adhesion.

12.3.11 Cement Paints

Chief constituent of cement paints is Portland cement preferably white cement (65 to 75%). Other constituents are :

Hydrated lime	10 to 20%
Hygroscopic salt;	
Calcium or sodium chloride	4 to 5%
Water repellent :	
Calcium or aluminium stearate	1%
White pigment :	
Titanium oxide or zinc sulphide	2 to 5%
Colouring pigments	upto 8%

Cement imparts strength, hardness, density and durability to the paint. Hydrated lime renders whiteness, opacity, and workability. Hygroscopic salt helps curing by retaining moisture. White pigment further increases whiteness and opacity. Colouring pigments give the mix the desired shade of colour. An inert filler (sand) is often added to make the paint more attractive and ensure better adhesion of successive coats. Quantity of filler (when used) is kept below 5% of the total.

In making the paint, cement, lime and colouring pigments are first intimately mixed, preferably in a ball mill. The other ingredients are then incorporated by thorough stirring in a suitable container. Batches of 50 kg or more can be conveniently mixed in a concrete mixer. Dry mix is prepared only a short time before gauging with water. If the mix is to be stored, it is to be kept in airtight container. The hygroscopic salt and water repellents are preferably mixed with gauging water. Only a part of estimated quantity of water is first added to any mix to make a stiff paste, which is then thinned to required consistency by adding more water and stirring up.

Applied to plastered brickwork masonry and concrete, a cement paint is known to last for 5 to 6 years without needing any renewal, even in exposed situations. It imparts beauty and elegance to a structure protecting the plaster below from damage by rain water. It is now used on an ever increasing scale, as a lasting finish for walls and ceilings of houses, offices, public buildings, cinema halls, hotels, factories, workshops, warehouses etc.

12.3.12 Tar and Bitumen Paints

Where appearance is not a governing consideration, tar and bitumen paints provide a useful alternative to other types of paints. In their simplest form, tar and bitumen paints consist of straight solutions of these products in solvent naphtha or white spirit. Protective properties of paints of this type are inferior to those of most of oil paints, both because they have no inhibitive properties and because, coat for coat; they produce an appreciably thinner film. Tar and bitumen paints. are therefore, most effective, when used for finishing rather than for priming coats, particularly when they are applied over red lead in linseed oil priming paint. Tar and bitumen paints can be materially improved by the incorporation of drying oils and the addition of pigments ; aluminium powder is a particularly useful pigment in this connection. Coatings of hot tar are also sometimes applied to protect structural steel work etc. This is prepared by adding tallow and slaked lime in the proportion of 0.1 kg of each per litre of tar. The mixture is heated to 200°C.

12.4 MATERIALS FOR STOPPING AND FILLING

(*a*) *For stopping.* Putty made from whiting, linseed oil and red lead.

(*b*) *For filling.* Putty made from whiting and white lead.

12.5 COVERING CAPACITY OF PAINTS

Covering capacity of a paint varies according to composition of a paint, surface to be covered, and the human element in application but the following may be taken as a general guide :

	Covering power
1 litre lead priming paint	
(a) on wood	9—11 sq m
(b) on metal	9—13 sq m
1 litre flat under coating	10—12 sq m
1 litre gloss paint	9—13 sq m
1 litre enamel	9—13 sq m
1 litre varnish (first coat)	11—13 sq m
1 litre varnish (second coat)	13—18 sq m

Water paints and oil bound distemper cover approximately 6—8 sq m per kg and usually one litre of petrifying liquid or water (as the case may be) to 4.5 kg of body colour.

12.6 PROCESS OF PAINTING

The process of painting involves the application of a priming coat followed by an undercoat and one or two finishing coats. All these coats are important for giving a desired protection.

12.6.1 Prime Coat

The prime coat has a double purpose, *i.e.*, it forms a layer which on one hand provides the adhesion of the paint to the given surface and on the other hand contains some reactive pigments which serve to protect the surface from the corrosive effect of weather. The selection of any primer for a given purpose depends upon its characteristics and the surface over which it has to be applied. For an absorbant surface, a primer should be of such a quality that a part of it should remain on the surface after absorption. This portion provides a key for adhesion. The proportions of oily and resinous materials are adjusted to suit the suction. Wood priming requires special consideration as it contains trapped moisture. In the case of steel, besides corrosion, another important factor which is to be considered is the expansion and contraction due to changes in the atmosphere and thus it has to stand this variation without any bad effect. For plastered surfaces, the presence of moisture must be taken into account while selecting a primer.

12.6.2 Under Coats

The main function of the undercoat relates to the filling properties, opacity, colour and support for the finishing coat. These are highly pigmented and thereby cover most of the irregularities. Heavy undercoats are generally avoided and hence suitable thinning agents must be mixed with a paint so that the painted surface is quite thin and there are not any excessive marks left after the application. In the case of old work, care has to be taken that there is no loss of adhesion. A good undercoat should be such that after its drying, no clogging marks should be produced when it is rubbed with a gloss paper. As this rubbing is essential before the application of any finishing coat, the tearing away of the film will result in a damage which is difficult to rectify.

12.6.3 Finishing Coat

The choice of the finishing materials depends upon the position where this is to be applied, *i.e.* whether exterior or interior surface ; the type of finish required, *i.e.*, glossy or matty ; atmospheric conditions, etc. For cold places, the paint should not become too brittle owing to the climatic conditions. In coastal areas or extremely wet places, growth of fungi presents problems in addition to corrosion.

The application of the paint on various surfaces is briefly described below :

12.7 PAINTING WOODWORK

Selection of the right type of paint and the correct method of applying it, keeping in view the surface to which the paint is to be applied, are important for good painting. The characteristics of wood which have to be considered in this connection are its cellulose or porous nature, its liability to contain oil substances, its liability to contain water soluble substances, its moisture content and its capacity of absorbing water thereby causing swelling or shrinkage. The internal structures of hard woods differ markedly from those of soft woods. In the case of hard woods, there are smaller cellular units but they contain pores which are wider and deeper than soft woods and thereby present a difficulty in painting. Ordinary paints applied by brush or spray will neither fill nor cover these pores. The pores in the soft woods are smaller and, in general, may be filled by painting brushes. But in the case of soft woods, the presence of cellular structures leads to the absorption of the paint and hence a proper proportion of oily contents and the pigments has to be maintained.

Hard woods do not contain resin and in general are free from objectionable oils. Teak and some other kinds of wood contain oils which tend to interfere with the drying, hardening and proper adhesion of a paint. They can, however, be painted successfully by the use of special primer. In soft woods, sometimes there are excessive amounts of resins present, particularly near knots, which lead to unsatisfactory painting.

With well seasoned timber, the effect of water soluble substances is relatively unimportant but some priming paints may be of slow drying type and may not harden on hard woods rich in tannins such as wall nuts, etc. In other circumstances, similar troubles arise with soft woods. Water soluble substances may cause discoloration of paints.

The timber which is to be primed should contain about 10 to 20 percent of moisture. The moisture content of dry seasoned timber may rise considerably higher merely by absorbing water vapours and may reach 25 to 30 percent in places where condensation occurs. The moisture in this case is not held in the wood cell but within the fibrous cell walls which consequently are in a swelling state.

Wood in contact with water would absorb more than 30 percent of moisture easily till the cavities are filled. The adverse effect of moisture present within the walls is due to shrinkage as the moisture is absorbed or dries out. The paints do not prevent entirely such drying or absorption. The changes in wood fibres will, therefore, weaken the adhesion of the paint film and thereby may peel off the wood surface.

Woodwork should generally be given four coats of paints, *i.e.*, one coat of primer, two undercoats and one finishing coat. For most soft woods of good varieties, white lead in linseed oil or similar material containing 10 percent of red lead may be used for priming coat. Higher percentage of red lead would produce a harder film which may need extra rubbing to ensure adequate adhesion of the next coat. This is more true if the prime coat has been applied a month or so before the application of the next coat. For lower grades of soft wood a highly water resistant varnish is sometimes used. For general interior woodwork, linseed oil primers are sometimes preferred to lead primers on account of their non-poisonous nature. They are also quite satisfactory. They should be made from litho-phone or titanium white with or without a proportion of pigment mixed with linseed oil. Quick drying primers should not be used on soft wood joinery.

Undercoats for use on wood should contain a more penetrative medium than the undercoats used on metals although the finishing coats may be the same in both the cases. This penetrative medium seeps into the primer coat and helps to restore any undue loss of oil brought about by absorption in wood. Good adhesion between the wood and the primer and between the primer and the undercoat is maintained, thereby preventing or delaying cracking, flaking and peeling. Proprietary undercoats usually contain a proper mixture of pigment, oil, varnish, thinner, etc. which enables it to be satisfactorily used for both coats. Undercoats of lead if properly constituted with respect to the particular finishing coat are quite satisfactory for external use.

The proprietary paint used as finishing coat on woodwork is described as hard gloss paints, synthetic gloss paints and enamel paints. These paints are intended to give a hard film having a smooth and clear surface. Finishing paints used for an exterior work usually differ from those on the interior woodwork.

12.8 PAINTING PROCESS ON NEW WOODWORK

If possible, the moisture content in timber should not be more than 15 percent at the time of painting. Before applying the priming coat, the new woodwork should be made ready for painting by removing or sealing off certain defects which are sources of trouble. The timber should be dried a little to reduce the moisture content. Large knots should be removed by cutting or otherwise replaced by sound wood. Projections should be smoothened with gloss paper. Knots should be given one or two coats of shellac knotting or aluminium paint. It should be ensured that the primer should work well into the nail, holes, etc. The primer should be applied firmly into the wood and sufficient time allowed for it to penetrate. Some of the primers may be good in this respect but should ensure good adhesion with wood.

Subsequent painting is done at site after erection. If the primer has been damaged, it should be touched. When the primer is dry and before applying the undercoat, nails, holes, etc. should be filled in by a stiff putty. This may consist of white lead and genuine linseed oil or glazing putty with or without some red lead. When it gets dry, it should be rubbed with a water-proof abrasive paper. Sometimes fillers are used to fill in the pores or to get smooth finished surface. They usually consist of a mixture of finely powdered slate and size. When dry, the filler is rubbed smooth with water-proof abrasive paper. Oil bound water paint is also sometimes used as filler ; but this is not so suitable as the other type mentioned earlier.

The subsequent coats should be applied carefully. Condition of the primer has to be observed before applying the first undercoat. The primer should be clean and not too hard as otherwise the undercoat well not stick will. Subsequent coats should be applied after about 24 hours' interval between each coat.

Repainting. The old paint should be completely removed if it is soft, or has cracked, blistered, peeled and shown loss of adhesion. Similarly it should be removed when it is highly discoloured or damaged by fungal growth. Small, loose patches of paint may be removed by a scraper. Interior paints should be rubbed with wet abrasive paper so as to provide a key and slightly rough surface to which the subsequent paint will adhere well. Exposed doors and windows should be touched with a primer and an undercoat as usual. Whenever it is desirable to remove old paint completely, blow lamp or other heating devices may be used. Alkaline removers should not be used on wood to be repainted. Whenever the use of a blow lamp is liable to cause damage to the paint, a paint remover of organic solvent type may be used. When the removal of the paints is complete, it should then be rubbed with wet abrasive paper and subsequent painting should be done in a manner similar to the one adopted for new woodwork.

Wood treated with tar oil as a preservative should be painted with dark coloured paints, or otherwise with light coloured paints if one or two sealing coats of good aluminium paints are given. A suitable type of paint should be used if the wood has been treated with other types of preservatives.

12.9 PAINTING OF IRON AND STEEL

After the steel has been rolled, it has an oily scale on the surface. It is sometimes considered that if a continuous layer of oil scale could be maintained, the surface of steel would get an effective protective layer. But this is not possible as once the break occurs, progressive loosening takes place and uncovers the steel. Similarly the oils and greases which may be present on the surface will not permit the paints to stick.

Mill scale can be removed with the aid of many methods. If the steel is stretched, the scale will flake away. This method is not easy to apply. Sand blasting may also be effective. Immersion in hot solutions of acids will remove the scale. Films of metallic phosphates can be created which are more resistant to rusting and form a good surface for painting. Thorough drying is essential beibre painting. Scrapers and grinding machines are used in bigger factories. Oxy-acetylene flame may be used along with the mechanical methods, but this method is not suitable for sections lesser than 6 mm in thickness.

Primers for painting steel should always contain pigments which resist the corrosion. Red lead and chromates of lead and zinc are commonly used. Red lead in linseed oil is the oldest primer still abundantly used, but this is unsuitable for dipping and spraying. Chromate based primers are more suitable for spraying. For dipping primers, red oxide primers containing a proportion of zinc chromate and metallic lead primer are useful.

Red oxide of iron and micaceous iron ore paints are cheap and satisfying for use as undercoats. Graphite paints and aluminium paints have a good resistance to moisture. Finishing coats of bitumen based aluminium paints can be applied advantageously. Paints based on silicons or silicon-alkyds are fairly suitable for resisting higher temperatures.

12.10 PAINTING PLASTERS, BRICKWORK, FLOORS AND OTHER CONCRETE SURFACES, ETC.

These surfaces present difficulties on account of the dampness or the presence of soluble salts. Usually efflorescence occurs on the surface of these materials which decrease the adhesion of the paint considerably and may lead to discoloration. When oil paints are to be used on alkaline surfaces, an alkali resisting primer must be applied.

For floor surfaces, enamel coatings are used. These surfaces have to resist alkali action, moisture and abrasion. The floor enamel must be tough to stand the impact of the foot traffic. Floor enamels should have gloss. They should be of quick drying type so as to use the floor in the least time. Oil modified phenols are occasionally used as floor finishes.

Areas subjected to the action of water vapours like bathrooms, kitchens etc. present special problems. Oil bound water paints do not have a longer life. They have a certain amount of porosity. While absorbing moisture they swell and contract when the moisture evaporates, this leading to disintegration. High-gloss paints are more resistant to moisture since they have an impermeable film, the effects of condensation are not much perceptible. Emulsion paints and synthetic rubber paints withstand washing and scrubbing and give a tougher film which makes them more suitable for such areas.

Cement paints is more suitable for application on concrete surfaces. The suction at the surface should be reduced to such an extent that the moisture is not removed quickly from the cement paint. The surface may slightly be wetted for this reason, but it should not present a damp surface. The paint available in a powder form is diluted with water to a creamy consistency. The paint is used within 2 to 3 hours after mixing. It is usually applied with coarse distemper brushes. Two coats with an interval of 24 hours are given. A fine spray of water is applied to the painted surface at intervals to cure the surface.

Silicate paints having a binding medium of sodium or potassium. Silicates can also be successfully used. They are sulphate resistant but may tend to promote efflorescence.

METAL CORROSION AND ITS PROTECTION

12.11 METAL CORROSION

Iron is very much affected by atmospheric agencies, especially when it comes in contact with water ; the slow but continuous oxidation of steel results in its corrosion or *rusting*, Non-Ferrous metals are also affected by atmospheric action, but to much less degree.

Iron, in the presence of water, combines with oxygen and carbon dioxide of the atmosphere to form soluble ferrous bicarbonate

$$Fe + 2CO_2 + O + H_2O = Fe(HCO_3)_2$$

Further oxidation converts bicarbonate to carbonate and finely to hydrated oxide, with the evolution of carbon dioxide

$$2Fe + 2(HCO_3)_2 + O = 2Fe(OH)CO_3 + 2CO_2 \uparrow + H_2O$$
$$Fe(OH)CO_3 + H_2O = Fe(OH)_3 + CO_2 \uparrow$$

12.12 CORROSION PROTECTION METHODS

Some of the methods adopted for protecting corrosion are given below :

(i) Painting

This is the most common method for protecting steel and other metals, but great care is needed in the selection of type of paint. Red lead and aluminium paints are mostly used for iron and steel works. The life of a paint is limited depending on the impurity in atmosphere. Where the action of salt and air is severe, the paint hardly lasts a year or two, while at places away from industrial town and the sea, repainting may not be required for even 10 to 20 years or more.

(ii) Galvanizing or Zinc Plating

Galvanizing is done by covering the steel or article of iron with a thin film of zinc. The latter is not affected by exposure to moisture. Air-zinc plating is carried out by immersing the article in a molten zinc bath. However, this is a crude method and does not produce a smooth surface. A better method is dry galvanizing or sherardizing. The pieces of iron, steel or copper is buried in zinc dust and is heated for several hours to a temperature of 300°C to 400°C. The surface of the buried metal is converted into a zinc alloy which has a very heavy percentage of zinc on the outer surface. The addition of zinc to metal is not in the form of a coat of paint, but is incorporated in the metal as an alloy and cannot be removed unless the top surface is filed off.

(iii) Electroplating :

On passing an electric current through a solution between the electrodes, the ions deposit on the negative electrode. The positive electrode consists of a piece of metal which is to be plated on article.

Nickel plating, chromium plating, and zinc plating are the common methods of electroplating.

(iv) Tin Plating :

Sheet of black iron or other base metals are plated with tin. The sheets are dipped in a 4% to 8% solution of sulphuric acid to remove rust, then washed and dipped in a flux of zinc chloride or NH_4C1. They are passed through a bath of molten tin and passed through rollers and a layer of palm oil.

The rollers spread out and squeeze off any excess of molten tin ; the oil prevents oxidation. The process is repeated, if needed.

(v) Terne Plating

It is similar to tin plating with the only difference that lead-tin alloy instead of pure tin is plated. Lead, if used alone, does not possess the required affinity with iron, while lead with tin makes a good soldering material. Roofing ventilators, non-food containers, metal furniture, etc. are terne plated.

(vi) Nickel Plating

Brass and copper articles are often nickel plated by immersing the articles in a solution of nickel cyanide and sulphuric acid which form a solution of nickel sulphate. Before nickel plating, the article, if not made of copper or brass, is first copper plated.

(vii) Chromium Plating

After carrying out nickel plating, the articles are usually immersed in a bath which gives a chromium plated surface which has a high finish.

(viii) Calorizing

It is the process of sherardizing except that a coating of aluminium is given instead of zinc on steel surfaces.

(ix) Spray Painting

It is a superior method to brush painting and is used especially for motor vehicles etc. The paint is sprayed on to the surface through a suitable nozzle under a compressed air pressure of about 1.5 kg/cm^2. Amount of paint required by spray method is $\frac{2}{3}$ of the quantity used by brush method. A gun for spray painting is shown in Fig. 12.1

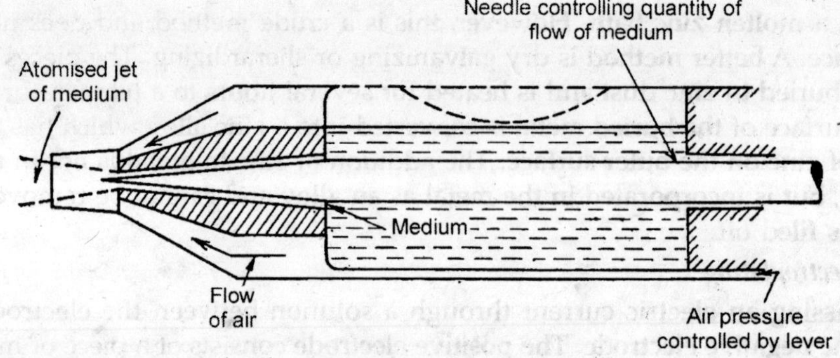

Fig. 12.1. Paint spraying gun.

QUESTIONS

1. List the ingredients of a paint. What functions are performed by the base?
2. What are the requisites of a good paint? State the functions performed by
   (*a*) Extenders
   (*b*) Vehicle
   (*c*) Thinners and driers.
3. List the commonly used
   (*a*) Extenders
   (*b*) Pigments
   (*c*) Thinners
   (*d*) Driers.
4. Name the types of paints. Give details of two of these.
5. Write short notes on :
   (*a*) Varnishes
   (*b*) Stains
   (*c*) Distempers

6. Give the constituents of cement paint alongwith their percentages. When and where are tar and bitumen paints used?
7. Explain the process of painting with special reference to prime coat, undercoats and finishing coat.
8. What considerations will influence the painting of following types of woods :
   (*a*) Soft wood
   (*b*) Hard wood
   (*c*) Wood in contact with water.
9. Explain painting process on a new wood-work.
10. Briefly give details of :
   (*a*) Painting iron and steel
   (*b*) Painting plasters and concrete.
11. Explain the process of iron corrosion. What methods are adopted to prevent corrosion of metals?

OBJECTIVE TYPE QUESTIONS

1. The solid ingredient in a paint is :
   (*a*) base (*b*) vehicle
   (*c*) drier (*d*) thinner
2. The liquid ingredient in a paint is :
   (*a*) pigment (*b*) extender
   (*c*) vehicle (*d*) base
3. Gypsum is used in a paint as :
   (*a*) medium (*b*) extender
   (*c*) base (*d*) drier
4. Stainer in a paint also serves the purpose of
   (*a*) extender (*b*) drier
   (*c*) thinner (*d*) base
5. The vehicle used in oil paints is
   (*a*) turpentine oil (*b*) kerosene oil
   (*c*) linseed oil (*d*) alcohol
6. The liquid medium used in varnish is
   (*a*) spirit (*b*) kerosene oil
   (*c*) water (*d*) linseed oil
7. The medium used in enamel is :
   (*a*) linseed oil (*b*) varnish
   (*c*) turpentine oil (*d*) castor oil

8. The solvent in lacquer is :
   (*a*) alcohol (*b*) castor oil
   (*c*) linseed oil (*d*) kerosene oil
9. The cheepest paint is :
   (*a*) oil paint (*b*) varnish
   (*c*) enamel (*d*) distemper
10. Distemper are suitable for
   (*a*) concrete surfaces (*b*) facework of brick
   (*c*) exposed plaster surfaces (*d*) interior plastered surface
11. 'Petrifying liquid' is a medium for :
   (*a*) lacquers (*b*) glue bound distempers
   (*c*) oil bound distempers (*d*) varnishes
12. The amount of hydrated lime in cement paint varies from
   (*a*) 0 to 10% (*b*) 10 to 20%
   (*c*) 20 to 30% (*d*) 30 to 40%
13. The chief extender used in a paint is
   (*a*) magnesium silicate (*b*) talc
   (*c*) barium sulphate (*d*) all of these
14. Which one of the following is not an oil drier
   (*a*) litharge (*b*) cobalt
   (*c*) magnesium dioxide (*d*) magnesium borate
15. The maximum amount of drier is limited upto :
   (*a*) 10% (*b*) 15%
   (*c*) 20% (*d*) 25%
16. The thinner used for distemper is :
   (*a*) Turpentine (*b*) White spirit
   (*c*) Water (*d*) Naphtha
17. ______ contains no pigment
   (*a*) oil paint (*b*) varnish
   (*c*) enamel (*d*) distemper
18. A cement paint applied on plaster or concrete has normally a life of ______ years
   (*a*) 2 to 3 (*b*) 3 to 4
   (*c*) 4 to 5 (*d*) 5 to 6
19. The coverage of water paint and oil bound distemper is approximately ______ sq. m/kg
   (*a*) 6 to 8 (*b*) 8 to 10
   (*c*) 10 to 12 (*d*) 12 to 14
20. For tar and bitumen paints, the solvent is :
   (*a*) turpentine (*b*) petrifying liquid
   (*c*) naphtha (*d*) none of these

21. The timber to be primed should contain about ______ percent of moisture
(*a*) 0 to 10 (*b*) 10 to 20
(*c*) 20 to 25 (*d*) 25 to 30

22. Wood in contact with water absorbs more than ______ percent of moisture
(*a*) 20 (*b*) 25
(*c*) 30 (*d*) 35

23. Finishing coat on woodwork is provided by
(*a*) hard gloss paint (*b*) synthetic gloss paint
(*c*) enamel (*d*) all of these

24. At the time of painting, the moisture content should not be more than :
(*a*) 15% (*b*) 20%
(*c*) 25% (*d*) 30%

25. Most common pigment used in primer to resist corrosion on steel is :
(*a*) Graphite (*b*) Red lead
(*c*) White lead (*d*) Sienna

26. The pigment which imparts earth colour is :
(*a*) Graphite (*b*) Chromium oxide
(*c*) iron oxide (*d*) lamp black

27. For floor finishes, the paint used is :
(*a*) oil paint (*b*) lacquer
(*c*) glaze (*d*) enamel

28. The paint most suitable for bath rooms is :
(*a*) emulsion (*b*) stucco
(*c*) cement (*d*) bitumen

29. Stains are used to
(*a*) change the colour of wood (*b*) bring out grains
(*c*) to improve texture (*d*) all of these

30. The paint which gives a very smooth surface of high tensile strength is :
(*a*) Enamel (*b*) Lacquer
(*c*) Cement paint (*d*) Varnish

Answers

1. (*a*)	2. (*c*)	3. (*b*)	4. (*d*)	5. (*c*)	6. (*a*)
7. (*b*)	8. (*a*)	9. (*d*)	10. (*d*)	11. (*c*)	12. (*b*)
13. (*d*)	14. (*b*)	15. (*a*)	16. (*c*)	17. (*b*)	18. (*d*)
19. (*a*)	20. (*c*)	21. (*b*)	22. (*c*)	23. (*d*)	24. (*a*)
25. (*b*)	26. (*c*)	27. (*d*)	28. (*a*)	29. (*d*)	30. (*b*)

GLASS

13.1 COMPOSITION OF GLASS

Glass is manufactured from silica (sand), chalk and potash or soda. In order to impart hardness, brilliance and colour ; other ingredients like borax, iron oxide, lead oxide etc. are sometimes added. Glass is produced by fusing all the ingredients in a kiln at a temperature of over 1000°C.

13.2 FUNCTIONS OF VARIOUS CONSTITUENTS

Silica, forms the main constituent of all kinds of glass. This has a very high melting point. Therefore, in order to lower its fusing point, alkaline material, *e.g.* sodium or potassium carbonate are added. These also help to make it viscous and workable.

Soda, causes the glass to fuse. If present in excess quantity, it is, therefore, harmful.

Potash, has opposite effect of soda. This makes glass infusible and imparts fire resisting property.

Lime, imparts toughness in glass.

Magnesia, gives whiteness, thereby reducing colour.

Oxides of metals, give colour to glass and, therefore, their presence in glass is not desirable.

Cullet, is broken glass pieces of the same quality as the one being manufactured. This is added to the raw materials to reduce cost.

13.3 MANUFACTURE OF GLASS

All the ingredients are thoroughly ground, sieved separately and mixed in required quantities. They are put in the furnace and fused at the proper temperature. Pot furnace is generally employed for small quantities whereas tank furnace of the reverberatory type is used for large quantities. The charge is fed at one end of the furnace and glass, in molten state, tapped at the other. Tank furnace is suitable for large scale manufacture of ordinary glass such as soda lime bottle glass, etc.

The molten glass can be given any desired shape.

13.4 TYPES OF GLASSES

13.4.1 Flat Transparent Sheet Glass (I.S. 2835 : 1987)

This is the most commonly type of glass used for engineering purposes. The sheet glass is flat, transparent and clear when judged by unaided eye but may possess a slight tint, when viewed edge wise. It should be free from any cracks. The range of nominal thicknesses vary from 1 mm to 32 mm (1, 1.5, 2, 3, 3.5, 4, 5, 5.5, 6.3, 8, 10, 12, 15, 19, 25, 32).

Various classes of flat transparent sheet glasses and their uses are :

(i) C Quality or Green house Quality (*GQ*) : There are used for green house glazing, production of frosted glass strips for flooring, etc.

(ii) B Quality or Ordinary Quality (*OQ*) : These are intended for use in glazing and framing purposes.

(iii) A Quality of Selected Quality (*SQ*) : Used for selected glazing, manufacture of mirrors, safety glass, etc.

(iv) AA Quality or Special Selected Quality (*SSQ*) : Intended for use where superior quality of safety glass, high quality mirrors, photographic plants, projection slides, etc.

Sheet glass should not have defects such as gaseous inclusion ; opaque gaseous inclusion ; knots, dirts and stones ; scratches, cuts and frush ; bow ; reams, strungs and lines ; waviness ; sulphur stains, corner breakage and chips more then those specified in IS code 2835 : 1987.

13.4.2 Flat Glass

Flat glass is a sheet of glass made by floating molten recycled glass, silica, sand, lime, potash and soda on a bed of molten metal, typically tin, although lead and various low melting point alloys were used in the past. The molten mass solidfies slowly while flowing over the bed of molten tin ; after which it is annealed to remove stresses induced during the cooling process. This method gives the sheet of glass uniform thickness and very flat surfaces.

Tinled float glasses are made by adding colouring agents during the melt process. Common colours include grey, green bronze and blue. Tinted glasses also provide materials with different properties including heat and light transmission (and/or reflectance) ultravoilet transmission and insulation properties. Reflective glasses are also available for applications where light transmission is an important design factor. These glasses have a metallic coating applied during manufacture.

Typical thicknesses range from 2 mm to 19 mm for commercial applications.

Main Properties

(*i*) High degree of light transmission

(*ii*) Ability to be produced in a range of colours

(*iii*) Ability to be produced in a range of opacities

(*iv*) Good chemical innertness

(*v*) Attacked by hydrofluoric acid.

Uses

Building Industry : Float glass is used for smaller windows in domestic housing. Large windows are made from toughened glasses. Glass is used for windows for aethetic and functional reasons, allowing the occupants to see out and at the same time allowing the light in.

Commercial Glazing : Float glass is becoming more and more popular in commercial applications as it allows structures to be constructed that give the impression of living outside with the benefits of being inside protected from the elements (with the exception of the sun). Float glass is also playing an increasing role in buildings where it provides an attractive and easy to maintain exterior surface. It should be noted that most glasses used for this application is subjected to a post heat treatment toughening process before use.

Displays : Based on its transparancy, hardness and ease of cleaning, float glass is often used for display cases in retail outlets and countertops. However many of these applications are being superseded by toughened glass due to its added strength.

13.4.5 Safety Glass : (I. S. 2553 (Part 1) : 1990)

Safety glass is made of 'AA' or 'A' qualities of transparent sheet glass. It may also be manufactured from float glass. Safety glass may be flat or curved and tinted/coloured itself or by providing coloured interlays.

Safety glasses are of two categories :

(*i*) Toughened safety glass ; and

(*ii*) Laminated safety glass

Toughened safety glass consists of a single piece of specially heat treated or chemically treated glass. This has such a stress pattern that the piece when fractured reduces to numerous granular fragments, with no large jagged edges.

Laminated safety glass comprise two or more pieces of glass held together by an interlocking layer or layers of plastic materials. This will crack and break under sufficient impact, but the pieces of glass tend to adhere to the plastic and do not fly. If a hole is produced, the edges are likely to be less jagged than that of ordinary glass. Laminated safety glasses are made in thicknesses of 3.5, 4, 5, 6, 8 and 10 mm.

Safety glass is of four types:

(*i*) Toughened safety (tempered) glass (TS)

(*ii*) Toughened-float safety glass (TF)

(*iii*) Laminated safety glass (LS) ; and

(*iv*) Laminated float safety glass (LF).

Toughned glass is subjected to the following tests :

(*i*) *Fragmentation test :* The glass is broken by giving a sudden punch and number of broken particles per unit area are counted.

(*ii*) *Resistance to shock test :* The glass is given a sudden shock with a smooth surface steel ball of 63.5 mm diameter from a height of one metre. Five samples are tested and not more than one sheet should break.

(*iii*) *Light stability test :* This test is carried out to ensure that the laminated safety glass is resistant to the effects of exposure to light.

(*iv*) *Boil test :* The specimens from light stability test are immersed vertically on the edge first in water at 65°C for three minutes and then immediately in boiling water for two hours and then removed. The specimen should not touch the bottom of the testing vessel. The glass may crack but no bubble or any other defect should appear within 15 mm from the edges or beyond 10 mm from the cracks.

(*v*) *Fracture and adhesion test* : The laminated safety glass is given a sudden punch and fragments from the under surface are collected and weighed.

Ten specimens are tested. The number of pierced samples should not exceed four. If the fragment from under side of unpierced specimens does not exceed 0.5 percent of the total weight of unpierced samples and if the yield of any fragment individually does not weigh more than 0.5 g, the specimens are considered to have passed the test.

13.4.4 Mirror Glass (I.S. 3438 : 1994)

A mirror consists of a transparent glass sheet coated with silver on one surface. The silver is protected by a metallic copper film, which is further covered by a suitable protective paint coating.

The transparent glass sheet should conform to 'AA' and 'A' qualities. Float glass may also be used. The back surface is given a coating of deposited silver by applying a solution of silver nitrate and ammonia. Silvering should be free from defects of blemishes in the reflecting surface such as lifting or separation of the silver from the glass, sulphide or other spots, haze or any other visible defects. A minimum amount of 0.8 g/m^2 silver is deposited. The silvered surface is protected by a film of deposited copper of not less than 0.4 g/m^2. A suitable paint coating is applied over the copper coating. The paint coating should not crack or peel off the copper or silver layer due to atmospheric temperature change or age-drying.

Reflectance of the mirrors should not be less than 80 percent for clear (untinted) glass mirrors.

The following tests are conducted on the mirror glasses as per IS 3438 ; 1994.

(*i*) Test for waviness

(*ii*) Salt spray test to assess the resistance of the paint coating

(*iii*) Hot water test

(*iv*) Tests fot copper and silver plating

(*v*) Reflectance test

13.4.5 Wired Glass (IS 5437 : 1994)

Wired glass is made with welded wire mesh embadded in the body of the glass to prevent if from shattering. It is manufactured in 6 mm thickness. The wire mesh used is made of steel wire 0.46 to 0.56 mm in diameter. The pattern of the mesh is either square or diamond with wires welded or hexagonal with wires twisted. The wire mesh is embedded completely in the glass sheet at least one mm from the surface and should not expose at any place. Wire mesh should not contain more than three broken wires per sq. m of the wired glass.

Figured wired glasses are also manufactured.

13.4.6 Obscured Glass

Many types of glasses are produced for use where light is to be transmitteed but vision obscured ; as in office doors, partitions and public toilets. The following are the main types :

(*i*) Patterned or figured glass (*ii*) Ground glass
(*iii*) Chipped glass, and (*iv*) Corrugated glass.

13.4.7 Coloured Glass

Coloured glass, generally known as stained glass is made by adding oxides of metals to molten glass.

13.4.8 Tinted Glass

It is generally used for decorative purpose. Some varieties have the property of excluding all the ultraviolet light and is used for show windows to prevent fading of materials.

13.4.9 Heat Absorbing Glass

It is blue green and has the property of absorbing the infra-red rays of sun. "Colorex" glass is of this type.

13.4.10 Tempered Plate Glass

It is produced by heating plate glass and then suddenly cooling it. It is much stronger than ordinary glass and is used for entrance doors, counter tops etc.

13.5 GLASS BLOCKS

Glass blocks are partially evacuated, completely sealed hollow units formed by fastening together two halves of pressed glass. Edges are coated with a grit bearing plastic material to ensure permanent and effective bond with the mortar. They are manufactured in 150 mm, 200 mm and 300 mm square sizes and approximate thickness is 100 mm.

In order to observe the glass and to difuse light ; ribs, flutes and prisms are cast on one or both the faces. In this manner not only translucence, but also diffusion, decorative effects and freedom from glare are obtained. The light may also be directed upwards. They have high insulating value.

Glass blocks are set in cement mortar or preferably in composite mortar 1 : 1 : 4.

Glass blocks are not intended to carry superimposed loads ; but they have adequate strength to carry their own weight upto a maximum height of 6 m and resist lateral wind pressure if individual panels do not exceed 11 square metres in area.

QUESTIONS

1. Give the composition of glass and the various functions performed by each constituent.
2. (*a*) What is flat transparent sheet glass and in what range of thickness it is manufactured?
   (*b*) State the classes of flat transparent sheet glass and their uses.

3. (*a*) What do you understand by float glass. How tinted float glasses are made?
(*b*) List the main properties of float glass and its uses in building and for commercial glazing.
4. (*a*) How is safety glass made and what are its types?
(*b*) Give brief details of the tests performed on toughened glass.
5. Describe mirror glass and its preparation. List the tests performed on mirror glasses.
6. Write short notes on :
(*a*) Wired glass
(*b*) Obscured glass
(*c*) Tempered plate glass
(*d*) Glass blocks
(*e*) Insulating glass

OBJECTIVE TYPE QUESTIONS

1. Which of the following is the main constituent of glass
(*a*) Chalk (*b*) Silica
(*c*) Potash (*d*) Lime
2. Which of the following constituents of glass has a very high melting point
(*a*) Magnesia (*b*) Soda
(*c*) Silica (*d*) Potash
3. ______ makes the glass infusible and imparts fire resisting property.
(*a*) Potash (*b*) Soda
(*c*) Silica (*d*) Oxides of metals
4. Glass is manufactured at a temperature of over :
(*a*) 800°C (*b*) 850°C
(*c*) 900°C (*d*) 1000°C
5. Which of the following is a special selected quality (SSQ)
(*a*) A quality (*b*) AA quality
(*c*) B quality (*d*) C quality
6. Float glass is made by floating the molten constituents on a bed of molten ______ .
(*a*) Copper (*b*) Zinc
(*c*) Tin (*d*) Iron
7. Float glass is mostly used in a building for
(*a*) Small windows (*b*) Large windows
(*c*) Both (*a*) and (*b*)
9. Laminated safety glass has an interlocking layer of :
(*a*) rubber (*b*) plastic
(*c*) bituminous felt (*d*) none of these
10. One side of a mirror has a coating of
(*a*) auminium (*b*) white lead
(*c*) siver (*d*) iron

11. The mirror coating is protected by a film of metallic
(*a*) copper (*b*) zinc
(*c*) tin (*d*) lead

12. Minimum amount of silver deposited on mirrors is :
(*a*) 0.6 g/m^2 (*b*) 0.8 g/m^2
(*c*) 1.0 g/m^2 (*d*) 1.2 g/m^2

13. The thickness of copper film on mirrors should not be less than
(*a*) 0.1 g/m^2 (*b*) 0.2 g/m^2
(*c*) 0.3 g/m^2 (*d*) 0.4 g/m^2

14. The reflectance of clear glass mirrors should not be less than
(*a*) 60% (*b*) 70%
(*c*) 80% (*d*) 90%

15. The diameter of mesh used for wired glass varies from
(*a*) 0.36 to 0.46 mm (*b*) 0.46 to 0.56 mm
(*c*) 0.56 to 0.66 mm (*d*) 0.66 to 0.76 mm

16. Wired glass should not contain broken wires per sq. m more than
(*a*) 3 (*b*) 4
(*c*) 5 (*d*) 6

17. The thickness of the wire glass in mm is
(*a*) 6 (*b*) 5
(*c*) 4 (*d*) 3

18. Which of the following types of glasses give obscured vision :
(*a*) figured glass (*b*) ground glass
(*c*) corrugated glass (*d*) all the above

19. ______ glass has the property of absorbing infra-red rays of sun :
(*a*) Stained glass (*b*) Colorex glass
(*c*) Chipped glass (*d*) Ground glass

20. Glass blocks have an approximate thickness of
(*a*) 250 mm (*b*) 250 mm
(*c*) 150 mm (*d*) 100 mm

Answers

1. (*b*)	**2.** (*c*)	**3.** (*a*)	**4.** (*d*)	**5.** (*b*)	**6.** (*c*)
7. (*a*)	**8.** (*d*)	**9.** (*b*)	**10.** (*c*)	**11.** (*a*)	**12.** (*b*)
13. (*d*)	**14.** (*c*)	**15.** (*b*)	**16.** (*a*)	**17.** (*a*)	**18.** (*d*)
19. (*b*)	**20.** (*d*)				

FURTHER REFERENCES

1. Bureau of Indian Standards, New Delhi
2. IS Codes 2553 (Part I), 2835, 3438, 5437

BITUMINOUS MATERIALS

Bitumen is defined as a viscous liquid or a solid consisting essentially of hydrocarbons and their derivatives, and which is soluble in carbon trichloroethylene. Bitumen is non-volatile and softens gradually when heated. It possesses water proofing and adhesive properties. American terminology uses the name 'asphalt' known as bitumen in India.

Bituminous materials include both bitumens and tars. Bitumens may be derived from the refining of petroleum or may occur as such in nature, either in a pure state or associated with varying quantities of mineral matter. Tars are bituminous condensates are produced by the destructive distillation of organic materials such as coal, wood, etc.

As far as Civil Engineering is concerned, bituminous materials are principally used as binders for road aggregates and as water-proofing materials. Both bitumens and tars are produced commercially in a number of different types and grades.

14.1. COMPARISON BETWEEN BITUMENS AND TARS

Although bitumens and tars are very much similar in appearance, being black to dark brown in colour, they are produced by vastly dissimilar processes and are quite different materials, both chemically and physically. They differ in certain properties which are of importance. Also, they have many properties in common, and the choice between the two is often a matter of availability.

Tars contain more free carbon which is insoluble in carbon trichloroethylene and they are also readily distinguished by odour.

Chemically speaking, the difference is concerned with the type of molecules present. The molecules in bitumens and tars are very large and complex and the chemistry of these substances is not well developed. In general, tar, especially derived from coal, is very resistant to the disintegrating action of water. For this reason, tar has greater utility of use below ground as water proofing material and for soil stabilization. However, tars are inferior in weather resistance to high quality bituminous materials when used in film form.

With regard to physical properties, the chief difference lies in their temperature susceptibility. Tar is more temperature susceptible than bitumen, becoming softer at summer temperatures and more brittle when cold. However, tars will usually wet and coat mineral aggregates with more facility than bituminous materials, especially when the aggregate is damp or dusty. Moreover, tar coating may be better retained in the presence of water.

The weathering depends on a number of external factors such as temperature, surface area exposed, action of light, possible leaching action of water and the advantages, and disadvantages of each of these materials should be considered for selection for a given use.

14.2 NATIVE BITUMENS

Natural deposits of bitumens are found in many widely separated sections of the world. These are usually designated as (*i*) Lake bitumens, (*ii*) Bitumenites, and (*iii*) Rock bitumens. All such natural bituminous materials are considered to have been derived from petroleum.

Lake Bitumens

These are found in depressions in earth's surface which have accumulated in lakes from the discharge of springs. Some of these deposits have been of commercial importance such as Bermudez lake bitumen and Trinidad lake bitumen. Trinidad lake, just off the north coast of South America has an area about 45 hectares and is 40 m deep in the centre. Bermudez lake extends over 450 hectares and varies in depth from 0.6 to 2.4 m.

Most lake bitumens contain more or less moisture, hygroscopic or in the form of emulsion and this water is removed by dehydration process by heating in a suitable open container. The refined bitumen has very low penetration (20 to 30 for Bermudez bitumen and 1.5 to 4.0 for Trinidad bitumen at 25°C) and this material is softened by fluxing with a petroleum flux to produce a material of desired consistency for paving purposes. Also, it may be blended with bitumens produced from petroleum to get blended bitumens.

Bitumenites

This is one of the natural bitumen like substances presumably derived from metamorphosis of petroleum. It is a hard, brittle, practically pure bitumen. Penetration at 25°C is 0 to 3. They are characterized by high fusing points. They are divided into three classes—gilsonite, glance pitch and grahamite. It is mined from seams and crevices in rock formations. Because of its higher cost and limited production, it plays a minor role in road works.They have been used in powdered form in road construction. They are principally used in the manufacture of varnishes, battery boxes, bitumen floor tiles, brake lining, printing inks, *etc.*

Rock Bitumens

These are deposits of limestone or sandstone naturally impregnated with bitumen and have been extensively developed in various parts of the world such as Italy, France, Germany, Russia and U.S. A. The bitumen content of rock may vary from trace to as much as 20 per cent and it may be very soft or very hard. These deposits are stripped, drilled and blasted as per ordinary quarrying operations. The excavated pieces are then crushed and pulverized and blended to make a uniform and satisfactory material for paving.

14.3 PETROLEUM BITUMENS

Petroleum bitumens are derived from petroleum crude oil taken from numerous oil wells throughout the world. The crude oil is transported by pipe lines, tank cars, or barges to the refineries where it is separated into its various components by continuous flow refining process.

Essentially the process is one of the fractional distillation wherein heat is used to boil off from the crude oil such distillates as bitumens or gasoline, kerosene, diesel oil, *etc.* and to leave behind bitumen as a residue. This residual bitumen may be of various grades depending upon the degree to which distillates are removed. These residual bitumens may be further processed by air blowing, solvent extraction, blending, compounding and admixing with other ingredients to produce a great variety of products used in road work and for other purposes such as sealing, roofing, water proofing, paints and various other industrial applications.

14.4 STRAIGHT RUN BITUMENS

The character of petroleum bitumen is dependent to some degree upon the nature of crude oil from which it is extracted. Crude petroleum may generally be divided into three classes (*i*) bituminous base, (*ii*) semi-bituminous base, and (*iii*) paraffinic base. In bituminous base crude oil, solid paraffins are either absent or are present only in traces. In paraffinic base or non-bituminous petroleums, solid paraffins are usually present. Since waxy materials are not desirable in road work, the most desirable bitumens for the road work are straight run materials produced by refining of bituminous base crudes.

The straight run bitumens are of two types; Type 1 and Type 2. Type 2 relates to bitumen derived from crude petroleum with excessive wax contents. Depending upon the penetration value, the straight run bitumens are further sub-divided into six grades; as follows

Penetration Test Value	*Type 1*	*Type 2*
30/40	S-35	A-35
40/50	S-45	-
50/60	S-55	A-55
60/70	S-65	A-65
80/100	S-90	A-90
175/225	S-200	

14.4.1. Dehydration of Petroleum

Nearly all crude petroleums carry more or less water, some being entrained mechanically and in other cases held in a state of emulsion. Crude oils containing less than 2 per cent of water are accepted by the pipe lines and refineries. Where it is necessary to remove water and bring it within the prescribed limit, usually the crude oil is held in storage for a long period for the water and other impurities, such as sand and salt to be removed by a simple process of sedimentation. If water is present in emulsified form, it may be necessary to add emulsion breaking agents to reduce water content. If the salt content of the crude oil reaching the refinery is so high as to cause corrosion problems, it may be necessary to remove it in a washing operation at the refinery before distillation begins.

14.4.2. Distillation

Petroleum is separated into various commercial products by means of distillation by which part of the hydrocarbon materials in petroleum are vaporized by heating them above their boiling points and then are condensed to liquids by cooling back to normal temperature.

Refining of crude oil is a continuous flow operation carried out by first pumping the crude through a tube still where its temperature is elevated and then immediately introducing the crude into a fractional tower. The fractional tower is an upright tall cylindrical vessel containing a series of platforms one above the other. As the hot crude is injected near the middle of the tower, lighter vapour or fractions gather on the uppermost platforms and are carried off to a condenser. Farther down from the top of the tower, heavier grades of crude are taken off for condensation until only the very heavy residue containing bitumen remains to be piped off from near the bottom of the tower. The temperature in such towers varies from top to bottom, decreasing upward.

In order to remove high-boiling point constituents such as heavy lubricating oils without changing them chemically, the pressure in the system must be reduced and/or steam distillation must be used. Action of steam is based on physical law that boiling point of a pair of non-mixible or slightly mixable liquids is lower than that of the pure components. Introduction of steam, therefore, serves to lower the boiling point of petroleum and produces the maximum yield of heavy lubricating oils. Bitumens produced by such means are said to be vacuum-reduced or steam-refined.

Increasing the temperature of the crude through the tube still and increasing the vacuum in the tower results in an end product of low-penetration-range bitumen cement; conversely, decreasing the temperature of the crude and decreasing the vacuum, produces a higher penetration or softer material.

When the residue is distilled to a definite consistency without further treatment, the distillation is known as *"straight running"* and the residue, is termed *"straight-run bitumen"*.

Cut-back bitumens in certain cases, a portion or fraction of the distillate is mixed with the residue at the close of distillation which is termed *"cutting back"* the object of which is to modify the properties of residual product or to dispose of economically a fraction which otherwise has little value commercially. They have the advantage, as compared to straight-run bitumens, that they can be used with cold aggregates and with a minimum of heat.

When a low-boiling solvent such as naphtha or gasoline is used to temporarily soften or cut-back the residual bitumen, the liquid bitumen produced is called rapid-curing (RC) cut-back bitumen.

When intermediate-boiling point solvent such as kerosene is used, the liquid bitumen is called medium-curing (MC) cut-back bitumen. When this type of liquid bitumen is used, the solvent will not evaporate as rapidly as the gasoline range solvent used in the manufacture of RC cut-backs.

Similarly liquid bitumens may be manufactured by either fluxing bitumen with a high-boiling point less volatile distillate such as gas oil or by controlling the rate of flow and

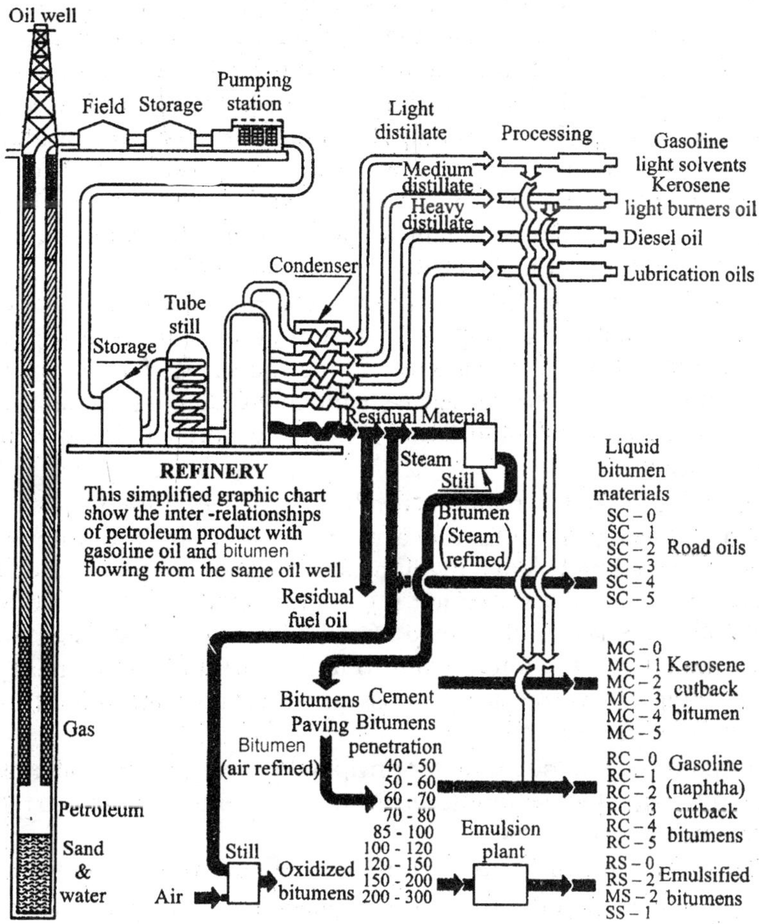

Fig. 14.1. Distillation of crude oil.

temperature of the crude during distillation process. These are slow-curing (SC) cut-back bitumens. Since the solvent in this case is a semi-volatile material, it requires much longer time for curing.

These three types of liquid bitumens are made in different grades which are distinguished on the basis of viscosity. In standard specifications there are six grades of RC and MC cut-backs, numbered from 0 to 5 (RC-0, RC-1, RC-2, RC-3, RC-4, and RC-5) and seven grades of SC material numbered from 0 to 6. Bitumen content and viscosity increase with the grade number, the 0 grade having the lowest viscosity and the 5 to 6 grade, the highest. For example, to make RC-0, the blend would be approximately 45 per cent bitumen whereas RC-5 grade would require approximately 15 per cent solvent and 85 per cent bitumen cement.

Emulsified bitumen. Emulsified bitumen is an intimate mixture of water, bitumens and emulsifying agent. This is another method of rendering a bitumen cement in liquid form. It has advantages over hot bitumen in that it can be used with cold as well as hot aggregate and with aggregate that is dry, damp, or wet. Its use with wet aggregate is an advantage for this material over the other type of liquid bitumens which are cut-backs. In addition, emulsified bitumen does not need any heat for its application.

Bitumen cement cannot dissolve in water, so it is necessary that bitumen and water exist in separate phases, one dispersed in the other in the form of minute droplets. In a properly made emulsion bitumen droplets range in size from 0.001 to 0.01 mm diameter. In order to prevent the bitumen droplets from coalescing, an emulsifying agent is necessary which is a material usually containing polar and nonpolar molecular structure, so that the molecules will concentrate at the interfacial layer between the two liquids and orient themselves to give the required action. Soaps are commonly used for this purpose. In addition to an emulsifying agent, the emulsion may contain a stabilizing agent, usually some form of protein, which will make it still more difficult for the particles to coalesce.

Bitumen emulsions do not depend upon evaporation of solvent for their setting characteristics but when emulsion is spread out thin on the surface of aggregate, it will set by virtue of coagulation of bitumen particles and the squeezing out of water between them. For this reason, emulsified bitumen may set faster and more completely than cut-back bitumens that cure over a long period of time. However, dry weather helps to remove the water quickly for road building.

Emulsified bitumen is usually manufactured by the use of a high speed disintegrator or colloid mill. Great shearing is applied as a mixture of water, bituminous materials and emulsifying agent is passed between a revolving and a stationary plate or two plates rotating in opposite directions which are separated by a very small space (0.05 to 0.075 mm). The bituminous material is reduced in the range of colloidal size and is dispersed in water. It is necessary to heat the bitumen before mixing to render it fluid. As temperature of finished emulsion cannot be high because of rapid loss of water, the hardness of bitumen used is usually limited up to 40 penetration.

Emulsified bitumens are made in three principal types-as per IS : 8887 rapid-setting (RS), medium-setting (MS), and slow-setting (SS). The type is controlled by the type and amount of emulsifying agent used.

Based on the type of emulsifier used, bituminous emulsion can be :

1. Anionic-bitumen globules being electro-negatively charged and the aqueous phase being alkaline.
2. Cationic-bitumen globules being electro-positively charged and the aqueous phase acidic.

The classification as per AASHTO/ASTM is more elaborate and has also been adopted in the Ministry's specifications. This is outlined below:

Anionic	*Cationic*
RS-1	CRS-1
RS-2	CRS-2
MS-1	
MS-2	CMS-2
MS-2h	CMS-2h
HFMS-1	
HFMS-2	
HFMS-2s	
SS-1	CSS-1
SS-1h	CSS-1h

Notes:

(*i*) The numbers 1, 2 relative to viscosity.

(*ii*) C stands for cationic, while absence of C denotes anionic.

(*iii*) *h* stands for harder

(*iv*) S stands for softer.

The disadvantage of emulsified bitumens is that freight charges for dispersing water have also to be paid which is an added expense, the usual amount of water contained in emulsion is in the range of 30 to 50 per cent.

Bitumen Felts for Water Proofing and Damp Proofing

Bitumen felts are comonly used for water-proofing and damp-proofing in buildings and other situations where penetration of moisture is to be stopped.

Classification

Bitumen felts are classified as given below depending on the type of base used in their manufacture and the use to which the felts are suited:

(*a*) *Fibre Base :*

Type 1—Saturated felt for underlay.

Type 2—Self-finished felt (for water-proofing) Grade 1.

Self-finished felt (for water-proofing) Grade 2.

Self-finished felt (for damp-proofing) Grade 3.

(*b*) *Hessian Base :*

Type 3—Self-finished felt (for water-proofing) Grade 1.

Self-finished felt (for damp-proofing) Grade 2.

Materials

(*a*) *Base Fabrics :*

1. *Fibre-Base Felt.* The base fabric for fibre base felts consists of a suitable blend of animal and vegetable fibres.

2. *Hessian Base Fabric.* The fabric is to be woven from good quality jute and shall have minimum weight of 225 gm per sq m.

(*b*) *Bituminous Saturant :*

Weight. The weight of the saturant shall not be less than 110 percent of the weight of the untreated felt for fibre-base felts and less than 60 percent of the weight of untreated fabric for hessian-base felts.

(*c*) *Bituminous Coatant :*

The bituminous coatant shall have a softening point not less than 105°C when measured by the Rind and Ball method, and a penetration of not less than 7 at 25°C.

(*d*) *Mineral Powder for Dusting :*

Composition. Mineral powder shall consist of powdered mineral matter, for example talc, mica, slate, etc., passing through I.S. sieve 60 micron.

Weight. The weight of the dusting powder incorporated shall be not less than 0.5 kg and not more than 1.8 kg per 10 sq m.

Dimensions

Unless specified otherwise, bitumen felts shall be supplied in widths of 90 mm and 100 mm. The length of felts shall be such that each roll shall be adequate to cover an area specified below, taking into account the necessary overlaps :

Type of Felt	*Area covered*
Type 1	40 sq m.
Type 2 —*Grade* 1 and 2 *Type 3*— Grade 1	20 sq m.
Type 2—Grade 3 *Type 3*— Grade 2	10 sq m.

Type of Felt	*Total weight per Roll in kg.*
Fibre Base :	
Type 1—Underlay	27.2
Type 2—Self-finished felt, Grade 1	35.6
Type 2—Self-finished felt, Grade 2	44.4
Type 2—Self-finished felt, Grade 3	31.8
Hessain Base :	
Type 3—Self-finished felt. Grade 1	45.4
Type 3—Self-finished felt. Grade 2	37.2

14.4.4. Bituminous Corrugated Sheets for Roofs

Bituminous roofing sheets are manufactured by saturating a dry fabric with hot bitumen and surfacing the final product with minerals in the shape of powder or granule and metals in the shape of thin sheets. The final product is pressed between corrugated rolls and sheets are cut to size.

Bituminous sheets are completely water proof, flexible, light, non-conducting and can be made to reflect more than 80 percent of solar light. They are not attacked by fungi or vermin. Being light and flexible, those can be easily transported and handled. There can be nailed directly to the wooden framework and do not require any nuts or bolts for assembly on the roof. These are not suspetible to corrosion and do not leak. The roof provides moderate protection against fire.

14.5. TARS

As already mentioned, tar generally is defined as a substance obtained by the condensation of distillates, resulting from the destructive distillation of organic and bituminous materials.

Many organic substances may be treated to yield tar. However, the following tar products are of importance.

14.5.1. Coke-oven Coal Tar

This is produced as a by-product in the manufacture of coke from bituminous coal for the steel industry and other uses. Tar is produced when bituminous coal is heated in the absence of air so that it decomposes into volatile substances and coke. Distillation is done continuously in series of ovens, each containing 8 tonnes or more of coal which is heated to temperatures of 1200°C or higher. The volatile substances include gases, light oils, tars and chemicals. Properties of tar vary with the coal used, design of oven, temperature used, and pressure in the system as it affects the time in which the volatile gases are exposed to incandescent coke.

The vapour is collected and passed through various forms of condensers and tar is removed by scrubbing with ammonia liquor in scrubbers. Ammonia liquor itself is a by-product of the operation. The tar and ammonia liquor are separated by decantation.

14.5.2. Water Gas Tar

The other main source of crude tar is from the cracking of petroleum in the carburated-water gas process in which water gas is produced for heating. Water gas is produced by allowing steam to pass through coke which is heated to a point of incandescence. This water gas (carbon monoxide and hydrogen gas) are too low in BTU value for heating purposes and it is enriched by the addition of hydrocarbon gases formed from the cracking of petroleum distillate (gas oil) as they are sprayed into a closed chamber carburator) maintained at a correct temperature where it is vaporized and combined with the gas produced in the heating chamber. The cracking of petroleum oils produces, in addition to hydrocarbon gases, a residue of heavy hydrocarbons which condenses to form tar.

14.5.3. Gas-work Coal Tar

In the process of eliminating gas from bituminous coal,, tar is produced as a byproduct. Coal is heated in ovens with water gas obtained by passing air and steam through incandescent coke. Vapours are driven off at temperatures ranging from 700 to 1000°C. These volatile materials are collected in a closed trough or hydraulic main which is partially filled with water. The temperature of vapour is reduced to about 65 °C in hydraulic main and a large percentage of tar condenses in water from which it is later removed. The remaining vapours are also converted into tar by passing them through a sequence of operations utilizing condensers, tar extractors, scrubbers and purifiers.

The crude tar, whether from coal or petroleum is then sent to the tar refining plant where it is dehydrated, if necessary and then fractionally distilled to produce residual products of various consistencies known as road tars and tar pitches. The crude tar contains materials such as benzole which are valuable as fuel and for the production of chemicals of many different kinds (*e.g.* dyestuffs, plastics, explosives, medicines). Moreover, crude tar being a thin material is quite unsuitable for use as road binder. It is therefore subjected to a distillation process whereby it is separated into a number of different fractions, some of which may later be blended together to form road tar and some of which are used in industry for other purposes. The distillation process of crude tar to produce various grades of road tars is similar to the distillation process for producing petroleum bitumens.

Tars are prepared in 14 grades designated as RT-1 to RT-12 and RTCB-5 and RTCB-6. Grades RT-1 to RT-5 and RTCB-6 are light tars which can be applied cold or at temperatures below 65°C. RT-7 to RT-12 are heavier grades of tar which require heating up to 120°C. The grades are progressively more viscous as the number designation increases. Only two cut-back tars RTCB-5 and RTCB-6 are used where low-temperature application and quick setting are desired. No emulsified tar materials are available in market at present for use as binders for road-building purposes.

Tar is also used for coating of wooden piles, paintings of both room walls and as a saturant for felts.

14.5.4. Coal Tar Pitch

Coal tar pitch is obtained by the direct distillation of crude coal tar or fluxing back pitch residues with high boiling coal tar distilletes to obtain products of the desired softening point. It is black or dark brown solid or semi-solid.

According to I.S. 216-2006 coal tar pitch is classified into four grades, namely :

(*a*) Soft pitch

(*b*) Soft medium pitch

(*c*) Hard medium pitch ; and

(*d*) Hard pitch.

Special coal tar pitch is manufactured those days under the following grades :

(*a*) Binder pitch

(*b*) Extra hard pitch ; and

(*c*) Super hard pitch

Coal tar pitch serves as a valuable ingredient in the production of a number of waterproofing, protective and binding compounds used in masonry, steel and timber structures. It is also used for water proofing concrete structures caulking of decks, as a saturant for roofing felts, for damp proof courses, flooring mastics and as a base for coal tar paints, as a binder for a variety of carbon products and coal briquettes.

Tests on coal tar pitch are conducted as per IS 13758 for soft to hard consistensies with softening points varying from 45°C to 95°C and for the range of three special coal tar grades with softening points varying from 86° to 130°C.

TESTING METHODS AND THEIR SIGNIFICANCE

Large number of different laboratory tests are performed upon bituminous materials for testing their properties for identification, checking the quality, uniformity and their suitability for any given use and also for checking compliance with the specifications. Most of these tests are conducted in accordance with methods of tests established by standards institutions such as BIS, ASTM, AASHTO. The usual laboratory tests of bituminous materials along with their designation are given in Table 14.1.

Brief description of test methods and their significance is given below. The reader may refer to the Indian standards for details of the test procedure in each case.

14.6 GENERAL TESTS

14.6.1. Specific Gravity Test

This is the ratio of the weight of a given volume of the material to the weight of an equal volume of water at a given temperature, usually 25°C. Specific gravity may be measured directly by means of hydrometer for the fluid grades and by displacement method for solid grades by weighing in air and weighing when completely immersed in distilled water at the standard temperature.

The test is usually performed with a pycnometer bottle. By filling the bottle only partially with bituminous material, the bitumen volume can be determined by the difference between the total volume of the bottle and the volume of water required to complete the filling; and thus specific gravity can be calculated.

The principal use of specific gravity is in converting from volume to weight measurements and vice versa for billing purposes and for mixture calculations.

The usual value of specific gravity for bituminous material varies from about 0.92 for thin liquid materials to about 1.06 for semi-solid materials. For road tars it varies from 1.08 to 1.24. This test has some value for identification purposes and for checking uniformity.

Table 14.1 Laboratory tests for bituminous materials

Name of test	*IS designation*	*AASTHO designation*	*Applicability*
General			
Specific gravity	IS : 2720 (Part 3) - 1980	T 43	Bitumens and tars
Flash point (open cup)	IS : 1209-1978	T 48	Bitumens only
Flash point (Tag open cup)		T 79	Bitumens only
Water content	IS : 2720 (Part 2) - 1973	T 55	Bitumens and tars
Consistency			
Penetration	IS : 1203-1978	T 49	Bitumens only
Viscosity (saybolt-furol)	IS : 1206-1978	T 72	Bitumens only
Specific Viscosity (Engler)		T 54	Tars only
Float test		T 50	Bitumens and tars
Softening point	IS : 1205-1978	T 53	Bitumens and tars
Solubility			
Solubility in carbon disulphide or	IS : 1216-1978	T 44	Bitumens and tars
Solubility in carbon tetrachlorate or trichloroethylene		T 45	Bitumens only
Spot test		T 102	Bitumens only

Ductility	IS:1208-1978			
Ductility test		T	51	Bitumens only
Volatility				
Loss on heating and penetration after loss on heating test	IS : 1212-1978	T	47	Bitumens only
Distillation				
Cut-back bituminous materials	IS : 217-1988	T	78	Bitumens only
Tar products	IS : 73-1992	T	52	Tars only
Bitumen residue of Specified penetration		T	56	Bitumens only

14.6.2. Flash Point Test

Two methods are in common use. For cut-back bitumens (RC and MC), flash point is generally determined by the use of Tag open cup apparatus in which heating takes place in a glass cup held in a water bath. For other bituminous materials Cleveland open cup is used. In this bitumen is heated in metal container suspended in air.

The test is performed by heating a sample of the material in an open cup at a specified rate and determining the temperature at which a small flame passed over the surface will cause the vapours to temporarily ignite or flash. The usual value of flash point for bitumen cement is above 175°C. For harder grades, it is above 230°C. Cut-back bitumens flash at low temperature, less than 40°C, RC cutback may flash at as low as 25°C.

Flash point requirements are included in the specifications for bituminous materials for safety reasons. Since the test is intended to determine the temperature to which the material may be heated safely, it is considered a good practice not to heat bitumen cements higher than 10°C below their flash point. Cut-back bitumens are commonly used at temperatures above their flash point and as such all of these materials present some danger in use and should be handled carefully. The more volatile the solvent in the liquid bitumen, the more hazardous is its use.

14.6.3. Water Content Test

The water content present in bituminous material is determined by mixing a specified weight of material with a water free petroleum distillate in a metal still or glass flask attached to water-cooled condenser with a glass graduated trap. Heating and fluxing is continued till all the water is transferred to the bottom of the glass trap where it is measured and expressed as a percentage by weight of the original material.

Determination of water content in bituminous material is necessary as low water contents are necessary to prevent foaming of the substance heated above the boiling point of water.

14.7 CONSISTENCY TESTS

The consistency of bituminous materials ranges from a very thin liquid for 0 grade of cut-back, to that of a stiff semi-solid resembling sealing wax in the case of blown bitumen cement. Because of this wide range, there is no one instrument which will satisfactorily measure the consistency of all bituminous materials. The following methods of measuring consistencies are now in common use:

1. Penetration
2. Furol viscosity
3. Engler specific viscosity
4. Float test
5. Softening point

14.7.1 Penetration

Penetration is the consistency test used to designate grades of bitumen cements. As bitumen cement being semi-solid at room temperature, it is not practicable to determine its consistency by Furol viscosity apparatus even when it is heated to decrease the viscosity.

Penetration is the distance in tenths of a millimeter that a standard needle can penetrate the sample under specified conditions of time, temperature, and load on the needle. Standard penetration, which is implied, unless other conditions are stated, is for a load of 100 g applied for 5 sec at a temperature of 25°C. Obviously, softer materials will have higher penetration values. Other test conditions used are (*i*) 0°C, 200 g, 60 sec; and (*ii*) 45°C, 50 g, 5 sec.

Penetration test is an identification test to identify a bitumen as to its hardness and grade. Lower the penetration, stiffer is the bitumen cement and more strongly it will bind the stones together in bitumen mix. For these practical considerations consistency is determined at 25°C which is about the average room temperature.

This test is not suitable for tars because of high surface tension exhibited by these materials and the fact that they contain relatively large amount of carbon.

14.7.2 Furol Viscosity (IS:3117)

Furol Viscosity (IS:3117) Viscosity is the broad general term used for consistency and is a measure of resistance to flow. Thus higher the viscosity of a liquid, the more nearly it approaches a semi-solid state in consistency. This test is applied to liquid bitumens.

In this test the time in seconds is measured for 60 cc of the material to flow through a standard orifice of 3.8 mm diameter under standard falling head conditions and at a standard temperature. It is not practicable to measure Furol viscosity at the same temperature on all liquid bitumens. The viscosity of bitumen is a function of the temperature, the viscosity decreasing with increased temperature. Because of this, the temperature at which each grade is tested is chosen so as to give a practical time of efflux. Standard temperatures used are 25, 50, 60, 80 and 100°C.

14.7.3 Engler Specific Viscosity

In this test the time in seconds for 50 cc of tar to flow through the standard orifice of the Engler viscosity meter at a specified temperature is divided by the time in seconds for 50 cc of water to flow through the same orifice at 25°C. The result is called specific viscosity. The test is similar to Furol viscosity except that the dimensions of the oil tubes are different. In

Saybolt apparatus, the column of oil above the orifice is relatively high in relation to its diameter while in Engler apparatus, the height of oil column is relatively low in relation to its diameter.

The viscosity tests are used to control the consistency, which in turn determines the grade of cutbacks and tars.

Viscosity is an important characteristic of a bituminous material because it determines in large measure how the material will function when used. For example, low viscosity materials (less than 150 sec viscosity) are recommended for priming. Similarly, materials with Furol viscosity less than 300 sec. at 60°C are recommended for road mixing with aggregates containing 5 to 15 per cent material passing 75 micron sieve. With coarser aggregates with less dust, more viscous grades can be used.

14.7.4 Float test

Float test is used to measure the consistency of materials too soft for the penetration test and too viscous at the desired test temperature for the viscosity test. It is also, a useful consistency test in those cases where the quantity of material tested is small. Usually bitumens which are more viscous than grade 5 liquid bitumens cannot be conveniently tested by the Furol viscosity test. Also, those which have a penetration higher than 300 cannot be conveniently tested by the penetration test. For this reason, the Float test is specified to the residue from the distillation of SC materials and heavier grades of tars.

The test is performed by filling a small brass collar with the material to be tested, cooling collar and contents to 5°C, screwing the collar to an aluminium float, and floating the assembly in water at a specified temperature (50° C for SC residues). The float test time in sec. is from placing the float in the water until the water displaces the plug of material in the collar and breaks through in the float.

14.7.5 Softening Point

The test is performed by forming a sample in a small brass ring (16 mm inside diameter and 6.2 mm high) and chilling it in a melting ice bath. After placing a 10 mm steel ball on the sample surface, the water bath temperature is raised at the rate of 4°C per minute. The temperature at which the sample sags under the weight of steel ball and touches the bottom of the container or other surface 25 mm below the sample is the softening point temperature.

This test may be classed as a consistency test in that it measures the temperature at which the bituminous material reaches a given consistency. It is used in specifications, of bitumens for crack filling, joint sealing of roofing, etc. where materials are used in thick films to ensure that they will not flow during application.

This is also a simple direct method of determining temperature susceptibility characteristics of bitumens. Bitumen having a higher softening point for a given penetration at 25°C is less susceptible to consistency change due to temperature. This is particularly so in case of blown bitumens as air blowing raises the softening point in relation to the hardness. The temperature susceptibility of bitumens also varies with the crude oil from which it was produced, but this variation is minor compared to one which exists between blown and paving bitumens.

Temperature susceptibility may also be measured by comparing consistency measurements such as penetration or viscosity at two or more temperatures.

The softening point for paving bitumens of penetration 40 to 300 ranges from about 55°C down to around 35°C. For blown bitumens, this test is highly significant because it is important that softening point of these materials be well above the temperature that they will reach when exposed to the sun which may be as high as 65°C.

14.8 SOLUBILITY TEST

The test is included in specifications for bituminous materials to determine the total bitumen present and to guard against the inclusion of undesirable amounts of foreign materials (salt, dirt, carbon and minerals). As the definition implies, it is necessary to use carbon disulphide as solvent to determine the bitumen content. However, carbon disulphide is highly inflammable and an unpleasant solvent to use. Carbon tetrachloride, because it is non-inflammable and is practically as good a solvent for bitumen as carbon disulphide is frequently substituted. For tars, carbon disulphide is always used.

In performing the test, a small sample of about 2 g of the bituminous material is dissolved in 100 ml of solvent and the solution filtered through an asbestos mat formed in the bottom of Gooch crucible. Insoluble material retained in the crucible is dried, cooled and the increase in weight of crucible is determined and expressed as a percentage of the weight of original sample.

As already stated, cracked bitumens weather much more rapidly than uncracked bitumens. Cracked bitumens, when subjected to solubility test in carbon tetrachloride, yield 0.5 per cent or more of a black carbonaceous residue (carbons-soluble in carbon disulphide but insoluble in carbon tetrachloride). Bitumens which are badly cracked may be easily identified. Uncracked bitumen has a glossy mirror-like surface. The one which is badly cracked has a dull surface.

Solubility test is used in specifications to ensure due care in refinery operations and to detect contamination in transit.

14.9 SPOT TEST

When the degree of cracking is slight, solubility test does not indicate these characteristics. For detection of overheating or cracking, a more sensitive means of detection is employed by means of spot test. This test measures whether a bitumen is hetrogeneous or homogeneous. Overheated or cracked bitumen will be colloidaly unstable.

In the standard test, 2 g of bitumen is dissolved in 10.2 ml of standard naphtha solvent. After one hour and again after 24 hr a drop of the solution is placed on a filter paper. If the stain on the paper is of a uniform colour, the test is negative and the bitumen is considered to be uncracked and is acceptable. If the spot forms a dark brown or black circle in the centre with a surrounding annular ring of lighter colour, the result is positive and the material is considered to be cracked.

14.10 DUCTILITY TEST

Ductility is the property which permits bitumens to undergo great deformation (elongation) without breaking.

The standard test is performed by determining the distance in mm that a briquette of material having a cross section of 1 mm sq will stretch without breaking when the rate of pull is 50 mm per rnin and the test temperature is 25°C.

High ductility is associated with materials of high temperature susceptibility. Also, materials with high ductility generally are adhesive and have good cementing properties. From the stand point of these two factors, it is felt that some reasonable value of ductility is desirable, but extremely high ductility is not necessarily good. Ductility may be required at two different temperatures to obtain some indication of the tendency of the material to change in temperature. Low temperature value may have more significance to indicate ductility of a material than standard test value at 25°C. Standard test method provides also for performing the test at 4°C and a rate of pull of 10 mm per minute. Some agencies also perform the test at 0°C with a rate of pull of 10 mm per minute.

14.11 VOLATILITY TESTS

14.11.1 Loss on Heating Test

This test measures the loss in weight when a 50 g sample of bitumen is placed in 85 gm metal can (vessel) and exposed to a temperature 163°C for 5 hr in a special oven with a revolving shelf. The loss in weight is expressed as a percentage of the original material. Also, the penetration of the bitumen after the test is determined, and expressed as a percentage of the original penetration.

This test is intended as a quality test as it measures the loss and hardening which the bitumen undergoes when heated. This is regarded as an accelerated test, the concept being that losses which occur on heating under the test conditions somewhat resemble the losses which occur in the construction process and during the service life of the pavement/building.

14.11.2 Distillation Test

This test is performed on cut-back bitumens, road oil and tars to determine the type and quantity of both the volatile distillate and the residue from distillation.

For cut-back bituminous materials, distillation is carried out by applying heat to 200 ml of the material in a glass flask. For tars, 100 g sample is distilled and the condenser is air - cooled rather than water-cooled as for bituminous materials. For tars, vapour (rather than liquid) temperature in the flask is measured, and the distillation result expressed as a percentage by weight of the total material rather than as a percentage by volume of the total distillate as applicable for bituminous materials.

Distillation test requirements are used to control the rate of curing of cut-back, road oil or tar.

14.11.3 Bitumen Residue of Specified Penetration

This test is intended to provide a method for determining the amount of 100 penetration bitumen in a SC bitumen or road oil. The test is conducted by heating a 100 g sample at a temperature between 250 and 260°C in a 170 gm metal can until the material reaches penetration at 25°C of 100 ± 15. The amount of residue at this point is weighed and expressed as a percentage of the original material. Heating is continued for indefinite time depending upon the operator's judgment and is stopped only when residue obtained is of desired penetration.

14.12 TEST FOR SULFONATION INDEX FOR TARS

Sulfonation index of road tar is the number of millimeters of unsulfonated residue per 100 gm of tar when determined in accordance with ASTM designation D 872 or AASHTO test T 108. Sulphuric acid (H_2SO_4) is added to a sample of the distillate that was taken at specified temperature or in a given temperature range. After repeated shaking to mix the liquids intimately, the sample is centrifuged at a speed of approximately 1000 rpm which separates the oils not acted upon by the acid.

Presence of high percentage of parafinic hydrocarbons causes distress in tar roads because either the residual tars become greasy and fail to hold the cover stone or priming materials do not set. This test is based on wide differences in reactivity of different hydrocarbons present in tars and bitumens. These are: (*i*) the parafinic or saturated straight-chain group, (*ii*) the naphthenic or cyclic saturated compounds, (*iii*) the olefinic or unsaturated straight-chain group, and (*iv*) the aromatics, which are the derivatives and homologs of benzene. Parafinic and naphthenic compounds are not attacked appreciably by sulphuric acid whereas the others are oxidized or converted into sulfonic acids.

Although AASTO specifications indicate that the test is applied only to tar grades RT-1 to RT-6, many agencies specify this control on all grades of tar.

14.13 TESTS OF EMULSIFIED BITUMENS

Emulsified bitumens are specified on the basis of the results of a complete series of tests. These are tested for composition, consistency, stability in mixing and in storage, and for character of residue. Composition tests include water content, residue by distillation, and residue by evaporation. Consistency is measured by Furol viscometer at 25°C, or at 50°C in the case of RS-2. Stability in mixing tests include demulsibility, cement mixing, and stone coating. Storage stability tests include sieve test, miscibility with water, modified miscibility with water, settlement and freezing. Tests on residue include penetration, ductility, solubility in carbon-trichloro-ethylene and ash content.

Standard methods of testing emulsified bitumens are designated as IS-8887. Brief description and their significance is given below.

14.13.1 Settlement

50 ml sample of emulsion is taken in two graduated cylinders and allowed to stand for 5 days. Samples are taken from the top and bottom parts of each cylinder. The samples are heated to allow the water to evaporate; leaving behind residue. The difference in weight of residue between bottom and top sample gives the settlement of bitumen globules. In case the percentage of bitumen residue at various levels in the container is considerably different, serious difficulty will be encountered in controlling the bitumen content of the finished pavement as the emulsion has to be stored for some duration before use.

14.13.2 Demulsibility

A 100 g sample is mixed with a dilute solution of calcium chloride. Two concentrations of calcium chloride are specified: 0.02 N if quick setting emulsion is being tested and 0.1 N solution if medium and slow-setting emulsions are being tested. Calcium chloride solution is added to accelerate breaking down of emulsion into bitumen and water. The bitumen will solidify and adhere to the vessel while the water and calcium chloride solution are poured off through 1.18 mm sieve. Demulsibility, which is stated as a percentage, is the ratio between the weight of bitumen separated from a sample by the demulsibility test and that obtained by distilling a sample of equal weight.

This test determines the relative resistance of emulsions to breaking down on contact with an aggregate surface.

14.13.3 Sieve Test

The percentage by weight of a 1000 g sample that fails to pass IS 850 micron (ASTM No. 20) sieve is determined. This test is used to determine whether or not an emulsion is of satisfactory character for application through pressure distributor and also whether or not the bitumen has properly emulsified. As a small amount of coarse sized globules will plug spray nozzles, the low limit of 0.10 per cent or less as specified is absolutely necessary.

14.13.4 Miscibility with Water

Approximately 50 ml of sample is mixed with 150 ml of water and the visible coagulation after 2 hr is observed. This test is used to demonstrate the stability of an emulsion when diluted with large quantities of water without the emulsion breaking down.

For slow-setting emulsions, a modified miscibility test is specified. After stirring 50 ml of emulsion diluted with 150 ml of water, the mixture is permitted to stand for 2 hr. The samples taken from three levels in the container are placed in an oven at 163°C for 2 hr to drive off water. Weighing before and after permits calculation of the percentage of bitumen residue at each level. Results are reported as the maximum numerical difference in percentage of bitumen content between any two of the three levels.

14.13.5 Cement Mixing

100 ml sample is mixed for one minute with 50 gm of high early strength cement. After this 150 ml of distilled water is added and mixing continued for 3 more minutes. The percentage by weight of the coagulated material that fails to pass 1.18 mm (No. 14 ASTM) sieve is dried at 163°C and weighed. The weight in grams of dried material retained is reported as the percentage emulsion broken. A maximum of 2 per cent is allowed.

This test is included for slow-setting emulsions to ensure an emulsion which will mix with all types of aggregates and this test has been found to approximate the severed condition of mixing emulsion with fine aggregate in field work.

14.13.6 Coating Test

The ability of a 35 gm sample to coat a specified stone when mixed therewith for 3 minutes is determined visually. This test is used for medium setting emulsions.

14.13.7 Freezing Test

A 400 gm sample is frozen three times at – 17.8°C for 12 hr each time, and the coagulation is observed visually. Samples that remain homogeneous or can be made so by stirring and pass the test are reported as *"homogeneous"*.

QUESTIONS

1. (*a*) Define bitumen and compare bitumen and tar.

   (*b*) What are the different types of native bitumens.
2. Explain the method of distillation of crude oil. What products are obtained as by-products.
3. List the classes of straight run bitumens and the different grades used in road work.
4. What do you understand by the term 'Cut-back bitumens? Give its different types and grades.
5. Define emulsified bitumen. How is it obtained? State the Indian and AASHTO classification of emilsion bitumens.
6. (*a*) State the different classifications of bituminous felts and the materials from which these are manufactured.

   (*b*) What are their normal dimensions and weights and how much area is covered by each type of felt?
7. State and describe three different types of tars. Distinguish between road tar and tar pitch.
8. Write short notes on :
   (*a*) Coal tar pitch
   (*b*) Bituminous corrugated sheets for roofs.
9. What is meant by the weathering of bituminous materials? What factors influence the deterioration action?
10. List in a tabular form the various laboratory tests carried out on bituminous materials and their applicability to bitumens and tars.
11. Briefly describe, how the following tests on bitumen are conducted :
    (*a*) Specific gravity test
    (*b*) Flash point test
    (*c*) Water content test
12. List the various methods of measuring the consistencies of bitumen. Describe these in brief.
13. Give details of the following tests on bitumenous materials :
    (*a*) Solubility test
    (*b*) Spot test
    (*c*) Ductility test

14. Explain the procedure for the following volatility tests :
 (*a*) Loss on heating test
 (*b*) Distillation test ; and
 (*c*) Bitumen residue test

15. Briefly describe the following tests :
 (*a*) Sulfonation index test for tars
 (*b*) Tests of emulsified bitumens.

16. What does the following types of bitumens represent :
 (*a*) S-200
 (*b*) A-90
 (*c*) RC-4
 (*d*) MS-2h

OBJECTIVE TYPE QUESTIONS

1. Bitumen and asphalt are synonymous :
 (*a*) True (*b*) False

2. Bitumen is :
 (*a*) Volatile (*b*) Hard on heating
 (*c*) A viscous liquid (*d*) Non-adhesive

3. Which of the following statement is incorrect :
 (*a*) Bitumens and tars are both chemically and physically different materials.
 (*b*) Tar has greater utility below ground as water proofing material.
 (*c*) Tars contain more free carbon.
 (*d*) Tars are superior in weather resistance to bitumens.

4. Which of the following is not a native bitumen :
 (*a*) Bitumenites (*b*) Petroleum bitumen
 (*c*) Lake bitumen (*d*) Rock bitumen

5. Which bitumen derived from the following classes of crude petroleum is most suited for road work
 (*a*) Bituminous base (*b*) Semi-bituminous base
 (*c*) Paraffinic base (*d*) All these

6. The bitumen derived from crude petroleum with excersive wax content is of :
 (*a*) Type 1 (*b*) Type 2
 (*c*) Both (*a*) and (*b*)

7. Crude oil containing less than ______ percent of water are accepted by refineries :
 (*a*) 5% (*b*) 4%
 (*c*) 3% (*d*) 2%

8. Cut-back bitumens can be used with :
 (*a*) Cold aggregates (*b*) Hot aggregates
 (*c*) Aggregates at normal temp erature (*d*) Very hot-aggregates

9. Rapid curing cut-back bitumens are obtained with the following solvent
(*a*) Kerosene oil (*b*) Naphtha
(*c*) Gas oil (*d*) Terpentine oil

10. RC-5 grade cut back bitumen would require about ______ percent bitumen :
(*a*) 75 (*b*) 80
(*c*) 85 (*d*) 90

11. Which of the following cut-back will require minimum solvent :
(*a*) RC-4 (*b*) MC-3
(*c*) RC-0 (*d*) SC-6

12. In emulsified bitumen, bitumen droplets range in size from :
(*a*) 001 to 0.01 mm (*b*) 0.01 to 0.1 mm
(*c*) 0.1 m to 1 mm (*d*) 1 mm to 2 mm

13. For emulsified bitumen, the hardness of bitumen used is restricted to ______ penetration :
(*a*) 35 (*b*) 40
(*c*) 45 (*d*) 50

14. Which emulsion bitumen has acidic acqueous phase?
(*a*) Cationic (*b*) Anionic
(*c*) Both (*a*) and (*b*) (*d*) Neither (*a*) nor (*b*)

15. The usual amount of water in emulsion is in the range of :
(*a*) 0 to 10% (*b*) 10 to 30%
(*c*) 30 to 50% (*d*) 50 to 70%

16. In gas-work coal tar process for tar, the vapours are formed at temperatures ranging from :
(*a*) 500 to 700°C (*b*) 700 to 1000°C
(*c*) 1000 to 1200°C (*d*) 1200 to 1500°C

17. The bituminous felt used for damp proofing is :
(*a*) Type 1 (*b*) Type 2 Grade 1
(*c*) Type 2 Grade 2 (*d*) Type 3 Grade 2

18. The maximum area covered by felt is by :
(*a*) Type 1 (*b*) Type 2 Grade 1
(*c*) Type 3 Grade 1 (*d*) Type 3 Grade 2

19. Tests on coal-tar pitch for soft to hard consistencies are done with softening points varying from :
(*a*) 0 to 45°C (*b*) 45 to 95°C
(*c*) 95 to 130°C (*d*) 130 to 170°C

20. The rate of oxidation and volatilization approx. doubles for each ______ °C rise of temperature.
(*a*) 10 (*b*) 50
(*c*) 100 (*d*) 200

21. The test applicable for bitumens only is:
(*a*) Float test (*b*) Softening point test
(*c*) Penetration test (*d*) Specific gravity test

22. The test prescribed for tars only is:
(*a*) Ductility test (*b*) Solubility test
(*c*) Flash point test (*d*) Engler specific viscosity test

23. The specific gravity test is carried out at a temperature usually:
(*a*) 20°C (*b*) 25°C
(*c*) 30°C (*d*) 35°C

24. The sp.cific gravity for road tars varies from
(*a*) 1.08 to 1.24 (*b*) 1.24 to 1.40
(*c*) 1.40 to 1,56 (*d*) 1.56 to 1.72

25. The flash point for ______ is lowest
(*a*) Bitumen cement (*b*) RC-5
(*c*) MC-4 (*d*) RC-cut back

26. Standard penetration test is done for the conditions:
(*a*) 0°C, 200 g, 60 sec (*b*) 25°C, 100 g, 5 sec
(*c*) 45°C, 50 g, 5 sec (*d*) None of these

27. The consistency test applied to liquid bitumens is:
(*a*) Float test (*b*) Engler specific viscosity test
(*c*) Penetration test (*d*) Furol viscosity test

28. In solubility test, about ______ gm of bituminous material is dissolved in a solvent.
(*a*) 1 gm (*b*) 2 gm
(*c*) 5 gm (*d*) 10 gm

29. In spot test, bitumen is dissolved in solvent
(*a*) Kerosene oil (*b*) Turpentine oil
(*c*) Naphtha (*d*) Coconut oil

30. In the dectility test, the rate of pull in mm/min. is:
(*a*) 50 (*b*) 100
(*c*) 150 (*d*) 200

31. In loss on heating test, 50 g of bitumen sample is placed in ______ gm of metal can (vessel):
(*a*) 25 (*b*) 50
(*c*) 75 (*d*) 100

Answers

1. (*a*)	**2.** (*c*)	**3.** (*d*)	**4.** (*b*)	**5.** (*a*)	**6.** (*b*)
7. (*d*)	**8.** (*a*)	**9.** (*b*)	**10.** (*c*)	**11.** (*d*)	**12.** (*a*)
13. (*b*)	**14.** (*a*)	**15.** (*c*)	**16.** (*b*)	**17.** (*d*)	**18.** (*a*)
19. (*b*)	**20.** (*a*)	**21.** (*c*)	**22.** (*d*)	**23.** (*b*)	**24.** (*a*)
25. (*d*)	**26.** (*b*)	**27.** (*d*)	**28.** (*b*)	**29.** (*c*)	**30.** (*a*)
31. (*b*)					

MISCELLANEOUS MATERIALS

ASBESTOS CEMENT PRODUCTS

Asbestos is a mineral fibre. It forms an excellent combination with cement. The percentage of asbestos varies from 8 to 15.

Asbestos cement is widely used in the manufacture of washers, fibre boards, roofing sheets and accessories, slates, ordinary rain water pipes, pressure pipes, soil pipes, gutters and sanitary fittings, flue pipes and fittings.

The products are light in weight, strong and durable, and can be finished to a rough or smooth surface as required. They are impervious to water, acid proof and non conductive, and can be sawn, nailed, screwed, veneered, ground and painted. Asbestos pipes are non corrodible and require no preservative treatment.

Asbestos fibre is used for paints also.

ADHESIVES

Adhesives are the substances used for bonding the surfaces of wood, metal, plastics and glass.

Adhesive may be classified into two categories:

(*a*) Organic adhesives, and

(*b*) Synthetic adhesives.

15.1 ORGANIC ADHESIVES

These are further divided into the following types :

1. Animal protein glues,
2. Blood albumin glues,
3. Casein adhesives, and
4. Starch adhesives.

15.1.1 Animal Protein Glues

These glues are obtained from hide trimming, bones and flashings by boiling them with hot water. Animal glues provide strong, tough, easily made joints; but they are affected by damp and moist conditions. Animal glue is supplied in the form of sheets, cakes, granules, pearls, flakes or powder or in a kibbled form.

According to IS : 852-1994 the average moisture content of animal glue shall not be greater than 14 percent and no individual value should be greater than 18 percent.

15.1.2 Blood Albumin Glues

This type of glue is made by drying raw blood. It is affected by damp and moist conditions.

15.1.3 Casein Adhesives

This type of adhesive is a strong water resistant adhesive from a type of milk protein called casein. The principal ingredients of a casein glue are casein, water, hydrated lime and sodium hydroxide. A properly proportioned mixture of these three substances will give a glue of high water resistance, but its working life will be very short. A glue can also be prepared of casein, water and sodium hydroxide. When properly prepared such a glue will have excellent dry strength and a long working life. Glues containing sodium hydroxide, hydrated lime and casein are suitable for wet-mix glues only.

According to IS : 850-1957, the casein should have following fineness characteristics.

(*a*) *For Wet mix glue.* It should wholly pass through IS Sieve, 600 micron when shaken in the sieve for 10 minutes.

(*b*) *For dry mix glue.* It should wholly pass through IS Sieve 250 micron. Dry casein and casein glue in powder form will keep for a long time, if stored under proper conditions. They should be kept in a cool, dry place because heat and moisture cause them to deteriorate. It is also advisable to store them in tight containers so that they cannot become infested with molds.

Casein glues when properly formulated provide highly moisture resistant glue joints although they cannot be called water-proof. They are not resistant to molds and bacteria and other micro-organisms. To improve resistance to deterioration caused by molds and other micro-organisms, preservatives such as copper salts, mercury salts or the chlorinated phenols or their sodium salts are added to casein glues. Chlorinated phenols, when added about 5 percent of the weight of the casein, results in marked improvement in mold resistance.

15.2 STARCH ADHESIVES

These glues are made from vegetable starch. They have good dry strength but are not moisture resistant.

Synthetic glues are based on synthetic resins. There are two main categories :—

(*a*) Thermo-setting glues, and

(*b*) Thermo-plastic glues.

These glues have, in recent years, revolutionized the method of gluing for plywood and other mass-produced woodwork. They may be applied in the liquid or dry form. Thermo-setting glues are permanent once they are set, but the thermo-plastic types can be made plastic again by reheating. All of them are strong, water-proof and fire-proof and the setting time can be regulated by varying the type (or kind) and amount of the hardener. They are resistant to fungal attack and possess non-staining qualities.

15.2.1 Types of Synthetic Glues

There are four main types of synthetic glues :

(*i*) Phenolic resins,

(*ii*) Resorcinol resins,

(*iii*) Urea resins, and

(*iv*) Melamine resins.

Other synthetic adhesives of lesser importance to the building industry include nitro-ellulose, ethyl cellulose, vinyl polymers and acrylic resins.

Phenolic glues are available in film, powder and liquid forms. In general the phenolic type of adhesives require heat and pressure to effect permanent strong bonding, and both Tego Film and phenolic glues are extensively used in the manufacture of resin-bonded plywood.

Resorcinol glues are dark viscous liquids. They cure at low temperatures in a relatively short time. They are resistant to heat, moisture, fungi and bacteria.

Urea formaldehyde resins are used extensively in joinery work where water-resistant glue joints are required. These adhesives are sold as syrups and powders for use with external catalyst or ready mixed with catalysts.

Melamine resins require heat and pressure for setting. They are used for plywood manufacture.

Acrylic and vinyl adhesives are used in the manufacture of safety :glass as interlayer.

According to IS : 851-1978 synthetic resin adhesives for construction work in wood are classified as below depending upon the degree of resistance to water and micro-organisms:

(*i*) BWP (Boiling water proof)

(*ii*) BWR (Boiling water resistance)

(*iii*) WWR (Warm water resistance)

(*iv*) CWR (Cold water resistance)

Gap filling adhesives and cold contact adhesives are distinguishhed by adding the symbols GF and CC respectively.

According to IS : 848-2006 synthetic resin adhesives for plywood are classified as below :

BWP-Boiling Water Proof,

BWR-Boiling Water Resistant,

MR-Moisture Resistant.

Their uses and cycle test are given in Table 15.1

Table 15.1: Characteristics of Different Types of Synthetic Resin Adhesives

S. No.	Type	Uses	Cycle Test
1.	BWP (Boiling water proof)	These have proved to make joints highly resistant to weather, cold and boiling water, steam and dry heat.	Six cycles , each cycle consisting of 8 h boiling in water and thereafter drying at 65 ± 2°C for 16 h.
2.	BWR (Boiling water resistant)	Joints will survive exposure to weather for only a few years. They will withstand cold water indifinitely and boiling water for a limited period.	Three cycles as above 8 h.
3.	MR (Moisture resistant)	Joints withstand cold water for a long period and hot water for a limited time, but fail in boiling water.	Three cycles, each cycle consisting of 3 h at 60 ± 2° C in water and there after drying at 65 ± 2°C for 8 h.

CORK

Cork is the outer layer of bark of an evergreen oak. Cork possesses a cellular structure, the specific gravity being 0.20.

Cork can be obtained in granulated form, in compressed slabs, and also in bitumen-bound slabs. It is one of the most efficient insulating materials and is used largely in cold storage construction and for insulation of flat roofs. The average density of compressed cork slab is about 210 kg/cm^2.

Compressed cork tiles form a resilient and almost soundless floor which is warm and of good appearance.

GYPSUM AND ALLIED PRODUCTS

Gypsum is a combination of calcium sulphate with water of crystallization and it is found in the form of rock in nature.

15.3 PLASTER OF PARIS AND HARD WALL PLASTER

When finely ground gypsum rock is heated to a temperature between 100° to 140°C, three fourths of the combined water passes off as steam. The remaining product is known as "Plaster of Paris", if pure gypsum has been used and "Hard Wall Plaster", if the gypsum rock contains impurities.

A large gypsum deposit at Montmartre in Paris led calcined gypsum to be commonly known as Plaster of Paris.

Plaster of Paris is used for cast ornamental plaster work and it is very valuable for this purpose. It produces hard surfaces, sharp contours and is sufficiently strong.

Neat Plaster of Paris will set in three or four minutes. This is too short a time for an operator to run mouldings or plaster walls and ceilings successfully, so small amount of retarders are added. They are glue, sodium citrate, gelatine and keratin. Gypsum plaster is rendered more plastic by the addition of clay or hydrated lime. The cohesiveness is increased by adding hair or sherded wood fibre.

15.4 GYPSUM ANHYDRITE

When gypsum rock is heated to 200°C, entire water of crystallization is driven off and the resultant product is known as gypsum anhydrite. This is also known as hard-burnt plaster. Hard-burnt plaster has a very slow set and to accelerate it certain substances such as alum or sulphate of potash are added.

15.5 PROPRIETARY PLASTERS

Some hard-burnt plasters are double burnt. The hard burnt plaster is impregnated with a chemical solution and burnt again. Som important varieties are the following :

(*a*) *Keene's Cement :* Gypsum calcined with alum,

(*b*) *Partina Cement :* Gypsum calcined with borax,

(*c*) *Mertin's Cement :* Gypsum calcined with pearl ash,

(*d*) *Sirapite :* Gypsum slaked in petroleum, and

(*e*) *Mack's Cement :* Gypsum calcined with Glauber's salt.

These plasters are widely used for corners arid angles, as their hard finish and mechanical resistance enable them to withstand hard wear. They are also valuable for repair work. They are also very much used for bathroom, kitchens and laundries or wherever a very hard waterproof coating is required on the walls. Such plasters may be polished to form a smooth surface and make a very satisfactory finish for interior walls. Often walls of these plasters are marked to immitate tiling with pleasing results.

15.6 PLASTER BOARD

Plaster board is a large sheet of gypsum plaster faced on both sides with stout paper as a reinforcement. Depending on the nature of the facing paper, the plaster board is classified into two types :

(*a*) *Gypsum Lath Board :* The facing paper is of rough nature to provide an adhesive grip for plaster finish.

(*b*) *Plaster Wall Board :* The facing paper is of "self finish" type, fit for decorating. Some boards have wood veneer applied on one side.

Plaster boards are usually 10.0 mm to 12.5 mm thick. They possess good insulation properties.

15.7 PYROCELL

It is a finely ground gypsum powder containing an admixture, which on being mixed with water, forms a gas and expands the mixture to 3 or 4 times its volume. This inflated paste hardens into a light, cellular, fire resistant mass possessing good acoustical and insulating properties.

15.8 OTHER GYPSUM PRODUCTS

Blocks and tiles made from wall plaster and suitable for floor and interior wall construction are now coming in the market. These forms are light, can be easily sawn to desired shape, possess sufficient strength for many types of construction and have high resistant to fire.

Mixed with fine cinders or wood chips and sufficient water to form a thin consistency, wall plaster has been used in making floors for buildings. Such floors can be much more rapidly constructed than concrete floors, owing to the rapidity with which the plaster hardens. However, they are neither so resistant to fire nor so strong as concrete floors.

LATH

When plaster is applied to masonry, some material must be provided to form a rigid backing or foundation for the plaster. This material has open spaces or corrugations in its surface to allow the soft plaster to form a key or grip to hold it in place. This is known as lath. The principal varieties of lath are as follows :

(*a*) *Wooden Lath :* Wood laths are thin strips of wood nailed to the under sides of ceiling joists and on the face of stud partitions. The laths are about 32 mm wide and the thickness varies from 5 mm to 8 mm. The spacing of laths varies from 6 mm to 9.5 mm.

(*b*) *Gypsum Lath Board :* This consists of a gypsum core with a facing of an absorbent paper which provides bond with the plaster by suction.

(*c*) *Metal Lath :* Metal lath may be woven from wire into a fabric with 50 to 60 mm square mesh called wire lath or it may be formed by cutting slits in metal sheets and by pulling the sheets transversely, opening out the slits and forming a perforated material called expanded metal lath.

LINOLEUM

Linoleum consists of a mixture of oxidised linseed oil, pulverised cork, wood flour, pigments and colours all spread in a uniform layer on canvas or burlap backing, the surface of which may be printed in different patterns. The surface is afterwards waterproofed with an oil paint.

Linoleum is available in the following gauges :

Gauges of Linoleum (From IS : 653 : 1992 (R 2006))

(Thickness in mm)

4.5, 3.2, 2.5 and 2.0

Linoleum is used for floor covering of hospitals, canteens, residences, offices and industrial buildings. It is easily cleaned with soap water and is usually wax polished. It is not affected by oils but slightly by acids. Linoleum is made in rolls as well as in tiles.

RUBBER

Natural Rubber

It is obtained from rubber trees which grow in hot moist climate. India is the 4th largest producer and 2nd largest consumer of natural rubber.

Synthetic Rubber

Synthetic rubber invariably a polymer is manufactured chemically. Some of the important varieties are as follows :

(*a*) Nitrile

(*b*) Neoprene—resistant to ozone and heat. It has got good mechanical properties.

(*c*) Butyl—It has got excellent resistance to acids, chemicals and abrasion.

(*d*) Polyurethane

(*e*) Polyisoprene

(*f*) Polyisobutylene.

Characteristics

The special field of usefulness of rubber as an engineering material depends on four salient characteristics :

1. Its value as an electrical insulator,
2. Its impermeability to water and gases,
3. Its ability to withstand great deformations without serious structural damage, and
4. Its ability to withstand energy of shock.

Uses

Rubber is used for belts, gaskets, lining of tanks and as a flooring material. It is also used for electrical insulation and sponge rubber is used for thermal insulation. Rubber door stoppers are used to prevent shutters slamming against the walls.

INSULATING MATERIALS

Insulating materials are of following types :

(*a*) Loose fills (*e.g.* slag wool, rock wool, glass wool, formed slag aggregate, exfoliated vermiculite, granulated cork and similar materials).

(*b*) Quilts, mats and rigid or semi-rigid slabs of any material in (*a*) above.

(*c*) Light weight concrete

(*d*) Reflective materials, *i.e.* products having one or both faces of some material of low emissivity (*e.g.* aluminium foil, foilfaced boards)

(*e*) Miscellaneous materials.

1. *Slag wool or Rock wool*

(*a*) Available in the form of loose fibres, quilts, mattresses, mats,, rigid or semi-rigid slabs or felts.

(*b*) *Density*

(*i*) *Loose fibres*—not more than 240 kg/cu m

(*ii*) *Quilts*—48 to 240 kg/cu m

(*iii*) *Mats*—not more than 240 kg/cu m excluding wire netting

(*iv*) *Slabs and felts*—48 to 240 kg/cu m

(*c*) *Thermal conductivity*—should not be greater than 0.037 k cal/sq m hr (deg C/m)

2. *Glass Wool*

(*a*) Available in the form of loose fibres, quilts, mats or rigid or semi-rigid slabs.

(*b*) *Density*

(*i*) *Loose fibres and felted fibres*—16 to 96 kg/cu m

(*ii*) *Quilts*—48 to 96 kg/cu m

(*iii*) *Mattresses and mats*—16 to 128 kg/cu m, excluding the wire netting

(*iv*) *Slabs*—48 to 192 kg/cu m.

(*c*) *Thermal conductivity*—should not be greater than 0.037 k cal/sq m hr (deg C/m)

3. *Hair Felt*

(*a*) *Density*—not greater than 192 kg/cu m

(*b*) *Thermal conductivity*—should not be greater than 0.037 k cal/sq m hr (deg C/m)

4. *Cork*

(*a*) Cork insulating material should be manufactured from pure natural cork. The material may be in the form of granulated cork, baked slab cork or regranulated baked cork.

(*b*) *Density*

(*i*) *Granulated cork*—not greater than 120 kg/cu m

(*ii*) *Regranulated baked cork*—not greater than 96 kg/cu m

(*iii*) *Slab cork*—not greater than 152 kg/cu m

(*c*) *Thermal conductivity*—should not be greater than 0.037 k cal/sq m hr (deg C/m)

5. *Exfoliated Vermiculite*

(*a*) Exfoliated vermiculite is prepared from naturally occurring vermiculite minerals by the action of heat.

(*b*) *Density*—not greater than 160 kg/cu m

(*c*) *Thermal Conductivity—not* greater than 0.062 k cal/sq m hr (deg C/m)

6. *Asbestos*

(*a*) Asbestos insulating materials should consist of natural asbestos fibres with or without a binding agent. It may be applied as loose fibre, either by handpacking or with a spray gun, or in slab form.

(*b*) *Density*

(*i*) *Loose fill*—not more than 240 kg/cu m

(*ii*) *Spray*—The weight per cu m of asbestos fibres applied by spray gun, measured after application should vary from 80 to 240 kg/cu m for applied thickness of 25 mm and more. For 12 mm thickness the weight per/cu m should not be more than 288 kg.

(*iii*) *Slabs*—from 128 kg/cu m to 240 kg/cu m

(*iv*) *Insulating boards*—not more than 800 kg/cu m

(*c*) *Thermal Conductivity*—should not exceed the following values :

(*i*) *Loosefill, hand packed*—0.074 k cal/sq m hr (deg C/m)

(*ii*) *Spray*—blue fibre 0.05 and white fibre 0.062 k cal/sq m hr (deg C/m)

(*iii*) *Slab*—0.05 k cal/sq m hr (deg C/m)

(*iv*) *Insulating boards*—0.105 k cal/sq m hr (deg C/m)

7. *Gypsum*

(*a*) *Density*—The weight of gypsum loosefill insulation should not be greater than 320 kg/cu m.

(*b*) *Thermal conductivity*—should not be greater than 0.062 k cal/sq m hr (deg C/m)

8. *Formed Slag Aggregate*

(*a*) *Density*—not greater than 672 kg/cu m.

(*b*) *Thermal conductivity*—should not be greater than 0.116 k cal/sq m hr (deg C/m)

9. *Light weight Concrete*

(*a*) Light weight concrete should consist of light weight aggregates with a suitable binder or a form of aerated concrete.

(*b*) *Density*—not more than 800 kg/cu m.

(*c*) *Thermal conductivity*—should not be greater than 0.124 k cal/sq m hr (deg C/m)

10. *Aluminium Foil*

Emissivity—The emissivity of the reflecting surfaces should not be greater than 0.08 at a temperature of 60°C.

11. *Insulating Gypsum Plaster Board*

This is similar to ordinary gypsum plaster board except that it is faced with aluminium foil on one face.

12. *Insulating Fibre Building Boards*

These boards are convenient to use, being light, easy to handle and fix but they are subject to some moisture movement, which causes "bellying" if fixing is not done properly. Thermal conductivity varies from 0.043 to 0.056 k cal/sq m hr (deg C/m) at 10 to 15 percent moisture content.

13. *Wood Wool and Compressed Straw Slabs*

They are used for roof insulation.

14. *Insulating Plasters*

Insulating plasters are those in which the aggregate consists of a material with low conductivity, for example, exfoliated vermiculite. The insulation varies with the amount of material used in the plaster. In addition, if the material is porous, the amount of the cementing agent absorbed affects the insulation.

15. *Magnesia*

Magnesia is very widely used as an insulator. The material consists of 85 percent of hydrated magnesium carbonate with 15 percent of asbestos fibre added as a binder. In the process of manufacture, innumerable small air pockets are formed in the crystalline structure. It is very fragile, and a hard setting finishing coat is used for appearance and where mechanical protection is necessary. It is usually applied in successive layers in plastic form, but can only be satisfactorily applied to surfaces at a temperature of at least 50°C. It is also made in block, sectional and mattress form, with paper or canvas backing.

15.9 METHODS OF INSULATION

(*a*) By replacing materials having a high rate of heat conduction with material having a low rate of heat conduction.

(*b*) By adding insulating materials to the construction, so as to partially or completely fill air spaces. Blankets, fills and baths are commonly used for this purpose. Except for very small air spaces, an insulating material has substantially greater heat resistance than the air space it fills.

(*c*) By replacing one or both of the surfaces of an air-space with a material of low emissivity, *i.e.* with a reflective insulation. The reflective insulation may be installed in such a manner as to divide the air space into two or more air spaces.

QUESTIONS

1. What are the different types of organic adhesives? Describe them briefly.
2. What are the main four types of synthetic glues? Give their details.
3. Define gypsum and explain how is it converted into the following forms :
   (*a*) Plaster of Paris
   (*b*) Gypsum anhydrite
   (*c*) Plaster board
   (*d*) Proprietary Plaster
   (*e*) Pyrocall
   (*f*) Gypsum blocks and tiles
4. (*a*) Name the different types of insulating materials. Give properties of five insulating materials.
   (*b*) What are the different methods of insulation.

5. Write short notes on :
   (*a*) Asbestos cement products
   (*b*) Lath
   (*c*) Linoleum
   (*d*) Rubber

OBJECTIVE TYPE QUESTIONS

1. Which of the following is not an organic adhesive?
   (*a*) Animal protein (*b*) Phenolic
   (*c*) Casein (*d*) Starch
2. The average moisture content of the animal glue should not be greater than ______ percent.
   (*a*) 14 (*b*) 16
   (*c*) 18 (*d*) 20
3. The individual value of the moisture content of animal glue should not be greater than ______ percent.
   (*a*) 14 (*b*) 16
   (*c*) 18 (*d*) 20
4. The presence of hydrated lime in casein glue makes its working life:
   (*a*) Short (*b*) Long
   (*c*) Very short (*d*) Very long
5. Glues containing sodium hydroxide, hydrated lime and casein are suitable for ______ mix glue only.
   (*a*) Wet (*b*) Dry
   (*c*) Both (*a*) and (*b*) (*d*) None of these
6. Addition of which of the folllowing preservative to casein glues results in marked improvement in mold resistance.
   (*a*) Copper sulphate (*b*) Mercury salts
   (*c*) Sodium salts (*d*) Chlorinated phenols
7. The BWP (Boiling water proof) synthetic resins should stand ______ cycles of boiling and drying.
   (*a*) 3 (*b*) 5
   (*c*) 7 (*d*) 9
8. The percentage of asbestos in asbestos cement products varies from :
   (*a*) 8 to 15 (*b*) 10 to 18
   (*c*) 12 to 20 (*d*) 14 to 24
9. The specific gravity of cork is
   (*a*) 0.1 (*b*) 0.15
   (*c*) 0.20 (*d*) 0.25

10. The average density of compressed cork slab is ______ kg/m^3.
(*a*) 150 (*b*) 170
(*c*) 190 (*d*) 210

11. 'Plaster of Paris' is obtained by heating finely ground gypsum rock to a temperature of
(*a*) 80 to 100°C (*b*) 100 to 140°C
(*c*) 140 to 200°C (*d*) 200 to 300°C

12. Neat plaster of Paris sets in
(*a*) 3 to 4 minutes (*b*) 5 to 7 minutes
(*c*) 7 to 9 minutes (*d*) 10 to 15 minutes

13. Entire water of cystallisation is driven off when gypsum rock is heated to
(*a*) 150°C (*b*) 175°C
(*c*) 200°C (*d*) 225°C

14. India is the ______ largest producer of natural rubber
(*a*) First (*b*) Second
(*c*) Third (*d*) Fourth

15. Which of the following insulating materials has the lowest thermal conductivity
(*a*) Glass wool (*b*) Asbestos
(*c*) Light weight concrete (*d*) Gypsum

Answers

1. (*b*)	**2.** (*a*)	**3.** (*c*)	**4.** (*c*)	**5.** (*a*)	**6.** (*d*)
7. (*b*)	**8.** (*a*)	**9.** (*c*)	**10.** (*d*)	**11.** (*b*)	**12.** (*a*)
13. (*c*)	**14.** (*d*)	**15.** (*a*)			

FURTHER REFERENCES

1. Bureau of Indian Standards, New Delhi.
2. IS Codes 653, 848, 850, 851, 852